Technological Change and the Environment

*Arnulf Grübler, Nebojsa Nakicenovic,
and William D. Nordhaus, editors*

Resources for the Future
Washington, DC, USA

International Institute
for Applied Systems Analysis
Laxenburg, Austria

A copublication of Resources for the Future (www.rff.org) and the International Institute for Applied Systems Analysis (www.iiasa.ac.at).

Printed in the United States of America

An RFF Press book
Published by Resources for the Future
1616 P Street, NW, Washington, DC 20036–1400

Library of Congress Cataloging-in-Publication Data

Technological change and the environment/ Arnulf Grübler, Nebojsa Nakicenovic, and William D. Nordhaus, editors.
     p. cm.
  Includes bibliographical references and index.
  ISBN 1-891853-46-5
  1. Technological innovations. 2. Environmental protection. I. Grübler, Arnulf, 1955- II. Nakicenovic, Nebojsa. III. Nordhaus, William D.

T173.8.T358 2002
628–dc21

2002017849

f e d c b a

The paper in this book meets the guidelines for permanence and durability of the Committee on Production Guidelines for Book Longevity of the Council on Library Resources.

This book was copyedited, typeset, and designed by the International Institute for Applied Systems Analysis. The cover was designed by Naylor Design.

ISBN 1-891853-46-5 (cloth)

About Resources for the Future

Resources for the Future (RFF) improves environmental and natural resource policymaking worldwide through independent social science research of the highest caliber.

Founded in 1952, RFF pioneered the application of economics as a tool to develop more effective policy about the use and conservation of natural resources. Its scholars continue to employ social science methods to analyze critical issues concerning pollution control, energy policy, land and water use, hazardous waste, climate change, biodiversity, and the environmental challenges of developing countries.

About the International Institute for Applied Systems Analysis

The International Institute for Applied Systems Analysis (IIASA) is an interdisciplinary, nongovernmental research institution founded in 1972 by leading scientific organizations in 12 countries. Situated near Vienna, in the center of Europe, IIASA has been producing valuable scientific research on economic, technological, and environmental issues for nearly three decades.

IIASA was one of the first international institutes to systematically study global issues of environment, technology, and development. IIASA conducts international and interdisciplinary scientific studies to provide timely and relevant information on critical issues of global environmental, economic, and social change for the benefit of the public, the scientific community, and national and international institutions. Research is organized around three central themes: energy and technology, environmental and natural resources, and population and society.

Contents

Chapter 1

Induced Technological Change and the Environment: An Introduction

Arnulf Grübler, Nebojsa Nakicenovic, and William D. Nordhaus

Technology is a double-edged sword for the environment: it can amplify as well as alleviate the impacts of human activities. The role of technology was largely ignored in the first round of global modeling efforts in the early 1970s, but it has recently moved to the forefront of both science and policy in addressing climate change. The long time span involved in climate change—from decades to centuries—puts technological change "in the driver's seat," because over long time periods technological systems are highly malleable, whereas in the short run they are largely inflexible.

A brief look at the history of technology illustrates the enormous scope for change: a century ago the industrialized countries relied on a coal-fueled steam-engine economy. Current technologies such as instantaneous global electronic communication or heavier-than-air flying machines were at most fantasies in the minds of science fiction writers. They were certainly not a source of inspiration for individuals in the newly founded research and development (R&D) laboratories. While it is easy to describe the enormous changes in technological hardware, software, and "orgware" that have characterized the past, and while it is comparatively easy to speculate about possible future developments, it is much harder to discern the factors that have caused all these changes. Why were certain technological options pursued while others were ignored? Why did some technologies gain widespread social acceptance and diffusion while others never moved beyond the status of a technological curiosity? And finally, what role was played by institutions and policies in triggering and promoting, or in obstructing and slowing down, change?

New and improved technologies do not "fall from heaven like autumn leaves." But do we have sufficient scientific knowledge about the sources and management of innovation to properly inform the policymaking process that affects technology-dependent domains such as energy or agriculture and their interactions with the environment? In light of current knowledge, we think the answer is, Not yet. We have theories and we have data, but we continue to lack a comprehensive conceptual framework for integration—a framework that can do justice to the "maze of ingenuity" surrounding technological innovation and the associated

1

uncertainties, to the complexity of factors that govern the incessant modifications and improvements of existing techniques, and to the diverse economic, social, and institutional factors driving their diffusion. Finally, we must recognize the intrinsic challenge of heterogeneity in technologies and agents for both innovation and diffusion.

A series of three workshops convened at the International Institute for Applied Systems Analysis (IIASA) in Laxenburg, Austria, from 1997 to 1999 focused on induced technological change (ITC), addressing the conceptual, empirical, and modeling challenges that an ITC perspective entails. By reviewing the state of the art of ITC theory, empirical case studies, and novel methodological approaches of modeling ITC, the workshops not only served to take stock, but also to generate new ideas and to frame a future research agenda. The current volume reports on the fruits of this collective endeavor. We wish to emphasize the collective nature of this exercise, to which all participants provided vital intellectual inputs. This volume necessarily contains only a limited number of contributions, which have been carefully chosen, reviewed, revised, and edited for the consideration of a wider readership.

Foremost, therefore, our sincere thanks go to the authors, the anonymous reviewers, and the workshop participants for their intellectual contributions and open minds, which enabled all of us to learn from one another and to report on the knowledge gained—as well as the knowledge gaps remaining. The contributors to this volume have been particularly generous with their time and efforts, not only in preparing draft manuscripts but also in revising them substantially in light of many fruitful discussions and the customary peer review process. Some contributions have been published in the meantime in a variety of scholarly journals. They are included here for the sake of completeness. The publishers—acknowledged in the relevant chapters—were kind enough to grant permission to reprint the articles so that the intellectual integrity of the contributions could be maintained. Transforming these individual contributions into a coherent volume is far from an easy task. Special thanks therefore go to the IIASA Publications Department for dealing with all technical production aspects so competently and expeditiously. Finally, we wish to acknowledge the financial support from IIASA, the United States National Science Foundation (which provided the resources through the Yale–NBER–IIASA program on international environmental economics), and the Austrian Federal Ministry for Education, Science and Culture. None of these institutions is responsible for the views expressed in this volume.

The individual chapters in this book are grouped into three main parts and are summarized in more detail below to guide the reader through this volume. Part One, including this overview (Chapter 1), provides historical background on the ITC debate (Vernon Ruttan, Chapter 2) and frames the issue from both a historical (Joel Mokyr, Chapter 3) and an international (Robert Evenson, Chapter 4) perspective. Part Two gives empirical insights into patterns of technological change either in response to price signals (Richard Newell and colleagues, Chapter 5) or technology policy (Chihiro Watanabe and colleagues, Chapter 6), or from a historical, evolutionary point of view (Nebojsa Nakicenovic, Chapter 7). Part Three presents recent ITC modeling approaches, embracing both "top-down" (William Nordhaus, Chapter 8; Lawrence Goulder and Koshy Mathai, Chapter 9) and "bottom-up" perspectives (Andrii Gritsevskyi and Nebojsa Nakicenovic, Chapter 10; Arnulf Grübler and Andrii Gritsevskyi, Chapter 11). It also provides an extensive literature review as well as reflections on the modeling research agenda ahead (Leon Clarke and John Weyant, Chapter 12). The final chapter (Vernon Ruttan, Chapter 13) moves the debate forward by extending the technological dimension into the realm of institutional innovations, a critical area when considering inducement mechanisms of technological change beyond the traditional dichotomy of "supply-push" versus "demand-pull" paradigms of research and policy.

Chapter 2, by Vernon Ruttan, reviews the historical roots of ITC as well as recent research streams. The chapter reviews the historical "demand-pull/supply-push" controversy and discusses the first ITC formulations in both their microeconomic and growth-theoretic foundations. The chapter then discusses two more recent research streams, namely, evolutionary theory and concepts of "path dependency" (often referred to also as technological "lock-in" phenomena). Of particular interest is Ruttan's careful assessment of the relative strengths and weaknesses of each research stream and his discussion of the need for a constructive dialogue between theory and empirical data. Ruttan reaches the conclusion that each individual research tradition—while having generated substantial insight into the generation and choice of new technology—is showing diminishing returns, but that, significantly, the different models of technological change (induced, evolutionary, path-dependent) can be seen as elements of a more general theory. The chapter makes concrete and useful suggestions toward the integration of different theoretical ITC streams. It concludes with a call for further integration of the insights gained from the theoretical and empirical research conducted within an ITC perspective and endogenous economic growth theory in order to gain new insights into the relationships between human capital, scale, and trade in the process of economic growth and development.

Chapter 3, by Joel Mokyr, provides an evolutionary interpretation of technological innovation in medical history. The chapter puts knowledge at the center of the debate. The history of medical technologies and techniques is indeed a most appropriate case study: nowhere is "demand-pull" as powerful as in the human desire to lead a long and healthy life. Yet, for all that demand-pull, improved medical technology will not emerge without adequate and improved knowledge. Quite appropriately, Mokyr refers to the "necessity is the mother of invention" theory of technological change as both a cliché and a historical fallacy. By developing a theoretical model of useful knowledge, Mokyr describes ITC as emerging from changes in knowledge as well as from changes in the selection environment of technological traits that can be realized with that knowledge. The evolution of knowledge in turn is seen as the net historical result of variation versus selective retention. From such a perspective, the chapter illustrates the limits of our capacity to steer or even to predict the evolution of new knowledge. Clearly, available resources and institutions are some of the factors that determine how conducive the existing knowledge base and society at large are to its expansion. But the uncertainties in ultimate outcomes remain enormous, with surprise and disappointment invariably accompanying the quest for new (technological) knowledge.

Chapter 4, by Robert E. Evenson, concludes Part One of this volume by explicitly addressing the international dimension of technological innovation and diffusion and the resulting impacts on productivity levels. Contrary to earlier expectations, empirical evidence suggests that productivity levels in developing countries have converged in the agricultural rather than the industrial sector. Evenson discusses two alternative mechanisms for productivity convergence to explain the puzzle: a mimicry mechanism and an induced adaptive invention mechanism with international invention recharge. The latter model requires both an endogenous capacity for innovation/adaptation in developing countries as well as a continuous "recharge" mechanism, by which innovation possibilities are replenished, for example, in the form of germplasm for the breeding of new plant varieties. This recharge mechanism, on which Evenson has pioneered research, highlights the importance of continuous investments in the knowledge base from which innovations can emerge. Evenson's tests of the model with empirical data (germplasm and diffusion of high-yield varieties in agriculture, international patent data in industry) clearly support the model. What seems even more important, however, is that the model of induced adaptive innovation with international recharge has been institutionally internalized in agriculture—leading to productivity convergence—whereas in industry no similar broad-based institutions facilitating local adaptation and assimilative capacity of innovations as well as an international recharge mechanism are available in developing countries. From that perspective, it is less surprising to see that convergence in industrial productivity levels is confined to the Organisation for Economic Co-operation and Development "club" and a small group of newly industrializing developing countries.

Chapter 5, by Richard G. Newell, Adam B. Jaffe, and Robert N. Stavins, opens Part Two of this volume by dealing with empirical analyses and findings from an ITC perspective. Using the novel approach of "characteristics transformation surfaces" for three consumer durables over several decades, they test a Hicksian inducement mechanism: how changing energy prices have affected the characteristics along the technological "possibility" frontier (otherwise frequently referred to as "best available technology," or BAT) as well as in the composition of product models available on the market (product substitution). Their analysis also includes the impacts of government regulations, including labeling and mandatory efficiency standards. The authors find that the overall rate of innovation was independent of prices and government regulation, but that the direction of technological change was responsive to price changes in a number of instances. Changes in energy prices in particular induced significant changes in the subset of models offered on the market, and this responsiveness increased substantially in the presence of labeling requirements. Finally, mandatory efficiency standards also had a significant impact on the average energy efficiency of the product menu available to consumers. Nonetheless, a sizable portion of energy-efficiency improvements was still found to be unrelated to changes in energy prices or regulation, pointing to the multitude of factors influencing the pace and direction of technological change.

Chapter 6, by Chihiro Watanabe, Charla Griffy-Brown, Bing Zhu, and Akira Nagamatsu, offers a unique glimpse into the workings of the Japanese system of innovation by examining in detail the spillover effects and positive feedback mechanisms at work in the development of photovoltaic (PV) technology. By drawing on a unique data set, the authors are not only able to shed light on knowledge (R&D) spillover effects among leading Japanese PV firms (a central area of concern regarding knowledge externalities inherent in technological innovation), but are able to relate it to the "virtuous cycle," most prominently reflected in the tremendous cost decline (of well over a factor of 40) in the economics of PVs over the past few decades. As such, the chapter breaks new ground in our understanding of the microeconomic foundations of technological "learning" phenomena, combining both supply- and demand-led inducement factors. Evidently, the ITC perspective offered in this chapter reports on a rather unique experience. The inducement effect of public R&D funding (by MITI, the Japanese Ministry for Trade and Industry) on private R&D, and its effect on improvements in PV technology, relies both on the specific coordination mechanisms that characterize the Japanese system of innovation as well as on a variety of demand-side factors. The latter include both a "moving target" of guaranteed PV purchase prices as well as the familiar impact of declining costs on market growth for a new technology. Perhaps the most important message to retain from this detailed analysis is the extreme importance of taking a long view in technology policy and the existence of an appropriate institutional setting.

Chapter 7, by Nebojsa Nakicenovic, explores the nature of the relationship between technological change, the costs and performance of new technologies, and their effect on reducing carbon dioxide (CO_2) emissions. The chapter shows that an important part of the secular decline of energy and carbon intensities is the result of technological change. Technologies that are more energy efficient have replaced less efficient ones, and technologies that are less carbon intensive have replaced those that are more carbon intensive. Further, it is argued that an important component of the dynamics of technological substitution is a cumulative process of learning by doing and that timely investment in new technologies with lower CO_2 emissions is a cost-effective strategy for reducing global carbon emissions. A number of implications are considered with reference to emission mitigation strategies. One is that there may be great leverage in policies that accelerate the accumulation of experience in new technologies with lower environmental impacts, for example, through early adoption and development of special niche markets. This leverage can be important, particularly if these policies can minimize the "deadweight" loss to society associated with the foregone exploitation of cheaper fossil fuels and possible reductions of research, development, and demonstration (RD&D) efforts in other parts of the economy.

Chapter 8, by William D. Nordhaus, opens Part Three of this volume with an analysis of the relative importance of ITC in the context of climate-change policy. Nordhaus notes that most studies of environmental and climate-change policy have ignored the thorny issue of induced innovation by assuming that technology is exogenous. He analyzes the impact of induced innovation by developing a model of induced innovation and incorporating this in an updated version of the globally aggregated DICE model called the R&DICE model. The approach specifies a model of induced innovation in which the stock of carbon-energy technological knowledge responds both to technological developments outside the carbon-energy sector and to profit-oriented R&D within it. The specification and parameters of the induced-innovation equations are determined from the extensive work available on the returns to R&D. The principal conclusion of the study is that, over the next century, induced innovation is likely to be a less powerful factor in implementing climate-change policies than substitution. The reductions in CO_2 concentrations and in global mean temperature due to induced innovation are estimated to be approximately one-half those due to substitution. If confirmed by further research, this study suggests that subsidizing R&D or energy technology is unlikely to be a fruitful approach to solving the climate-change problem.

Chapter 9, by Lawrence H. Goulder, and Koshy Mathai, examines optimal CO_2 abatement profiles under various model specifications of ITC. The model considers two specifications for technological progress: one relying on an R&D-based regime of knowledge accumulation, and the other relying on a learning-by-doing formulation. These two formulations are then tested for their impact on both the timing and extent of emission abatement as well as on optimal carbon tax levels under both a cost-effectiveness and a benefit-cost criterion using a by now classic formulation first proposed by Nordhaus where abatement costs are contrasted with a nonlinear climate-change damage function. The particular appeal of the Goulder/Mathai approach lies, first, in the analytical solutions to the problem formulation, and, second, in the extensive sensitivity analyses performed. As expected, the presence (or absence) of ITC has its most significant impacts on lowering (or increasing) optimal carbon tax profiles, assuming of course that the shape and parameters of the climate damage function are known with certainty. Conversely, the impacts of ITC on the timing of emission abatement are more ambiguous and depend on the assumed regime of knowledge accumulation (R&D versus learning by doing). The Goulder/Mathai model and the numerical simulations discussed in the chapter provide valuable guidance for future research, proposing, first, a more detailed examination of the microeconomic foundations and complexities of different forms of knowledge accumulation, and, second, representing the numerous uncertainties involved in knowledge accumulation and in the abatement cost and climate damage functions.

Chapter 10, by Andrii Gritsevskyi and Nebojsa Nakicenovic, presents a new method for modeling induced technological learning and uncertainty in energy systems. Three related features are introduced simultaneously: (1) increasing returns to scale for the costs of new technologies; (2) clusters of linked technologies that induce learning "spillovers" depending on their technological "proximity," in addition to the technology relations through the structure (and connections) of the energy system; and (3) uncertain costs of all technologies and energy sources. One of the results of the analysis is that there is a large diversity across alternative energy technology strategies. The strategies are path dependent, and it is not possible to choose *a priori* the "optimal" direction of energy systems development. Another result of the analysis is that endogenous technology learning with uncertainty and spillover effects will have the greatest impact on the emerging structures of energy system during the first few decades of the next century. Finally, the results imply that fundamentally different future energy system structures might be reachable with similar overall costs. Thus, future energy systems with low CO_2 emissions need not be associated with costs higher than those of systems with high emissions.

Chapter 11, by Arnulf Grübler and Andrii Gritsevskyi, extends the traditional social planner perspective of models of technology choice in the direction of heterogeneity of actors. These actors operate under deep uncertainty with respect to both economic returns from their innovation efforts and the market environment (demand, resource availability, environmental regulations) under which their decisions take place. Using a Pareto-optimal formulation and a risk function term that integrates uncertainties directly into the objective function, the model illustrates substantial computational progress. Workable formulations and algorithms are now available that can simultaneously deal with intermediate levels of technological complexity, heterogeneous actors, and a full range of uncertainties, including uncertain increasing returns. Among the important conclusions for policy is that technology hedging and portfolio strategies are an economically rational response in view of persistent uncertainties, even if these strategies are initially much more cautious and gradual than our immediate concern for the environment suggests they should be. Second, there is no single "magic bullet" among the many inducement factors of technological change. Radical departures in long-term emission trends appear to be possible only when factoring into the analysis a combination of supply and demand-led inducement factors combined with environmental regulation (taxes), once these are treated in models as they are in reality, that is, as inherently uncertain.

Chapter 12, by Leon E. Clarke and John P. Weyant, concludes the chapters dealing with modeling issues of Part Three of this volume. The chapter offers a thorough review of the modeling work in the literature and in this volume, as well as much useful food for thought on further developments in modeling ITC. The authors single out four important areas of extensions of ITC modeling: first, explicitly recognizing knowledge externalities in technological innovation that can lead to underinvestments—even if such externalities may only be representable heuristically in models; second, explicitly considering technological heterogeneity—and thus representing a minimum degree of technological detail; third, dealing with uncertainty—especially coming to grips with how to represent the ways uncertainties may gradually evolve over time; and fourth, explicitly representing technological diffusion in both time and space, along with the substantial lags such diffusion processes entail and the benefits that would accrue if diffusion could be accelerated. The chapter also stresses the need for improved communication and interpretation of model structures and results—a useful reminder for both modelers as well as decision makers. Models, after all, are simplified abstractions of our perceptions of reality, and as such, they are an important input to decision making but are not a substitute for it.

The concluding chapter, Chapter 13 by Vernon Ruttan, extends the ITC debate by considering institutional innovations. This perspective is important because resource endowments, technologies, institutions, and culture (e.g., as expressed in consumer preferences) cannot be considered as given and static, but rather must be treated as dynamic, interacting variables. Ruttan's model of induced institutional innovation considers that changes in these noninstitutional ("environmental") variables are an important endogenous mechanism for inducing institutional innovation. Using agricultural history in preindustrial England, nineteenth-century Thailand, and twentieth-century Philippines, Ruttan illustrates how changes in technology and relative resource endowments (e.g., costs of labor versus land costs) have induced important institutional changes in the form of changing property rights and organization of work. As another example, Ruttan describes the "constructed market" for sulfur dioxide emission permits in the United States to illustrate an important institutional innovation beyond the traditional "command-and-control" paradigm. The institutional innovation consists, in this case, of inventing new institutions for property rights to manage formerly open-access resources. The available evidence suggests not only that emissions trading has been more cost-effective than initially anticipated, but also that the performance-based emission trading system has induced powerful changes in technology, substantially lowering sulfur abatement costs. Even if the sulfur-trading example is orders of magnitude simpler than an equivalent global regime for greenhouse-gas emissions, the example provides a powerful message: in the seamless web of institutional and technological change, departures from "business as usual" approaches are needed on all fronts to induce the changes required to respond to the climate-change challenge.

Chapter 2

Sources of Technical Change: Induced Innovation, Evolutionary Theory, and Path Dependence

Vernon W. Ruttan

2.1 Introduction

This is an appropriate time to take stock, as economists, of our understanding of the determinants of the rate and direction of technical change. The 1960s through the 1980s produced considerable new theory and empirical insight into the process of technical change. In the 1960s and 1970s, major attention was focused on the implications of changes in demand and in relative factor prices. In the late 1970s and early 1980s, attention shifted to evolutionary models inspired by a revival of interest in Schumpeter's insight into the sources of economic development. Since the early 1980s, these models have been complemented by the development of historically grounded "path-dependent" models of technical change.

Each of these models has contributed substantial insight into the generation and choice of new technology. It appears to me, however, that each research agenda is approaching a dead end. In this chapter I argue that the three models—induced, evolutionary, and path-dependent—represent elements of a more general theory. The purpose of this chapter is to review the development of the three models to identify their complementarity and to suggest how they might be incorporated into a more general theory.

2.2 Induced Technical Change

Three major traditions of research have attempted to confront the impact of change in the economic environment on the rate and direction of technical change. The "demand-pull" tradition has emphasized the relative importance of market demand on the supply of knowledge in inducing advances in technology. A microeconomic approach was built directly on an early observation by Sir John Hicks

---

Throughout this volume the terms "technical change" and "technological change" are generally used interchangeably. In Chapters 2 and 3, the use of the term "technical change" reflects the long-standing terminological tradition in the field of the economics of technical change. We have not attempted to standardize the usage across chapters, however, and instead chose to respect the terminological choices of the authors. Eds.

(1932:124–125) that a change in the relative prices of factors of production is a spur to invention. A macroeconomic growth theoretic tradition stems from attempts by economic theorists to understand the apparent stability in factor shares in the American economy during the twentieth century in spite of the very large substitution of capital for labor. In the language of the "new growth economics," each of these approaches could be described as an attempt to interpret the process of technical change as at least partially endogenous.

2.3   Demand Pull and the Rate of Technical Change

Schumpeter, whose writings have been exceptionally important in influencing the way economists think about technical change, made a sharp distinction between invention (and the inventor) and innovation (and the innovator): "Innovation is possible without anything we should identify as invention, and invention does not necessarily induce innovation but produces itself . . . no economically relevant effect at all" (Schumpeter 1934:84). The Chicago sociologist S.C. Gilfillan viewed invention as proceeding under the stress of necessity, with the individual innovator being an instrument of luck and process (Gilfillan 1935).

In his now classic study of the invention and diffusion of hybrid maize, Griliches demonstrated the role of demand in determining the timing and location of invention (Griliches 1957). Schmookler, in a massive study of patent statistics for inventions in four industries (railroads, agricultural machinery, paper, and petroleum), concluded that demand was more important in stimulating inventive activity than advances in the state of knowledge (Schmookler 1962, 1966). The Griliches–Schmookler demand-induced model received further support from papers by Lucas (1967) and Benzion and Ruttan (1975, 1978) that showed technical change to be responsive to aggregate demand. In the mid-1960s, Vernon (1966, 1979) introduced a demand-pull model to interpret the initial invention and diffusion of consumer durable technologies (e.g., automobiles, television, refrigerators, and washing machines) in the United States versus other developed countries. His interpretation came just as the United States was about to lose its dominance in several of these technologies to Japan.

Arguments about the relative importance of the role of demand-side forces and supply-side forces, such as advances in knowledge, in inducing advances in technology were intensified in the late 1960s. A study conducted by the Office of the Director of Defense Research and Engineering (1969) purported to show that the significant "research events" contributing to the development of major weapons systems were predominantly motivated by military need rather than disinterested scientific inquiry. This view was challenged in studies commissioned by the National Science Foundation that, not unexpectedly, found that science events were of much greater importance as a source of technical change (Thirtle and Ruttan 1987:6–11).

In a review of the "demand-pull/supply-push" controversy, Mowery and Rosenberg argue that much of the research purporting to show that technical change has been demand induced is seriously flawed (Mowery and Rosenberg 1979). They insist that the concept of demand employed in many of the studies has been so broad or imprecise as to embrace virtually all possible determinants. Rosenberg also insists that the demand-pull perspective has ignored "the whole thrust of modern science and the manner in which the growth of specialized knowledge has shaped and enlarged man's technological capacities" (Rosenberg 1974). Research conducted from a demand-pull perspective appears to have atrophied since the late 1970s, partly as a result of the Rosenberg criticism.

Careful industry studies, such as the study of innovation in the chemical industry by Walsh, suggest that both "supply and demand factors play an important role in innovation and in the life cycles of industries, but the relationship between the two varies with time and the maturity of the industrial sector concerned" (Walsh 1984:233). A rigorous econometrics study by Scherer (1982) that simultaneously tests both the demand-induced and supply-push hypotheses across a broad range of industries confirms the earlier Schmookler finding of a strong association between capital goods investment and invention. However, Scherer found a weaker association between demand pull and industrial materials inventions. He also found that the introduction of an index of technological opportunity based on the richness of an industry's knowledge base added significantly to the power of his model to explain differences in the level of inventive activity among industries.

It should no longer be necessary to insist that basic research is the cornucopia from which all inventive activity must flow to conclude that investment in the generation of scientific and technical knowledge can open up new possibilities for technical change. Nor should it be necessary to demonstrate that advances in knowledge, inventive activity, and technical change flow automatically from changes in demand to conclude that changes in demand represent a powerful inducement for the allocation of research resources.

2.4 Factor Endowments and the Direction of Technical Change

Modern interest in the effect of factor endowments on the direction of technical change dates to the early 1960s. Hicks had previously suggested:

> The real reason for the predominance of labor saving inventions is surely that which was hinted at in our discussion of substitution. A change in the relative prices of the factors of production is itself a spur to invention and to inventions of a particular kind—directed at economizing the use of a factor which has become relatively expensive. (Hicks 1932:124–125)

Hicks' suggestion received implicit assent but little attention until the early 1960s. In his work on the theory of wages, Rothschild repeated the Hicks argument (Rothschild 1956:118, 176). In a book on economic growth, Fellner argued that firms with some degree of monopsony power had an incentive to make "improvements" that economized on the progressively more expensive factors of production, and that expectations of future changes in relative factor prices would be sufficient to induce even firms operating in a purely competitive environment to seek improvements that would save the more expensive factors (Fellner 1956:220–222; see also Fellner 1961, 1962).

An intense dialogue concerning the issue of induced innovation by economic theorists in the 1960s and early 1970s was triggered by Salter's explicit criticism of Hicks' induced technical change hypothesis. Salter insisted that "at competitive equilibrium each factor is being paid its marginal value product; therefore all factors are equally expensive to firms" (Salter 1960:16). He went on to argue that "the entrepreneur is interested in reducing costs in total, not particular costs or capital costs. When labor costs rise any advance that reduces total cost is welcome, and whether this is achieved by saving labor or saving capital is irrelevant" (Salter 1960:43–44; see also Blaug 1963). It is difficult to understand why Salter's criticism attracted so much attention except that students of economic growth were increasingly puzzled about why, in the presence of substantial capital deepening in the US economy, factor shares to labor and capital appeared to remain relatively stable. The differential growth rates of labor and capital in the US economy were regarded as too large to be explained by simple substitution along a neoclassical production function. Below, I discuss a series of papers published in the 1960s that successfully refuted Salter's argument and established a solid theoretical and empirical foundation for the induced technical change hypothesis.

2.4.1   The growth theoretic model

The debates about induced technical change centered on two alternative models— one a growth theoretic approach and the other a microeconomic version. The most formally developed version was the growth theoretic approach introduced by Kennedy (1964). The Kennedy article initiated an extended debate on the theoretical foundations and the implications of incorporating the process of induced technical change into the theory of economic growth (Samuelson 1965, 1966; Kennedy 1966; Drandakis and Phelps 1966; Wan 1971).

In the Kennedy model the initial conditions include (1) given factor prices, (2) an exogenously given budget for research and development (R&D), and (3) a fundamental trade-off (a transformation function) between the rate of reduction in labor requirements and the reduction of capital requirements. The model assumes a production function with factor-augmenting technical change. Kennedy cast his analysis in terms of the effect of changes in relative factor shares rather than changes in relative factor prices on bias in invention because of the growth theory implications.

The following example from Binswanger represents an intuitive interpretation of the Kennedy model:

> Suppose it is equally expensive to develop either a new technology that will reduce labor requirements by 10 percent or one that will reduce capital requirements by 10 percent. If the capital share is equal to the labor share, entrepreneurs will be indifferent between the two courses of action. ... The outcomes of both choices will be neutral technical change. If, however, the labor share is 60 percent, all entrepreneurs will choose the labor reducing version. If the elasticity of substitution is less than one, this will go on until the labor and capital shares again become equal, provided the induced bias

in technical change does not alter the (fundamental) trade-off relationship between technical changes that reduce labor requirements on the one hand, or capital requirements on the other. (Binswanger 1973, 1978a:32)

The Kennedy variant of induced innovation was subsequently incorporated into neoclassical growth theory (Wan 1971). Nordhaus notes

> Until recently, only Harrod-neutral (or purely labor-augmenting) technological change could be introduced into neoclassical growth without leading to bizarre results. Neoclassical growth models were "saved" from such restrictiveness by the introduction of the theory of induced innovation. Under the usual neoclassical assumptions and, in addition, when the innovation possibility curve takes the form assumed by Kennedy and Samuelson the system settles down into a balanced growth path exactly like that of the labor-augmenting case. (1973:209)

By the early 1970s the growth theoretic approach to induced technical change was under severe attack (Wan 1971; Nordhaus 1969:93–115, 1973; David 1975:44–57). Nordhaus notes that in the Kennedy model no resources are allocated to inventive activity. A valid theory "of induced innovation requires at least two productive activities: production and invention. If there is no invention then the theory of induced innovation is just a disguised case of growth theory with exogenous technological change" (Nordhaus 1973:210). He further notes that the Kennedy innovation-possibility frontier (IPF) implies that the rate of capital-augmenting technological change is independent of the level of labor augmentation. Thus, as technological change accumulates there is no effect on the trade-off between labor- and capital-augmenting technological change (Nordhaus 1973:215). He insisted that the model is "too defective to be used in serious economic analysis" (Nordhaus 1973:208). The growth theoretic version of induced innovation has never recovered from the criticism of its inadequate microeconomic foundation.[1]

2.4.2   The microeconomic model

A microeconomic approach to induced innovation, built directly on Hicksian foundations, was developed by Ahmad (1966). His criticism of the growth theoretic approach initiated a vigorous exchange (Fellner 1967; Ahmad 1967a, 1967b; Kennedy 1967). In his 1973 critique, Nordhaus mentioned that Ahmad was the only person to attempt to formulate the theory of induced technical change along microeconomic lines, but he did not comment explicitly on the Ahmad paper or on the subsequent exchange.[2]

In his model, Ahmad employed the concept of a historic *innovation-possibility curve* (IPC). At a given time there exists a set of potential production processes, determined by the basic state of knowledge, available to be developed. Each process in the set is characterized by an isoquant with rather narrow possibilities for substitution. Each process in the set requires that resources be devoted to R&D before the process can be actively employed in production. The IPC is the

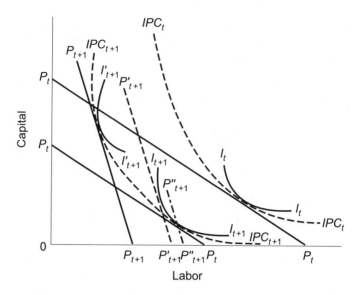

Figure 2.1. Ahmad's Induced Innovation Model.

Adapted from Ahmad (1966), Figure 1 amended.

envelope of all unit isoquants of the subset of those potential processes which the entrepreneur might develop with a given amount of R&D expenditure.

Assume that I_t is the unit isoquant describing a technological process available in time t and that IPC_t is the corresponding IPC (Figure 2.1). Given the relative factor prices described by line P_tP_t, I_t is the cost-minimizing technology. Once I_t is developed, the remainder of the IPC becomes irrelevant because, for period $t + 1$, the IPC shifts inward to some IPC_{t+1}. This occurs because it would take the same R&D resources to go from I_t to any other technique on IPC_t as to go from I_t to any technique on IPC_{t+1}. If factor prices remain unchanged and technical change is neutral, the new unit isoquant will be I_{t+1} on IPC_{t+1}. If, however, factor prices change to $P_{t+1}P_{t+1}$, then it is no longer optimal to develop I_{t+1}. Instead, a technological process corresponding to some I'_{t+1} becomes optimal. In the graph, $P_{t+1}P_{t+1}$ corresponds to a rise in the relative price of labor. If the IPC has shifted neutrally, I'_{t+1} will be relatively labor saving in comparison to I_t.

Ahmad's graphical exposition is a useful illustration of the induced innovation process of a one-period microeconomic model in which a firm or a research institute has a fixed exogenous budget constraint. When research budgets are no longer fixed, a mathematical exposition is more convenient (Binswanger 1978a:26–27).[3] In a multiple-period model the shift from I_t to I'_{t+1} would occur in a series of steps in response to incremental shifts from P_t to P_{t+1}. One way of describing such an incremental process would be to appeal to "learning-by-doing" and "learning-by-using" concepts (Arrow 1962; Rosenberg 1982).

2.5 Dialogue with Data

The initial dialogues about the logic of the Kennedy–Samuelson–Weizsäcker growth theoretic and the Hicks–Ahmad microeconomic approaches to induced technical change were conducted within the confines of the standard two-factor (labor and capital) neoclassical model. Among economic historians there has been an ongoing debate about the role of land abundance on the direction of technical change in the industrial sector. Among agricultural economists there has emerged a large literature on the bias of technical change along mechanical (labor-saving) and biological (land-saving) directions.

Habakkuk (1962) argued that the ratio of land to labor, which was higher in the United States than in Britain, raised real wages in American agriculture and thereby increased the cost of labor to manufacturers. He argued, in effect, that in the nineteenth century, the higher US wage rates resulted not only in the substitution of capital for labor (more capital) but in induced technical changes (better capital) biased in a labor-saving direction as well. The issue became controversial among economic historians even before they became fully sensitive to the emerging theoretical debates of the 1960s concerning the issue of induced technical change or the earlier empirical work by Hayami and Ruttan (1970, 1971).

The criticisms of the Rothbard–Habakkuk labor scarcity theses by Temin (1966) and the debates that his criticism engendered (Fogel 1967; Ames and Rosenberg 1968; David 1973, 1975:24–30) focused primarily on the issue of the impact of land abundance on the substitution of capital for labor— the "more capital" rather than the "better capital" part of the thesis. David argued that economic historians "steered away from serious re-evaluation of the proposition about the rate and bias of innovation, precisely because standard economic analysis was thought to offer less reliable guidance there than on questions of the choice of alternative known techniques of production" (David 1975:31).

David insisted that the argument could not be resolved without a more intensive mining of the historical evidence. But recourse to measurement could not be expected to get very far without a theoretically grounded definition of an operational concept that distinguishes between choice of technology and technical change and between bias in the direction of technical change and the rate of technical change. David argued that this can be done by embracing "the concept of a concave, downward sloping 'innovation-possibility frontier' ... along the lines of the neoclassical theory of induced technical progress due to Kennedy, Weizsäcker and Samuelson" (David 1975:32). He then went on to argue along the same lines as Wan (1971) and Nordhaus (1973) that the particular pattern of changes in macro-production relationships observed in the United States could not be rationalized within the framework of a stable IPF: "While shifts of the innovation-possibility frontier are entirely conceivable, the necessity of accepting their occurrence in this context signifies a practical failure of the underlying theoretical construct. For the latter treats the position of the frontier as established autonomously for each economy, and has no explanation to offer for it" (David 1975:33).

David also insisted that bias in the direction of technical change could only be understood by building a theory of induced innovation on microeconomic foundations consistent with engineering and agronomic practice. To David this also meant abandoning both neoclassical growth theory and the neoclassical theory of the firm. Furthermore, it would be necessary to incorporate the intimate evolutionary connection "between factor prices, the choice of technique and the rate and direction of global technical change" (David 1975:61).

In attempting to develop a non-neoclassical "evolutionary and historical" approach to induced innovation, David introduced the concepts of (1) linear fixed-coefficient processes or techniques from activity analyses (which he credits to Chenery) and (2) a latent set of potential processes that could be designed with the currently existing state of knowledge (which he attributed to Salter). He then added (3) localized learning, which directs technical change toward the origin along a specific process ray (which he attributes to Stiglitz), and (4) a probablistic learning process that is bounded by transition probabilities that depend on the firm's initial technical state (David 1975:57–86).

The model of the search process appears to have been inspired by the Nelson–Winter evolutionary model (see Section 2.6). David insisted, however, that his transition probabilities, in which past states—the firm's initial myopic selection of a technical process—influence the future course of development, are "clearly non-Markovian" (David 1975:81). He also distinguished the mechanism that accounts for the evolutionary nature of technical change from the form employed by Nelson and Winter.

David differentiated his approach from neoclassical production theory by suggesting that substitution may involve an element of innovation. This is similar to the mechanism that Ahmad (1966) and Hayami and Ruttan (1970) had earlier employed to account for the shift in the IPC (or in David's terms, the frontier production function). It should be viewed as an extension rather than an alternative to the neoclassical model.

When he turned to the technical relationships among natural resources, labor, and capital, drawing on the work of Ames and Rosenberg (1968) and his own earlier work (David 1966), David argued that in the mid-nineteenth century mechanical technology and land were complements: "The relevant fundamental production functions for the various branches of industry and in agriculture did not possess the property of being separable in the raw materials and natural resource inputs; instead the relatively capital-intensive techniques ... were also relatively resource using" (David 1975:88). Greater availability of natural resources facilitates the substitution of capital for labor. "Thus, even if the same labor/capital price ratios had faced producers in Britain and America, the comparatively greater availability of natural resources would have suggested to some American producers the design and to others the selection of more capital-intensive methods. ... In America the on-going capital formation spurred by the greater possibilities of jointly substituting natural resources and capital for labor may well have been responsible for driving up the price of labor from the demand side" (David 1975:89–90). The formal introduction of the role of relative resource abundance (or scarcity) clearly represents an important extension as compared to the traditional two-factor (labor and capital) neoclassical models. But the primary significance is that David opened the door to, and identified most of the elements of, what has since become known as the path-dependent model of technical change (David 1975:65, 66).

There are substantial differences in the extent to which the several induced technical change models have been tested against empirical data. The demand-induced model was developed in close association with empirical studies and was not subjected to formal modeling or theoretical critique until fairly late (Lucas 1967; Mowery and Rosenberg 1979).[4] The growth theoretic version of factor-induced technical change has produced very little empirical research. The only test against empirical data seems to have been that of Fellner. Fellner interpreted his results as indicating that, except during periods of very rapid increase in rising capital intensity, and hence rapidly rising demand for labor, the induced labor-saving bias was sufficient to prevent the labor share from rising (Fellner 1961).

In contrast, the microeconomic version of factor-induced technical change has stimulated a wide body of applied research. The first formal test based directly on microeconomic foundations was the Hayami–Ruttan test against the historical experience of agricultural development in the United States and Japan (Hayami and Ruttan 1970). It seemed apparent that neither the enormous differences in land/labor ratios between the two countries nor the changes in each country over time could be explained by simple factor substitution. Hayami and Ruttan employed a four-factor model in which (1) land and mechanical power were regarded as complements and land and labor as substitutes, and (2) fertilizer and land infrastructure were regarded as complements and fertilizer and land as substitutes.

The processes of advance in mechanical technology in the Hayami–Ruttan model are illustrated in Figure 2.2a. i_0^* represents the IPC at time zero; it is the envelope of less elastic unit isoquants that correspond, for example, to different types of harvesting machinery. The relationship between land and power is complementary. Land-cum-power is substituted for labor in response to a change in the wage rate relative to an index of land and power prices. The change in the price ratio from BB to CC induces the invention of labor-saving machinery—say, a combine for a reaper.

The process of advance in biological technology is illustrated in Figure 2.2b. Here, i_0^* represents an IPC that is an envelope of relatively inelastic land-fertilizer isoquants such as L_0. When the fertilizer/land price ratio declines from bb to cc, a new technology—a more fertilizer-responsive crop variety—represented by cc is developed along i_0^*. Since the substitution of fertilizer for land is facilitated by investment in land and water development, the relationship between new fertilizer-responsive varieties and land infrastructure is complementary.

In Figure 2.2 the impact of advances in mechanical and biological technology on factor ratios is treated as if the advances were completely separable. This is clearly an oversimplification. It is not essential to the Hayami–Ruttan induced technical change model that changes in the land/labor ratio be a direct response to the price of land relative to the wage rate (Thirtle and Ruttan 1987:30, 31).

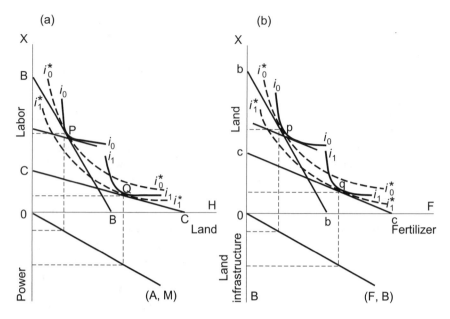

Figure 2.2. Induced Technical Change in Agriculture.
Source: Hayami and Ruttan (1985:91).

The econometric tests conducted by Hayami and Ruttan suggested that the enormous changes in factor proportions that occurred during the process of agricultural development in the two countries represent "a process of dynamic factor substitutions accompanying changes in the production function induced by changes in relative factor prices" (Hayami and Ruttan 1970:1135).

The initial Hayami–Ruttan article and the further exposition in their book *Agricultural Development* (1971) became the inspiration for a large number of empirical tests of the microeconomic version of the induced technical change hypothesis in the agricultural and natural resource sectors. Binswanger advanced the methodology for measuring technical change bias with many factors of production (1974a, 1974b). In a 1987 literature review, Thirtle and Ruttan (1987) listed 29 empirical studies of induced technical change in agriculture. Most of the studies drew their inspiration from the initial study by Hayami and Ruttan (1970). Thirtle and Ruttan also listed 38 empirical studies in the industrial sector. The initial studies of biased technical progress change in industry typically did not involve direct tests of the induced technical change hypotheses. By the late 1970s and early 1980s, however, a substantial number of studies, some stimulated by the rise in energy prices in the 1970s, involved direct tests of the induced technical change hypotheses. Within the industrial sector the evidence is strongest in those industries using natural resources and raw materials (Jorgenson and Fraumeni 1981; Wright 1990; Jorgenson and Wilcoxen 1993). As of the mid-1980s, the evidence of tests of the induced technical change hypotheses in agriculture, both in the United States and abroad, was sufficient to support the view that changes (and sometimes differences) in relative factor endowments and prices exert a substantial impact on the direction of technical change.[5]

2.6 Evolutionary Theory

The modern revival of interest by economists in an evolutionary theory of technical change derives largely from a series of articles by Nelson and Winter in the mid-1970s (Nelson and Winter 1973, 1974, 1975, 1977; Nelson *et al*. 1976).[6] These articles in turn served as the basis for the highly acclaimed book *An Evolutionary Theory of Economic Change* (Nelson and Winter 1982). The theory advanced by Nelson and Winter has been identified by the authors as "Schumpeterian" in its interpretation of the process of economic change. In much of the literature that has drawn its inspiration from Nelson and Winter, "evolutionary" and "Schumpeterian" have been used interchangeably.[7] The second cornerstone of the Nelson–Winter model is the behavioral theory of the firm, in which profit-maximizing behavior is replaced by decision rules that are applied routinely over an extended period of time (Simon 1955, 1959; Cyert and March 1963).

The Nelson–Winter evolutionary model, particularly Chapters 9–11, jettisons much of what they consider to be the excess baggage of the neoclassical microeconomic theory: "the global objective function, the well-defined choice set, and the maximizing choice rationalization of firm's actions. And we see 'decision rules' as very close conceptual relatives of production 'techniques' whereas orthodoxy sees these things as very different" (Nelson and Winter 1982:14). The production function and all other regular and predictable behavior patterns of the firm are replaced by the concept of "routine": "... a term that includes characteristics that range from well-specified technical routines for producing things, procedures for hiring and firing, ordering new inventory, or stepping up production of items in high demand to policies regarding investment, research and development (R&D), or advertising, and business strategies about product diversification and overseas investment" (Nelson and Winter 1982:14). The distinction between factor substitution and shifts in the production function is also abandoned. The two fundamental mechanisms in the Nelson–Winter model are the *search* for better techniques and the *selection* of firms by the market (Elster 1983:14). In their models the microeconomics of innovation is represented as "a stochastic process dependent on the search routines of individual firms" (Dosi *et al*. 1992:10). The activities leading to technical changes are characterized by (1) local search for technical innovations, (2) imitation of the practices of other firms, and (3) satisfying economic behavior.

In their initial models, a firm's search for new technology—whether generated internally by R&D or transferred from suppliers or competitors—is set in motion when profits fall below a certain threshold. The models assume that in this search the firms draw samples from a distribution of input–output coefficients (Figure 2.3). If A is the present input combination, then potential input coefficients are distributed around it such that there is a much greater probability

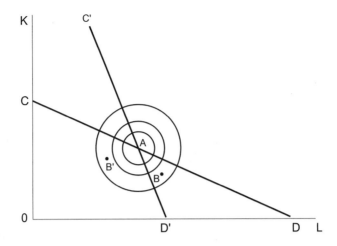

Figure 2.3. Sampling and Selection of New Input–Output Coefficients.
Adapted from Nelson and Winter (1975:472).

of finding a point close to A than finding one far from it. The search is local. Once
the firm finds a point B it makes a profitability check. If costs are lower at B than
at A, the firm adopts the technology represented by point B and stops searching.
Otherwise, the search continues. Thus, the technology described by the point B,
input–output and factor ratios will be accepted if labor is relatively inexpensive,
that is, if relative prices are described by line CD. However, if labor is relatively
expensive, as described by line C'D', the firm will reject the technology at point
B and continue to search for another technology until it finds another point, say
B'. The technology at point B' will be labor saving relative to that at B.

The stochastic technology search process is built into a model with many
competing firms. All profits above a "normal" dividend—investors are satisfi-
cers rather than optimizers—are reinvested so that successful firms grow faster
than unsuccessful ones. The capital stock of the economy is determined by the
total investment by all firms. Labor supply is elastic to the firm.

Simulation runs rather than formal analysis or tests against historical experience are employed to demonstrate the plausibility of the models. The simulations
start from an initial point where all firms are equal. The model endogenously
determines the output of the economy, the wage-rental rate, and the capital accumulation rates. Nelson and Winter have used a series of variations of their basic
model to explain how changes in market structure influence the rate of technical
change, the direction of technical change, and the importance of imitation and
innovation.

When firms check the profitability of the alternative techniques that their
search processes uncover, a higher wage rate will cause certain techniques to fail
profitability tests they would have passed at a lower wage rate and enable others to pass tests they would have failed at a lower wage rate. The latter will be

capital intensive relative to the former. Thus a higher wage rate nudges firms in a direction that is more capital intensive than that in which they would have gone. Also, the effect of a higher wage rate is to make all technologies less profitable (assuming, as in their model, a constant cost of capital), but the cost increase is proportionately greatest for those that involve a low capital/labor ratio. Since firms with high capital/labor ratios are less adversely affected by high wage rates then those with low capital/labor ratios, capital-intensive firms will tend to expand relative to labor-intensive ones. For both these reasons, a higher wage rate will tend to increase capital intensity relative to what otherwise would have been obtained (Nelson and Winter 1974:900). The responsiveness of the capital/labor ratio to changes in relative factor prices is rather striking because, except for the profitability check, search (or research) outcomes are random (Nelson and Winter 1982:175–184), and the inducement mechanism comes about through competition, survival, and growth rather than through efforts to maximize profits.

The early Nelson–Winter models were criticized for the "dumb manager" assumption in which the search (or research) process is triggered only when profits fall below a threshold level. For example, "here we assume that firms with positive capacity do not search if they are making positive or zero profits; they satisfice on their prevailing routines" (Nelson and Winter 1982:149). An implication is that an increase in demand for an industry's product can lead to a reduction in research effort. This was hardly consistent with either historical evidence (Schmookler 1966) or with the Schumpeterian perspective. The restriction was relaxed in the second round of Nelson–Winter models by the explicit introduction of directed research. As the wage/rental ratio rises, research effort is allocated, as in the microeconomic induced technical change model, to sampling the spectrum of potential capital-intensive techniques (Nelson and Winter 1975, 1977).

Winter has devoted considerable attention to extensions of the initial Nelson–Winter models. In a 1984 article, for example, he abandons the assumption of the level playing field in which the initial conditions are the same for all firms. The basic model is augmented by a model that includes entirely new firms. Winter uses this expanded model to explore the growth path of two industrial regimes. One is an "entrepreneurial regime," which he identifies with the early Schumpeter of *The Theory of Economic Development* (1934; originally published in German, 1911). The second is a "routinized regime," which he identifies with the Schumpeter of *Capitalism, Socialism and Democracy* (1950). The entrepreneurial regime model is designed so that innovations are primarily associated with the entry of new firms. In the routinized regime, innovations are primarily the result of internal R&D by established firms. Several suggestions for further extension of the Nelson–Winter models to include the creation of new industries, interaction among industries, and product innovation and imitation, for example, have been summarized and extended by Andersen (1994:118–131).

It is important to clarify the role of historical process in the Nelson–Winter evolutionary models. The condition of the industry in each time period shapes its condition in the following period:

Some economic processes are conceived as working very fast, driving some of the model variables to (temporary) equilibrium values within a single period (or in a continuous time model, instantaneously). In both the entrepreneurial and routinized Schumpeterian models, for example, a short-run equilibrium price of output is established in every time period. Slower working processes of investment and of technological and organizational change operate to modify the data of the short-run equilibrium system from period to period (or from instant to instant). The directions taken by these slower processes of change are directly influenced by the values taken by the subset of variables that are equilibrated in the individual period or instant. (Winter 1984:290)

I find it difficult to resolve the question of why there have been so few efforts by other scholars to advance the Nelson–Winter methodology,[8] or to test the correspondence between the plausible results of the Nelson–Winter simulations and the historical experience of particular firms or industries.[9] Simulation is capable of generating a wide range of plausible behavior. But the hypotheses generated by the simulations have seldom been subjected to rigorous empirical tests. The closest Nelson and Winter come to empirical testing is the demonstration that it is possible to generate plausible economywide growth paths or changes in market share.

2.7   Path Dependence

The argument that technical change is "path dependent" was vigorously advanced by Arthur and several colleagues in the late 1970s and early 1980s (Arthur 1989, 1990, 1994; Arthur et al. 1987).[10] In the mid- and late 1980s, David presented the results of a series of historical studies—of the typewriter keyboard, the electric light and power supply industries, and others—that served to buttress the plausibility of the path-dependence perspective (David 1985, 1986, 1993; David and Bunn 1988). As noted previously, the emphasis on path dependence in David's more recent work represents an extension of his earlier research on the relationship between labor scarcity and modernization in nineteenth-century America (David 1975). This earlier work was strongly influenced by Arrow's article on learning by doing (Arrow 1962) and by Habakkuk's historical research on British and American technology in the nineteenth century (Habakkuk 1962).

The work by Arthur and his colleagues has emphasized the importance of increasing returns to scale as a source of technological "lock-in." In some nonlinear dynamic systems, positive feedbacks (Polya processes) may cause certain patterns or structures that emerge to be self-reinforcing. Such systems tend to be sensitive to early dynamical fluctuations (Figure 2.4). Often there is a multiplicity of patterns that are candidates for long-term self-reinforcement; the accumulation of small events early on "pushes" the dynamics of technical choice into the orbit of one of these and thus "selects" the structure that the system eventually "locks into" (Arthur et al. 1987:294).[11]

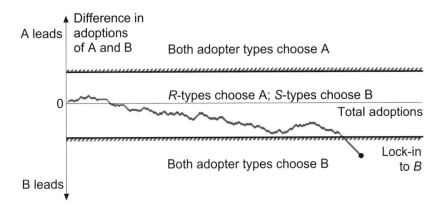

Figure 2.4. Increasing Returns Adoption: A Random Walk with Absorbing Barriers.
Source: Arthur (1989:120).

The authors provide an intuitive example. Think of an urn of an infinite capacity:

> Starting with one red and one white ball in the urn, add a ball each time, indefinitely, according to the following rule. Choose a ball in the urn at random and replace it; if it is red, add a red; if it is white, add a white. Obviously this process has increments that are path dependent—at any time the probabilities that the next ball added is red exactly equals the proportion red. ... Polya proved in 1931 that in a scheme like this the proportion of red balls does tend to a limit X_1 and with probability one. But X is a random variable uniformly distributed between 0 and 1. (Arthur *et al.* 1987:259)

Thus in an industry characterized by increasing returns, small historical or chance events that give one of several technologies an initial advantage can (but need not) "drive the adoption process into developing a technology that has inferior long-run potential" (Arthur 1989:117).

The historical small events that result in path dependence are "outside the *ex ante* knowledge of the observer—beyond the resolving power of his model or abstraction of the situation" (Arthur 1989:118). Arthur employs a series of progressively complex models to simulate situations in which several technologies compete for adoption by a large number of economic agents. Agents have full knowledge of the technology and returns functions but not of the events that determine entry and choice of technology by other agents. His analysis is carried out for three technological regimes (constant, increasing, and diminishing returns) with respect to four properties of the paths of technical change (predictable, flexible, ergodic, and path efficient).[12] The only unknown is the set of historical events that determine the sequence in which the agents make their choices. The question he attempts to answer is whether the fluctuations in the order of choice will make a difference in final adoption shares. Arthur's simulations emphasize the importance of increasing returns as a necessary condition for technological lock-in:

> Under constant and diminishing returns the evolution of the market reflects only *a priori* endowments, preferences, and transformation possibilities; small events cannot sway the outcome. ... Under increasing returns, by contrast, many outcomes are possible. Insignificant circumstances become magnified by positive feedbacks to "tip" the system into the actual outcome "selected." The small events in history become important. ... (Arthur 1989:127)

The network externalities are important not only because of their impact on the direction or path of technology development, but also because they represent a source of market failure—welfare losses that cannot be resolved by normal market processes—and hence call for public intervention (Arthur 1994:9–10). Other factors that may contribute to lock-in include adjustment costs, switching costs, and the costs of maintaining parallel technologies (David 2001).

In *Technical Choice, Innovation and Economic Growth*, David (1975) characterizes his work as an evolutionary alternative to neoclassical theory. As noted previously, he explicitly rejected the Fellner and Kennedy versions of the induced technical change approach to the analysis of factor bias. He also rejected the early work of Nelson and Winter as being "fundamentally neoclassical-inspired" (David 1975:76).[13] But he shares the Nelson and Winter view that the neoclassical model is excessively restrictive, as factor substitution typically involves not simply a movement along a given production function but an element of innovation leading to a shift in the function itself. He does assume that the firm has knowledge of available (or potentially available) alternative technologies and chooses rationally from among them.

David's early analysis of factor bias was remarkably similar to the Hicks–Ahmad–Hayami–Ruttan interpretation of the process of induced technical change. And, in spite of his assertion in *Technical Choice* (1975) that the future development of the system depends not only on the present state but also on the way the present state evolved, I find his research on path dependence in the 1980s to be a distinct departure from his research on factor bias in the 1970s.

In his research from the mid- and late 1980s, David employs historical analysis of a series of technical changes—the typewriter keyboard, the electric light, and power supply industries—to buttress the plausibility of the path-dependence perspective. His now classic paper on the economics of QWERTY (the first six letters on the left side, topmost row of the typewriter and now the computer keyboard) explored why an inefficient (from today's perspective) typewriter keyboard was introduced and why it has persisted.[14] David's answer is that an innovation in typing method, touch typing, gave rise to three features "which were crucially important in causing QWERTY to become 'locked in' as the dominant keyboard arrangement. These features were *technical inter-relatedness, economies of scale*, and *quasi-irreversibility* of investment" (David 1985:334). Technical interrelatedness refers to the need for system compatibility—in this case, the linkage between the design of the typewriter keyboard and typists' memory of a particular keyboard arrangement. Scale economics refers to the decline in user cost of the QWERTY system (or any other system) as it gained acceptance relative to other systems. The quasi-irreversibility of investments is the result of the acquisition of specific touch-typing skills (the "software"). These characteristics are sometimes bundled under the rubric of positive "network externalities."

As David has drawn increasingly on Arthur's path-dependence model, his research has moved even further in the direction of interpreting the QWERTY-like phenomenon in dynamic systems characterized by network externalities and path-dependent technical change as a dominant paradigm for the history of technology (David 1993:208–231).[15] This paradigm would seem particularly apt at a time when the impact of scale economies on productivity growth has been rediscovered and embodied in a "new growth economics" literature (Romer 1986; Lucas 1988; Barro and Sala-i-Martin 1995).[16] But Arthur's results suggest some caution is necessary: "Increasing returns, if they are bounded, are in general not sufficient to guarantee eventual monopoly by a single technology" (Arthur 1989:126). And there is substantial empirical evidence that scale economies, which often depend on prior technical change, are typically bounded by the state of technology (Levin 1977:208–221).

Both induced innovation and evolutionary theory suggest that as scale economies are exhausted (and profits decline) the pressure of growth in demand will focus scientific and technical effort on breaking the new technological barriers. Superior technologies that lost out as a result of chance events in the first round of technical competition have frequently turned out to be successful as the industry developed.[17] And induced technical change theory suggests that research effort will be directed to removing the constraints on growth resulting from technological constraints or inelastic (or scarce) factor supplies.[18]

The transition from coal to petroleum-based feedstocks in the heavy organic chemical industry is a particularly dramatic example. From the 1870s through the 1930s, German leadership in the organic chemical industry was based on coal-based technology. Beginning in the 1920s with the rapid growth in demand for gasoline for automobiles and trucks in the United States, a large and inexpensive supply of olefins became available as a by-product of petroleum refining. By the end of World War II, the US chemical industry had shifted rapidly to petroleum-based feedstocks. In Germany this transition—impeded by skills, education, and attitudes that had been developed under a coal-based industrial regime—was delayed by more than a decade. By the 1960s, however, Germany was making a rapid transition to the petroleum-based feedstock path of technical change in heavy organic chemicals (Grant *et al.* 1988; Stokes 1994).

2.8   Toward a More General Theory?

In this section I first summarize my assessment of the strengths and limitations of each of the three models of technical innovation. I then outline the elements of a more general theory. I would like to make clear to the reader my particular historical and epistemological bias: Departures from neoclassical microeconomic theory, when successful, are eventually seen as extensions and become incorporated into neoclassical theory. Thus, for example, the microeconomic version of induced technical change can now be viewed as an extension of, rather than a departure from, the neoclassical theory of the firm.[19]

2.8.1   Assessment

One common theme pervading the three approaches to understanding sources of technical change is the disagreement with the assumption in neoclassical growth models that a common production function is available to all countries regardless of human capital, resource, or institutional endowments. It should now be obvious that differences in productivity levels and rates of growth cannot be overcome by the simple transfer of capital and technology. The asymmetries between firms and between countries in resource endowments and in scientific and technological capabilities are not easily overcome. The technologies that are capable of becoming the most productive sources of growth are often location specific (Kenney and von Burg 2001). A second common theme is an emphasis on microfoundations. This emphasis on microfoundations is common to the approaches that have abandoned neoclassical microeconomics as well as to those that have attempted to extend neoclassical theory (Dosi 1995). This stands in sharp contrast to the limited attention to microfoundations in neoclassical growth theory (Ruttan 1998, 2001).

The primary limitation of the growth theoretic version of the induced innovation model is the implausibility of the innovation-possibility function. The shape of the function is independent of the bias in the path of technical change. As technical change progresses, there is no effect on the "fundamental" trade-off between labor- and capital-augmenting technical change. Thus, as Nordhaus notes, the growth theoretic approach to induced innovation fails to rescue growth theory from treating technical change as exogenous. It has not produced empirical research and is no longer viewed as an important contribution to growth theory.

The primary limitation of the microeconomic version is that its internal mechanism—the learning, search, and formal R&D processes—remains inside a black box. The model is driven by exogenous changes in the economic environment in which the firm (or public research agency) finds itself. But the process of technical change is, itself, not entirely endogenous. Exogenous advances in scientific and technological knowledge can, for example, open up new opportunities for technical change. The microeconomic model has, nevertheless, produced a substantial body of empirical research and has helped to clarify the historical process of technical change, particularly at the industry and sector levels both within and across countries.

The strength of the evolutionary model lies precisely in the area where the microeconomic induced innovation model is weakest. It builds on the behavioral theory of the firm in an attempt to provide a more realistic description of the internal workings of the black box. The Nelson–Winter evolutionary approach has not, however, become a productive source of empirical research. The results of the various simulations are defended as plausible in terms of the stylized facts of industrial organization and of firm, sector, and macroeconomic growth. It is possible that the reason for the lack of empirical testing is that the simulation methodology lends itself to the easy proliferation of plausible results. At present, the evolutionary approach must be regarded as a "point of view" rather than a theory (Arrow 1995).

The strength of the path-dependent model lies in the insistence of its practitioners on the importance of the sequence of specific micro-level historical events.[20] In this view, current choices of techniques become the link through which prevailing economic conditions may influence the future dimensions of technology and knowledge (David 1975:39, 57). However, the concept of technological lock-in, at least in the hands of its more rigorous practitioners, applies only to network technologies characterized by increasing returns to scale. In industries with constant or decreasing returns to scale, historical lock-in does not apply.

There can be no question that technical change is path dependent in the sense that it evolves from earlier technological development. In spite of somewhat similar motivation, the path-dependent literature has not consciously drawn on the Nelson–Winter work for inspiration (Arthur 1996). It is necessary to go beyond the present path-dependent models, however, to examine the forces responsible for changes in the rate and direction of technical change. But there is little discussion of how firms or industries escape from lock-in. What happens when the scale economies resulting from an earlier change in technology have been exhausted and the industry enters a constant- or decreasing-returns stage? At this point in time it seems apparent that changes in relative factor prices would, with some lag, have the effect of bending or biasing the path of technical change along the lines suggested by the theory of induced technical change. Similarly, a new radical innovation may, at this stage, both increase the rate and modify the direction of technical change.

The study of technical change in the semiconductor industry by Dosi (1984) represents a useful illustration of the potential value of a more general model. The Dosi study is particularly rich in its depth of technical insight. At a rhetorical level, Dosi identifies his methodology with the Nelson–Winter evolutionary approach. In practice, however, he utilizes an eclectic combination of induced innovation, evolutionary, and path-dependence interpretations of the process of semi-conductor technology development. A more rigorous approach to the development of a general theory of the sources of technical change will be required to bridge the three "island empires."

2.9   Integration of Models

A first step toward developing a more general theory of technical change is to integrate the "factor-induced" and the "demand-induced" models (Ruttan and Hayami 1994:180). Binswanger, drawing on Nordhaus (1969:105–109) and Kamien and Schwartz (1969), has sketched outlines of how a more general model can make both the rate and direction of technical change endogenous (Binswanger 1978b:104–110).

If one assumes decreasing marginal productivity of research resources in applied research and technology development and, in addition, incorporates the effects of changes in product demand, then growth (decline) in product demand would increase (decrease) the optimum level of search and research expenditure. The larger research budget, induced by growth in product demand, increases the rate at which the meta-production function shifts inward toward the origin. Even when the initial path of technological development is generated by "technology push," factor market forces often act to modify the path of technical change. Differential elasticities of factor supply result in changes in relative factor prices and direct research effort to save increasingly scarce factor supplies. The result is a non-neutral shift in both the neoclassical and the meta-production functions.[21] More recently, Christian has elaborated the Binswanger model and analyzed more formally the innovators' decision to conduct R&D directed toward process innovation (Christian 1993). As yet, however, there has been no attempt to implement empirically an integrated factor- and demand-induced innovation model.

A second step would be the integration of the induced technical change and the path-dependent models. As noted earlier, David has pointed to the persistent failure to replace the inefficient QWERTY layout of the typewriter and computer keyboards with the more efficient DSK keyboard. Wright (1990) has suggested that the historical resource intensity of American industry, based on domestic resource abundance, has been an important factor in weakening the capacity of American industry to adapt to a world in which lower transportation costs and more open trading systems have reduced the traditional advantage of US-based firms. If this perspective is correct, Japan's industrial success may be attributed to its historical resource scarcity.

The difference in perspective seems to hinge on how the elasticity of substitution changes over time in response to changes in resource endowments or relative factor prices. David has shown how localized induced technical change can lead to path-dependent technical change (1975:65–68). The effect of localization is to lower the elasticity of substitution and lock in the trajectory of technical change. As relative factor prices continue to shift, however, it is difficult to believe technological competition would not result in a "bending" of the path of technical change in the direction implied by changing factor endowments. The path-dependent and the induced innovation models are appropriately viewed as complementary rather than as alternative interpretations of the forces that influence the direction of technical change.

The path-dependent model will remain incomplete, however, until it is more fully integrated with the microeconomic version of the induced technical change model and with the Nelson–Winter evolutionary model. Development of an industry seldom proceeds indefinitely along an initially selected process ray (Landes 1994). As technical progress slows down or scale economies erode, a shift in relative factor prices can be expected to induce an intensified search for technologies along a ray that is more consistent with contemporary factor prices.[22] At the theoretical level, research should be directed toward the development of a succession of more fully integrated models of the sources of technical change. At the empirical level, research should be directed toward more comprehensive testing of the induced, evolutionary, and path-dependent models. There is a substantial empirical literature on induced technical change in agricultural economics and economic history, but only limited efforts have been made to test the evolutionary and path-dependent models against historical experience. Metaphor is not enough!

2.10 Induced Technical Change and Endogenous Economic Growth

Since the late 1980s, students of economic growth have been engaged in a re-evaluation of neoclassical growth models. The re-examination was stimulated by concern that the neoclassical growth theory was inconsistent with the evidence of a lack of convergence of growth rates among rich and poor countries (Baumol 1986; Dollar and Wolff 1993; Ruttan 1998). One result of this re-examination has been the emergence of a new generation of endogenous growth models.

The primary focus of the new "macroeconomic endogenous" growth models is to attribute differences in growth performance among countries to factors such as investment in human capital, learning by doing, scale economies, and technical change (Romer 1986, 1994; Lucas 1988; Ruttan 1998). In the initial Romer–Lucas framework, the accumulation of human capital adds to the productivity of the person in whom it is embodied.[23] But the general level of productivity rises by more than can be accounted for or captured by the person or firm that makes each particular investment. Gains in scale economies are enhanced by the integration into multinational trading systems of economies that are human-capital intensive (Grossman and Helpman 1992).

There has also been a renewed interest in the theory of induced technical change. This renewed interest is reflected in the subsequent chapters of this book. It has been stimulated, at least in part, by the focus on endogenous technical change in the new growth literature. It has also been stimulated by a growing concern with a series of natural resource issues such as the effects of global climate change discussed in this book. The new endogenous growth literature has yet, however, to incorporate the richness and depth of understanding of the sources of technical change that the three traditions reviewed in this chapter have achieved (Bardhan 1995; Ruttan 1998). Like the older neoclassical growth literature, its focus is on the proximate sources of growth rather than the sources of technical change. A major challenge for the future is to integrate the insights about endogenous growth gained from the theoretical and empirical research conducted within the induced technical change, the evolutionary, and the path-dependence theories with new insights into the relationship between human capital, scale, and trade opened up by the macro-endogenous growth models.

Acknowledgments

The author is indebted to Esben Sloth-Anderson, W. Brian Arthur, Erhard Bruderer, Jason E. Christian, Paul A. David, Jerry Donato, Giovanni Dosi, Laura McCann, Richard Nelson, Nathan Rosenberg, Tugrul Temel, Michael A. Trueblood, Andrew Van de Ven, and Sidney Winter for comments on an earlier draft of this paper. Earlier versions of this paper have been presented in seminars at the International Institute for Applied Systems Analysis (IIASA), at the University of Minnesota Economic Development Center, and at the Hong Kong University of Science and Technology. The research on which the paper is based was supported, in part, by a grant from the Alfred P. Sloan Foundation. Earlier versions of this paper have appeared in Ruttan (1996, 1997, 2001). For critical reviews, see Dosi (1997) and Wright (1997).

Notes

1. Zvi Griliches pointed out to me (in conversation) that another reason for the decline in interest among economic theorists was the difficulty, noted by Diamond *et al.* (1978:125–147), of simultaneously measuring the bias of technical change and the elasticity of substitution among factors. This problem had, however, already been solved (Binswanger 1974c; Binswanger and Ruttan 1978:73–80, 215–242). For a more recent discussion see Haltmaier (1986).

2. It is interesting to speculate on what the course of induced innovation theory might have been if the Ahmad article had, as it might have, appeared first. The initial drafts of the articles were written while Kennedy was teaching at the University of the West Indies (Kingston) and Ahmad was teaching at the University of Khartoum (Sudan). Ahmad submitted his article to the *Economic Journal* in 1963. Kennedy served as a reviewer of the Ahmad article. His article, which was published in 1964, was originally written as a comment on the Ahmad article. Ahmad's article was rewritten, resubmitted, and published in 1966 (Ruttan and Hayami 1994:24).

3. Binswanger authored a series of articles in the mid-1970s that were fundamental in the formalization of the theoretical foundations of the microeconomic model (1974a, 1974b). The material in these two articles is incorporated in Binswanger and Ruttan (1978). See also Kamien and Schwartz (1969, 1971).

4. At the time the article was written, Hayami and Ruttan were familiar with the growth theoretic literature by Fellner, Kennedy, and Samuelson, but not with the Ahmad article and his subsequent exchange with Fellner and Kennedy. The inspiration for the 1970 Hayami–Ruttan paper was the historical observations about the development of British and American technology by Habakkuk (1962). See Ruttan and Hayami (1994).

5. Olmstead and Rhode (1993) have criticized the Hayami and Ruttan work on both conceptual and empirical grounds. At the conceptual level they find confusion between the relative factor "change variant" used in explaining productivity growth over time within a given country and the "level variant" of the model used in analysis of international productivity differences. They also argue, on the basis of regional tests in the United States, that the induced technical change model holds only for the central grain-growing regions. In a later paper (Olmstead and Rhode 1995) using state-level data, they found somewhat stronger support for the induced technical change hypothesis. For further criticism and a defense, see Koppel (1995).

6. Nelson and Winter identify Alchian (1950) and Penrose (1952) as representing direct intellectual antecedents of their work. For the theoretical foundations of the Nelson–Winter collaboration, see Winter (1971). For the historical and philosophical foundations, see Elster (1983:131–158) and Langolis and Everett (1994:11–47). Witt has assembled many of the most important articles in the field of evolutionary economics in a collection of readings, *Evolutionary Economics* (1993). For a review of recent evolutionary thought about economic change, see Nelson (1995:48–90).

7. The Nelson–Winter model departs in its treatment of the linkage between invention and innovation. For Schumpeter there was no necessary link between invention and innovation (Ruttan 1959). Nelson and Winter employ the term "evolutionary" metaphorically: "We emphatically disavow any intention to pursue biological analogies for their own sake" (1982:11). Nelson and Winter regard their approach as being closer to Lamankianism than Mendelianism. Yet their description of the evolutionary process of firm behavior and technical change as a Markov process and their use of the Markov mechanism in their simulation are analogous to the Mendelian model.

8. For a useful interpretation and extension, see Andersen (1994). Andersen's work is particularly helpful in clarifying the "poorly documented" computational steps of the Nelson–Winter models. Andersen supplements the mathematical notation employed by Nelson and Winter with an algorithmically oriented programming notation. An appendix, "Algorithmic Nelson and Winter Models" (pp. 198–219), is particularly useful.

9. A large body of empirical research on technical change that can be categorized as broadly Schumpeterian or evolutionary in inspiration has emerged since the mid-1970s [see the review by Freeman (1994)]. The point I am making, however, is quite different. There has been very little effort to use the simulation models to generate hypotheses about the process of technology development and then to either identify historical counterparts or test the outcomes against historical experience in a rigorous manner. The one exception with which I am familiar is the Evenson–Kislev (1975:140–155) stochastic model of technological discovery. The model was used to interpret the stages in sugarcane varietal development. The Evenson–Kislev model did not, however, draw directly on the Nelson–Winter stochastic model.

10. Arthur encountered unusual delay before his work was accepted in a leading economics journal. His 1986 *Economic Journal* paper was initially submitted to the *American Economic Review* in 1983. It was rejected by the *American Economic Review* twice and by the *Quarterly Journal of Economics* twice and accepted by the *Economic Journal* only after an appeal. By the time the paper was finally accepted in the *Economic Journal*, referees were noting that the path-dependence idea was already recognized in the literature (Gans and Sheperd 1994:173).

11. There has been considerable confusion regarding the interpretations of path dependence and the meaning of "lock-in" in the literature. David has recently provided explicit definitions: "A path dependent stochastic process is one whose asymptotic distribution evolves as a consequence (function of) the process's own history" (David 2001:5). "The term lock-in is a vivid way to describe the entry of a system into a ... region ... that surrounds a locally (or globally) stable equilibrium. When a dynamic system enters such a region it cannot escape except through the intervention of external force, or shock" (David 2001:10).

12. "A process is *predictable* if the small degree of uncertainty built in 'averages away' so that the observer has enough knowledge to predetermine market shares accurately in the long run; *flexible* if a subsidy or tax adjustment to one of the technologies' returns can always influence future market choice; *ergodic* (not path dependent) if different sequences

of historical events lead to the same market outcome with probability one; ... and *path efficient* if at all times equal development (equal adoption) of the technology that is behind in adoption would not have paid off better" (Arthur 1989:118, 199).

13.   For a further comparison of the David and Nelson–Winter evolutionary approaches, see Elster (1983:150–157). Elster notes that David regards the Nelson–Winter model as evolutionary, but ahistorical. In his view, it differs from the neoclassical model only in its conception of microeconomic behavior. It is ahistorical since, in David's view, the Markovian-like transition probabilities depend only on the current state and not on earlier states of the system. Elster rejects David's criticism of Nelson and Winter on the basis that, for the past to have a causal influence on the present, it must be "mediated by a chain of locally causal links." Thus, since all the history that is relevant to the prediction of the future is contained in the state description, if the present state is known, prediction cannot be improved by considering the past history of the system (Elster 1983:157).

14.   Liebowitz and Margolis (1990, 1994) argue that David's version of the history of the market's rejection of the supposedly more efficient Dvorak keyboard represents bad history. Given the available knowledge and experience at the time QWERTY became dominant, it represented a rational choice of technology. For a vigorous response to the Liebowitz and Margolis criticism, see David (2001).

15.   Both Arthur and David emphasize the role of network externalities in locking in inferior technological trajectories. In their review of the creation and growth of Silicon Valley, Kenney and von Burg note that while the evolution of Silicon Valley venture capital has a path-dependent history, "it is difficult to imagine a more efficient system for formulating high technology startups" (2001:144).

16.   Scale economics have become the new "black box" of contemporary growth theory. It is difficult to believe that much of the productivity growth that is presumably accounted for by scale economies is not the disequilibrium effect of prior technical change (Landau and Rosenberg 1992:93; Liebowitz and Margolis 1994:139).

17.   See, for example, the exceedingly careful study of technological substitution in the case of Cochlear implants by Van de Ven and Garud (1993) and Garud and Rappa (1994). The Cochlear implant is a biomedical invention that enables hearing by profoundly deaf people. The industry is characterized by the conditions that David and Arthur identify with technological lock-in. Yet in spite of initial commercial dominance, the "single-channel" technology was completely replaced by the "multiple-channel" technology. For other cases, see Foray and Grübler (1990), Cheng and Van de Ven (1994), and Liebowitz and Margolis (1992, 1995).

18.   The development of semiconductor technology as a replacement for vacuum tubes for amplifying, rectifying, and modulating electrical signals is an example of a shift in technological trajectories induced by technological constraints (Dosi 1984:26–45). The development of fertilizer-responsive crop varieties represents an example of a shift in technological trajectories induced by changes in resource endowments (Hayami and Ruttan 1985:163–198). The gas turbine's emergence from a niche technology to become an important source of electric power generation since the early 1980s was induced, in part, by the exhaustion of scale in steam turbine generation (Islas 1997:49–66).

19.   Nelson and Winter attempt to confront this problem by arguing that there are two alternative views of neoclassical theory. One is the more rigorous "literal" view. The other is termed the "tendency" view. Applied economists with a primary interest in interpreting economic history or behavior tend to employ the "tendency" view, while theorists who are more concerned with the formal properties employ a more literal interpretation. They identify evolutionary theory with the "tendency" view (Nelson and Winter 1975:467).

20. This emphasis creates an important opportunity to incorporate the contributions of noneconomists (particularly historians and other social scientists) along with those of economists into a more comprehensive understanding of the sources of technical change (Kenney and von Burg 2001).

21. In Binswanger, both the production function and the meta-production function are neoclassical. David (1997) noted that if the advances in technology are highly "localized," in the sense suggested by Atkinson and Stiglitz (1969), the neoclassical assumption is inappropriate. See also David (1975:65–68) and Antonelli (1995:1–18).

22. See, for example, the patterns of factor substitution in the transition in primary energy sources and transportation infrastructure (Grübler and Nakicenovic 1988:13–44; Nakicenovic 1991).

23. The initial models are frequently referred to as AK models after the assumed production function ($Y = AK$). In expanded versions of the model, K can be thought of as "a proxy for a composite of capital goods that includes physical and human components" (Barro and Sala-i-Martin 1995:146). In a retrospective assessment, Romer notes, "My interpretation ... was that investments in physical capital tended to be accompanied by investments in new ideas. Looking back ... it has pushed the discussion away from knowledge and ideas and toward a more narrow focus on the marginal productivity of capital" (Romer 1994:558).

References

Ahmad, S., 1966, On the theory of induced innovation, *Economic Journal*, **76**:344–357.

Ahmad, S., 1967a, Reply to Professor Fellner, *Economic Journal*, **77**:662–664.

Ahmad, S., 1967b, A rejoinder to Professor Kennedy, *Economic Journal*, **77**:960–963.

Alchian, A.A., 1950, Uncertainty, evolution and economic theory, *Journal of Political Economy*, **58**:211–222.

Ames, E., and Rosenberg, N., 1968, The Enfield arsenal in theory and history, *Economic Journal*, **78**:730–733.

Andersen, E.S., 1994, *Evolutionary Economics: Post-Schumpeterian Contributions*, Pinter Publishers, London, UK.

Antonelli, C., 1995, *The Economics of Localized Technological Changes and Industrial Dynamics*, Kluwer Academic Publishers, Dordrecht, Netherlands.

Arrow, K., 1962, The economic implications of learning by doing, *Review of Economic Studies*, **29**:155–173.

Arrow, K., 1995, Viewpoint, *Science*, **267**(March 17):1617.

Arthur, W.B., 1989, Competing technologies, increasing returns, and lock-in by historical events, *The Economic Journal*, **99**(March):116–131.

Arthur, W.B., 1990, Positive feedbacks in the economy, *Scientific American*, **262**(February):92–99.

Arthur, W.B., 1994, *Increasing Returns and Path Dependence in the Economy*, The University of Michigan Press, Ann Arbor, MI, USA.

Arthur, W.B., 1996, Letter, October 17.

Arthur, W. B., Ermoliev, Y.M., and Kaniovski, Y.M., 1987, Path dependence processes and the emergence of macro-structure, *European Journal of Operational Research*, **30**(June):294–303.

Atkinson, A.B., and Stiglitz, J.E., 1969, A new view of technological change, *Economic Journal*, **79**(September):573–578.

Bardhan, P., 1995, The contribution of endogenous growth theory to the analysis of development problems: An assessment, in J. Behrman and T.N. Srinivasan, eds, *Handbook of Development Economics*, Vol. III, Elsevier, Amsterdam, Netherlands.

Barro, R.J., and Sala-i-Martin, X., 1995, *Economic Growth*, McGraw Hill, New York, NY, USA.

Baumol, W., 1986, Productivity growth, convergence and welfare: What the long run data show, *American Economic Review*, **76**:1072–1085.

Benzion, U., and Ruttan, V.W., 1975, Money in the production function: An interpretation of empirical results, *Review of Economics and Statistics*, **57**:246–247.

Benzion, U., and Ruttan, V.W., 1978, Aggregate demand and the rate of technical change, in H.P. Binswanger, and V.W. Ruttan, eds, *Induced Innovation: Technology Institutions and Development*, The Johns Hopkins University Press, Baltimore, MD, USA.

Binswanger, H.P., 1973, The Measurement of Biased Efficiency Gains in US and Japanese Agriculture to Test the Induced Innovation Hypothesis, Ph.D. dissertation, North Carolina State University, Raleigh, NC, USA.

Binswanger, H.P., 1974a, A microeconomic approach to induced innovation, *Economic Journal*, **84**:940–958.

Binswanger, H.P., 1974b, A cost function approach to the measurement of elasticities of factor demand and elasticities of substitution, *American Journal of Agricultural Economics*, **56**:377–386.

Binswanger, H.P., 1974c, The measurement of technical change biases with many factors of production, *American Economic Review*, **64**:964–976.

Binswanger, H.P., 1978a, Induced technical change: Evolution of thought, in H.P. Binswanger and V.W. Ruttan, eds, *Induced Innovation: Technology, Institutions and Development*, The Johns Hopkins University Press, Baltimore, MD, USA.

Binswanger, H.P., 1978b, The microeconomics of induced technical change, in H.P. Binswanger and V.W. Ruttan, eds, *Induced Innovation: Technology, Institutions and Development*, The Johns Hopkins University Press, Baltimore, MD, USA.

Binswanger, H.P., and Ruttan, V.W., 1978, *Induced Innovation: Technology, Institutions and Development*, The Johns Hopkins University Press, Baltimore, MD, USA.

Blaug, M., 1963, A survey of the theory of process-innovation, *Economica*, **63**:13–32.

Cheng, Y.-T., and Van de Ven, A.H., 1994, Learning the Innovation Journey: Order out of Chaos?, Discussion Paper 208, October, University of Minnesota Strategic Management Center, Minneapolis, MN, USA.

Christian, J.E., 1993, The Simple Microeconomics of Induced Innovation, Department of Agricultural Economics, University of California, December 23, Davis, CA, USA (mimeo).

Cyert, R.M., and March, J.G., 1963, *A Behavioral Theory of the Firm*, Prentice Hall, Englewood Cliffs, NJ, USA.

David, P.A., 1966, Mechanization of reaping in the antebellum Midwest, in H. Rosovsky, ed., *Industrialization in Two Systems: Essays in Honor of Alexander Gerschenkron*, John Wiley, New York, NY, USA.

David, P.A., 1973, Labor Scarcity and the Problem of Technological Practice and Progress in the Nineteenth Century, Research Paper 297, Harvard Institute of Economic Research, May, Cambridge, MA, USA.

David, P.A., 1975, *Technical Choice, Innovation and Economic Growth*, Cambridge University Press, Cambridge, UK.

David, P.A., 1985, Clio and the economics of QWERTY, *American Economic Review*, **76**:332–337.

David, P.A., 1986, Understanding the economics of QWERTY: The necessity of history, in W.N. Parker, ed., *Economic History and the Modern Economist*, Basil Blackwell, New York, NY, USA.

David, P.A., 1993, Path dependence predictability in dynamic systems with local network externalities: A paradigm for historical economics, in D. Foray and C. Freeman, eds, *Technology and the Wealth of Nations: The Dynamics of Contracted Advantage*, Pinter Publishers, London, UK.

David, P.A., 1997, Letter, October 24.

David, P.A., 2001, Path dependence: Its critics and the quest for historical economics, in P. Garrouste and S. Ioannides, eds, *Evolution and Path Dependence in Economic Ideas: Past and Present*, Edward Elgar Publishing, Cheltenham, UK.

David, P.A., and Bunn, J.A., 1988, The economics of gateway technologies and network evolution: Lessons from electricity supply history, *Information Economics and Policy*, **3**:165–202.

Diamond, P., McFadden, D., and Rodriguez, M., 1978, Measurement of factor substitution and bias of technological change, in M. Fuss and D. McFadden, eds, *Production Economics: A Dual Approach to Theory and Applications*, Vol. 2, North Holland, Amsterdam, Netherlands.

Dollar, D., and Wolff, E.N., 1993, *Competitiveness, Convergence and International Specialization*, The MIT Press, Cambridge, MA, USA.

Dosi, G., 1984, *Technical Change and Industrial Transformation*, St. Martins Press, New York, NY, USA, pp. 26–45.

Dosi, G., 1988, Sources, procedures, and microeconomic effects of innovation, *Journal of Economic Literature*, **26**:1120–1171.

Dosi, G., 1995, Letter, February 15.

Dosi, G., 1997, Opportunities, incentives and the collective patterns of technological change, *Economic Journal*, **107**:1530–1547.

Dosi, G., Giannetti, R., and Toninelli, P.A., eds, 1992, *Technology and Enterprise in a Historical Perspective*, Oxford University Press, Oxford, UK.

Drandakis, E.M., and Phelps, E.S., 1966, A model of induced invention, growth and distribution, *Economic Journal*, **76**:823–840.

Elster, J., 1983, *Explaining Technical Change*, Cambridge University Press, New York, NY, USA.

Evenson, R.E., and Kislev, Y., 1975, *Agricultural Research and Productivity*, Yale University Press, New Haven, CT, USA.

Fellner, W., 1956, *Trends and Cycles in Economic Activity*, Henry Holt, New York, NY, USA.

Fellner, W., 1961, Two propositions in the theory of induced innovations, *Economic Journal*, **71**:305–308.

Fellner, W., 1962, Does the market direct the relative factor saving effects of technological progress?, in R.R. Nelson, ed., *The Rate and Direction of Inventive Activity: Economic and Social Factors*, Princeton University Press for the National Bureau of Economic Research, Princeton, NJ, USA.

Fellner, W., 1967, Comment on the induced bias, *Economic Journal*, **77**:664–665.

Fogel, R.W., 1967, The specification problem in economic history, *Journal of Economic History*, **27**:283–308.

Foray, D., and Grübler, A., 1990, Morphological analysis, diffusion and lock-out of technologies: Ferrous casting in France and the FRG, *Research Policy*, **19**:535–550.

Freeman, C., 1994, The economics of technical change, *Cambridge Journal of Economics*, **18**:463–514.

Gans, J.S., and Shepherd, G.B., 1994, How are the mighty fallen: Rejected classic articles by leading economics, *Journal of Economic Perspectives*, **8**:165–179.

Garud, R., and Rappa, M., 1994, A socio-cognitive model of technology evolution: The case of Cochlear implants, *Organization Science*, **5**:344–362.

Gilfillan, S.C., 1935, *The Sociology of Invention*, Folliett, Chicago, IL, USA.

Grant, W., Patterson, W., and Whitston, C., 1988, *Government and the Chemical Industry: A Comparative Study of Britain and West Germany*, Oxford University Press, Oxford, UK.

Griliches, Z., 1957, Hybrid corn: An exploration in the economics of technological change, *Econometrica*, **25**:501–522.

Grossman, G., and Helpman, E., 1992, *Innovation and Growth in the Global Economy*, The MIT Press, Cambridge, MA, USA.

Grübler, A., and Nakicenovic, N., 1988, The dynamic evolution of methane technologies, in T.H. Lee, H.R. Linden, D.A. Dryfus, and T. Vasco, eds, *The Methane Age*, Kluwer Academic Publishers, Dordrecht, Netherlands.

Habakkuk, H.J., 1962, *American and British Technology in the Nineteenth Century*, Cambridge University Press, Cambridge, UK.

Haltmaier, J., 1986, Induced Innovation and Productivity Growth: An Empirical Analysis, Federal Reserve Board Special Studies Paper 220, February, Washington, DC, USA.

Hayami, Y., and Ruttan, V.W., 1970, Factor prices and technical change in agricultural development: The United States and Japan, 1880–1960, *Journal of Political Economy*, **78** (September–October):1115–1141.

Hayami, Y., and Ruttan, V.W., 1971 and 1985, *Agricultural Development: An International Perspective*, The Johns Hopkins University Press, Baltimore, MD, USA.

Hicks, J., 1932, *The Theory of Wages*, 1st ed., Macmillan, London, UK (2nd ed., 1963).

Islas, J., 1997, Getting round the lock-in in electricity generating systems: The examples of the gas turbine, *Research Policy*, **26**:49–66.

Jorgenson, D., and Fraumeni, B.M., 1981, Relative prices and technical change, in E.R. Bendt and B. Fields, eds, *Modeling and Measuring National Resource Substitution*, MIT Press, Cambridge, MA, USA.

Jorgenson, D., and Wilcoxen, P.J., 1993, Energy, the environment, and economic growth, in A.V. Kneese and J.L. Sweeney, eds, *Handbook of Resource and Energy Economics*, Vol. III, Elsevier Science Publishers, Amsterdam, Netherlands.

Kamien, M., and Schwartz, N., 1969, Optimal induced technical change, *Econometrica*, **36**(1):1–17.

Kamien, M., and Schwartz, N., 1971, The theory of the firm with induced technical change, *Metroeconomica*, **23**:233–256.

Kennedy, C., 1964, Induced bias in innovation and the theory of distribution, *Economic Journal*, **74**:541–547.

Kennedy, C., 1966, Samuelson on induced innovation, *Review of Economics and Statistics*, **48**:442–444.

Kennedy, C., 1967, On the theory of induced innovation—A reply, *Economic Journal*, **77**:958–960.

Kenney, M., and von Burg, U., 2001, Paths and regions: The creation and growth of Silicon Valley, in R. Garud and P. Karnoe, eds, *Path Dependence and Path Creation*, Lawrence Earlbaum and Associates, New York, NY, USA.

Koppel, B.M., ed., 1995, *Induced Innovation Theory and International Agricultural Development*, The Johns Hopkins University Press, Baltimore, MD, USA.

Landau, R., and Rosenberg, N., 1992, Successful commercialization in the chemical process industries, in N. Rosenberg, R. Landau, and D.C. Mowery, eds, *Technology and the Wealth of Nations*, Stanford University Press, Stanford, CA, USA.

Landes, D.S., 1994, What room for accident in history? Explaining big changes by small events, *Economic History Review*, **47**:637–656.

Langolis, R.N., and Everett. M.J., 1994, What is evolutionary economics?, in L. Magnusson, ed., *Evolutionary and Neo-Schumpeterian Approaches to Economics*, Kluwer, Dordrecht, Netherlands.

Levin, R.C., 1977, Technical change and optimal scale: Some evidence and implications, *Southern Economic Journal*, **44**(October):208–221.

Liebowitz, S.J., and Margolis, S.E., 1990, The fable of the keys, *Journal of Law and Economics*, **33**(April):1–325.

Liebowitz, S.J., and Margolis, S.E., 1992, Market Processes and the Selection of Standards, March, Working Paper, University of Texas at Dallas School of Management, Dallas, TX, USA.

Liebowitz, S.J., and Margolis, S.E., 1994, Network externality: An uncommon tragedy, *Journal of Economic Perspectives*, **8**(2):133–150.

Liebowitz, S.J., and Margolis, S.E., 1995, Path dependence, lock-in, and history, *Journal of Law, Economics, and Organization*, **11** (April):205–226.

Lucas, R.E. Jr., 1967, Tests of a capital-theoretic model of technological change, *Review of Economic Studies*, **34**:175–180.

Lucas, R.E. Jr., 1988, On the mechanics of economic development, *Journal of Monetary Economics*, **22**:3–42.

Mowery, D.C., and Rosenberg, N., 1979, The influence of market demand upon innovation: A critical review of some recent empirical studies, *Research Policy*, **8**:103–153.

Nakicenovic, N., 1991, Diffusion of pervasive systems: A case of transport infrastructures, *Technological Forecasting and Social Change*, **39**(March/April):181–219.

Nelson, R.R., 1995, Recent evolutionary theorizing about economic change, *Journal of Economic Literature*, **33**(March): 48–90.

Nelson, R.R., and Winter, S.G., 1973, Toward an evolutionary theory of economic capabilities, *American Economic Review*, **63**:440–449.

Nelson, R.R., and Winter, S.G., 1974, Neoclassical vs. evolutionary theories of economic growth: Critique and prospects, *Economic Journal*, **84**:886–905.

Nelson, R.R., and Winter, S.G., 1975, Factor price changes and factor substitution in an evolutionary model, *Bell Journal of Economics*, **6**:466–486.

Nelson, R.R., and Winter, S.G., 1977, Simulation of Schumpeterian competition, *American Economic Review*, **67**:271–276.

Nelson, R.R., and Winter, S.G., 1982, *An Evolutionary Theory of Economic Change*, Harvard University Press, Cambridge, MA, USA.

Nelson, R.R., Winter, S.G., and Schuette, H.L., 1976, Technical change in an evolutionary model, *Quarterly Journal of Economics*, **40**:90–118.

Nordhaus, W.D., 1969, *Invention, Growth and Welfare: A Theoretical Treatment of Technical Change*, The MIT Press, Cambridge, MA, USA.

Nordhaus, W.D., 1973, Some skeptical thoughts on the theory of induced innovation, *Quarterly Journal of Economics*, **87**:208–219.

Office of the Director of Defense Research and Engineering, 1969, Project Hindsight: Final Report (HINDSIGHT), Washington, DC, USA.

Olmstead, A.L., and Rhode, P., 1993, Induced innovation in American agricultures: A reconsideration, *Journal of Political Economy*, **101**:100–118.

Olmstead, A.L., and Rhode, P., 1995, Induced Innovation in American Agriculture: An Econometric Analysis, Institute of Government Affairs, University of California, Davis, CA, USA (mimeo).

Penrose, E.T., 1952, Biological analogies to the theory of the firm, *American Economic Review*, **42**:804–819.

Romer, P., 1986, Increasing returns and long-run growth, *Journal of Political Economy*, **94**:1002–1037.

Romer, P., 1994, Idea gaps and object gaps in economic development, *Journal of Monetary Economics*, **8**:3–22.

Rosenberg, N., 1974, Science, invention and economic growth, *Economic Journal*, **84**(March):90–108.

Rosenberg, N., 1982, Learning by using, in N. Rosenberg, ed., *Inside the Black Box: Technology and Economics*, Cambridge University Press, Cambridge, UK.

Rothschild, K., 1956, *The Theory of Wages*, Blackwell Publishers, Oxford, UK.

Ruttan, V.W., 1959, Usher and Schumpeter on invention, innovation, and technological change, *Quarterly Journal of Economics*, **73**:596–606.

Ruttan, V.W., 1996, Induced innovation and path dependence: A reassessment with respect to agricultural development and the environment, *Technological Forecasting and Social Change*, **53**(1):41–59.

Ruttan, V.W., 1997, Induced innovation, evolutionary theory and path dependence: Sources of technical change, *Economic Journal*, **107**:1520–1529.

Ruttan, V.W., 1998, The new growth theory and development economics: A survey, *Journal of Development Studies*, **35**:1–26.

Ruttan, V.W., 2001, *Technology, Growth and Development: An Induced Innovation Perspective*, Oxford University Press, New York, NY, USA.

Ruttan, V.W., and Hayami, Y., 1994, Induced innovation theory and agricultural development: A personal account, in B. Koppel, ed., *Induced Innovation Theory and International Agricultural Development: A Reassessment*, The Johns Hopkins University Press, Baltimore, MD, USA.

Salter, W.E.G., 1960, *Productivity and Technical Change*, 2nd ed., Cambridge University Press, New York, NY, USA.

Samuelson, P.A., 1965, A theory of induced innovation along Kennedy–Weizsäcker lines, *Review of Economics and Statistics*, **47**:343–356.

Samuelson, P.A., 1966, Rejoinder: Agreements, disagreements, doubts and the case of induced Harrod-neutral technical change, *Review of Economics and Statistics*, **48**:444–448.

Scherer, F.M., 1982, Demand pull and technological inventions: Schmookler revisited, *Journal of Industrial Economics*, **30**:225–237.

Schmookler, J., 1962, Determinants of industrial invention, in R.R. Nelson, ed., *The Rate of Direction of Inventive Activity: Economic and Social Factors*, Princeton University Press, Princeton, NJ, USA.

Schmookler, J., 1966, *Invention and Economic Growth*, Harvard University Press, Cambridge, UK.

Schumpeter, J.A., 1934, *The Theory of Economic Development*, Harvard University Press, Cambridge, MA, USA.

Schumpeter, J.A., 1950, *Capitalism, Socialism and Democracy*, Harper, New York, NY, USA.

Simon, H.A., 1955, A behavioral model of rational choice, *Quarterly Journal of Economics*, **69**:99–118.

Simon, H.A., 1959, Theories of decision making in economics, *American Economic Review*, **49**:253–283.

Stokes, R.G., 1994, *Opting for Oil: The Political Economy of Technical Change in the West German Chemical Industry, 1945–1961*, Cambridge University Press, Cambridge, UK.

Temin, P., 1966, Labor scarcity and the problems of American industrial efficiency, *Journal of Economic History*, **26**:277–298.

Thirtle, C.G., and Ruttan, V.W., 1987, *The Role of Demand and Supply in the Generation and Diffusion of Technical Change*, Harwood Academic Publishers, London, UK.

Van de Ven, A., and Garud, R., 1993, Innovation and industry development: The case of Cochlear implants, in R. Burgelman and R. Rosenbloom, eds, *Research on Technological Innovation, Management and Policy*, Vol. 5, JAI Press, Greenwich, CT, USA.

Vernon, R., 1966, International investment and international trade in the product cycle, *Quarterly Journal of Economics*, **80**:190–207.

Vernon, R., 1979, The product cycle hypothesis in a new international environment, *Oxford Bulletin of Economics and Statistics*, **40**:255–267.

Walsh, V., 1984, Invention and innovation in the chemical industry: Demand-pull or discovery-push?, *Research Policy*, **13**:211–234.

Wan, H.Y., Jr., 1971, *Economic Growth*, Harcourt Brace Javonovich, New York, NY, USA, pp. 215–226.

Winter, S.G., 1971, Satisficing, selection and the innovating remnant, *Quarterly Journal of Economics*, **85**(May):237–261.

Winter, S.G., 1984, Schumpeterian competition in alternative technological regimes, *Journal of Economic Behavior and Organization*, **5**:287–320.

Witt, U., 1993, *Evolutionary Economics*, Edward Elgar Publishing, Aldershot, UK.

Wright, G., 1990, The origins of American industrial success, 1889–1940, *The American Economic Review*, **80**(September):651–667.

Wright, G., 1997, Toward a historical approach to technological change, *Economic Journal*, **107**:1560–1566.

Chapter 3

Induced Technical Innovation and Medical History: An Evolutionary Approach

Joel Mokyr

3.1 Introduction

The motivation for this project is derived from my amazement that changes in human knowledge have been so little analyzed in the economic history literature. For most relevant problems, we tend to assume that knowledge is given and should be regarded, insofar as it is considered at all, a constraint on the maximization problem to be solved. In that approach, knowledge is much like income: for a one-period optimization problem, it is quite warranted to consider income as given and a binding constraint, but nobody would recommend the same for a study of changes in long-term economic growth. Whereas studies of changes in income are now as numerous as ever, little is being done in the history of knowledge. In part, this is because human knowledge is such a slippery concept, and most economists—including the present author—do not have the background in philosophy to understand the finer points of epistemology. In part, it is because, even if we could agree on a definition of what knowledge is, the economics of its creation and historical development violate every axiom of economic goods: it is usually non-rival, often non-excludable, never trades at marginal cost, is often lumpy, nonlinear, nonconvex, non-differentiable, and externality laden, and at times even totally fails to obey the laws of arithmetic. Neoclassical approaches to knowledge growth are therefore unlikely to make much progress. Yet economic history is unthinkable without relating it to what people knew or thought they knew, and the topic is simply too important to be left to the historians of science and technology, especially since so many of them have recently lost interest in knowledge as such and are focusing increasingly on the social context and political construction of knowledge rather than the thing itself.

This chapter was originally published in the *Journal of Evolutionary Economics*, Volume 8:119–137, 1998. ©Springer–Verlag.

One possible avenue to take is to adopt a Darwinian paradigm which regards the evolution of knowledge as the net historical result of blind variation and selective retention. Such an approach has enormous promise and enormous danger. At its worst, it provides empty epistemological boxes to regurgitate old concepts and well-worn facts and observations without adding much insight. Yet the evolutionary approach, when practiced by experts such as David Hull and Robert Richards, has shed considerable light on the history of scientific and engineering knowledge, and whereas it has yet to find much application in economic history, it could become a fruitful approach to a hitherto poorly developed area.

3.2   Induced Technological Change and the History of Medicine

The argument of this paper is simple. In its barest version, it just says that the more we know about a particular subject, the more likely it is that techniques of any kind will be able to adjust to environmental changes and thus generate induced technological change. Many years ago, Rosenberg (1976) pointed out that for a demand-induced mechanism to work in technological change, the technological capabilities have to exist. In a paper published subsequently, I relied on Rosenberg to question the importance of demand factors in bringing about major episodes of technological progress such as the Industrial Revolution (Mokyr 1985). The point was not so much that demand played no role as that, as a general phenomenon, human preferences for a higher standard of living of some kind were given, and thus wide-ranging episodes of technological progress in which many diverse areas of production were affected seem unlikely to have been the consequences of exogenous demand changes or even the effects of sharp changes in relative prices for whatever reason. In its crudest form—the "necessity is the mother of invention" theory of technological change—this approach succeeds in being at once a cliché and a historical fallacy. Economists and historians alike have treated this folk wisdom with contempt (Mokyr 1990:151, n. 1).[1]

A useful example of this kind of logic is to be found in the history of medicine. The idea that changes in medical technology should be regarded as a special case of technological change seems obvious enough. By medical technology I mean the techniques that prevent, cure, and alleviate the symptoms of disease. Medical technology provides an unusually fruitful ground to study induced technological progress. First, frequent exogenous changes in the environment occurred due to exogenous changes in pathogenic agents and new contacts between people and societies. Human health clearly exists in the least stable environment of any comparable variable. Its history was riddled with autonomous shocks. Throughout recorded history, new diseases appeared apparently *ex nihilo* and old diseases changed or vanished inexplicably. Until recently, adaptive responses were extremely slow in coming, ineffective, or altogether absent. Second, the demand side was in part biologically and not socially determined. The desire to survive, be disease-free and pain-free, and have one's children and relatives enjoy the same seems at first glance to be more or less constant over history.[2] It is therefore perhaps surprising that the history of medicine, as viewed from the point of view of the technological historian, shows remarkably little progress of any significance before 1800. Indeed, it could be argued that the ability of mankind to understand, avoid, let alone cure diseases by 1850 was little better than it had been at the time of Galen. The previous century had witnessed huge changes in the deployment of energy, the manipulation of materials, the transportation of goods and people, the transmission and communication of information, and the raising of crops and animals. Yet while the centuries since Vesalius (1514–1564) and Paracelsus (1493–1541) did witness major improvements in the understanding of the human body, these developments had little or no practical medicinal significance. If ever there should have been "demand-induced" innovation, it would have been in the avoidance of physical sufferance. Yet the supply side, at least until 1850, budged but little, particularly as far as infectious disease was concerned.

Moreover, what few improvements there were before then seem to have been not so much induced adaptations to changing circumstances as much as fortuitous events not based on systematic knowledge. Serendipitous discoveries were rarely fully exploited, did not lead to further developments, and often ended up being badly applied or forgotten. For instance, Roman physicians discovered a crude form of antibiotics when they applied a mixture of rotting wood and flowers to wounds to prevent infection (Galdston 1958). One successful medical advance of the more recent premodern age was the discovery of the Cinchona bark (quinine) as an effective cure for malaria ("ague" as it was known at the time), which became widely used in Europe in the last third of the seventeenth century. Yet the medication was applied to other fevers where it was of course ineffective. The smallpox vaccination process was discovered by Edward Jenner in 1798, but no other disease was conquered the same way for almost a century, and even today many diseases have escaped effective immunization. The successful war waged against bubonic plague through tough public policies prevented the spread of the dreaded disease. By the time of the British Industrial Revolution it had entirely vanished from Europe (Biraben 1975–1976; Cipolla 1981). All the same, until the closing decades of the nineteenth century, European medical technology remained as ignorant as it was powerless against the bulk of infectious diseases which killed people in the West. Measures effective against the plague failed to produce results for influenza, pneumonia, typhus, or cholera. Mortality rates rose and fell more with exogenous changes in the disease environment than with medical knowledge (Goldstone 1991).

To make any progress in the understanding of subsequent advances, we need a clearer theory of useful knowledge and its role in economic and social change. Such a theory does not exist, and the economic history of technological change has been written largely in a neoclassical competitive market paradigm or a theoretical vacuum. What I propose to do below is to sketch the bare bones of a framework in which such a theory could one day be constructed, and then show how "induced" innovation can be defined in such a theory. I will then return to the issue of medical technology as a case study of such a theory and try to show how the theoretical concepts can be made operative.

3.3   An Evolutionary Theory of Useful Knowledge

The idea that human knowledge can be analyzed using an evolutionary epistemology based on blind variation and selective retention was proposed first by Campbell and has since been restated by a number of scholars in a wide variety of disciplines.[3] In previous work, I have outlined the potential of the use of evolutionary biology in the economic history of technological change.[4] A reasonable criticism of such arguments has been that, while models of blind variation with selective retention are a useful way to look at innovations, they add little direct insight that cannot be gained from standard models. The example of induced innovation in medical technology should be regarded as an attempt to show the potential usefulness of such models. We should not think of such models as written in *analogy with* models in evolutionary biology. Instead, as I have argued elsewhere, both biological and cultural evolution are special cases of a larger class of dynamic models that share certain well-understood properties (Mokyr 2000).

The fundamental unit at which selection takes place is not a living being or a species as in Darwin's theory, but an epistemological one, the *technique*.[5] The technique is in its bare essentials nothing but a set of instructions, if-then statements (often nested) that describe how to manipulate nature for our benefit, that is to say, production widely defined (including medical and domestic technology).[6] In the case of medicine, such instructions are reasonably straightforward whether they deal with preventive medicine ("boil your water before drinking it") or curative practice ("stay in bed and drink lots of liquids until the fever has passed"). How are we to understand the Darwinian dynamics proposed by Campbell in such a model?

One element in this theory is the notion of the relation between an *underlying structure* that constrains but does not entirely determine a *manifested entity*. In biology, the underlying structure is the genotype, which does not respond to the environment, whereas the manifested entity is the phenotype, which does. The relation between the two is more or less understood, although there is still an endless dispute of their respective contributions of the environment and the underlying structure to the phenotype. In the history of technology, I submit, the underlying structure is the set of *useful knowledge* that exists in a society.[7] This set contains but is not confined to scientific knowledge. It also contains traditions and other strongly autocorrelated knowledge systems which may not get down to the principles of *why* something works but all the same codify it.

The set of useful knowledge needs to be defined with some care. Useful knowledge is defined as the union of all the knowledge possessed by individuals that can conceivably be applied to production in its widest sense (including household activities) in a given society. This knowledge is confined to the natural world: knowledge about the epistemological philosophy in Biblical texts, say, does not count, while knowledge of the planets of Jupiter does. There is a certain arbitrariness about this, and in some gray areas the line may not be as sharp as we would like. All the same, because techniques always and everywhere involve the manipulation of natural regularities, this seems a natural definition. Such useful knowledge does not have to be actually applied or even be, in some definable sense, correct. Knowledge could well be a set of untested beliefs and prejudices that posterity will eventually reject.

Knowledge can reside in people's minds and in storage devices with greater or lesser accessibility. Leaving out non-human storage devices, let there be n members of society and let each individual in a society possess technical knowledge S_i. Let $N_i(S_i)$ define the number of "useful" pieces of information contained in the set S_i. We can then define the total knowledge of society as the union of all the individual knowledge of members of society:

$$\Omega = \bigcup S_i = S_i \cup S_j \ for \ \forall i \neq j \tag{3.1}$$

and

$$\Phi = N(\Omega) \ . \tag{3.2}$$

Φ represents the total number of pieces of useful knowledge possessed by society. When a single individual produces an innovation (that is, discovers something hitherto unknown about nature), we observe unequivocally an increase in Ω. In a biological sense, Ω can be thought of as the gene pool. Ω is a union over n members, although it is very likely that the knowledge of many individuals is redundant in that their knowledge is wholly subsumed in that of others (so that their removal from society does not reduce Ω).[8] To define diffusion, we simply look at the intersection of S_i and S_j, which is the amount of knowledge that two individuals share, $S_i \cap S_j$.

The set Ω maps into a second set, the manifested entity, which I will call the feasible techniques set λ. This set defines what society *can* do, but not what it *will* do. The mapping function, in essence, translates the knowledge about natural facts and regularities into "how to" blueprints that actually can manipulate nature into improving the human condition in some way (that is, produce a good or service). The outcome is then evaluated by a set of selection criteria that determine whether this particular technique will be actually used or not, in a fashion similar to selection criteria that pick living specimens and decide which will be selected for survival and reproduction and thus continue to exist in the gene pool. The analogy is inexact and to some extent forced: while genes are a mechanism of inheritance and will vanish as soon as the species is extinct, knowledge can continue to exist even if the techniques it implies are no longer chosen. All the same, the bare outline is quite similar in that the dualism between the underlying structure and the manifest entity is maintained. Above all, selection can only pick entities from existing material; variants that cannot be constructed from existing potential will not appear, no matter how beneficial or desirable.

To stick with the example chosen for this chapter, an example of an underlying structure in medical technology is the humoral theory of disease, which viewed all diseases as resulting from imbalances between the four basic bodily fluids: blood, yellow bile, black bile, and phlegm. This theory, propounded by the Hippocratic school of medicine and part of the Galenian canon, implied that certain techniques be used by physicians on their patients, the best-known of them being bleeding and purging patients suffering from fever. The technique was then examined against alternatives and chosen by physicians as their main weapon against infectious disease for many centuries. Note that there are a number of distinct stages in the translation of knowledge into practices. One of them is the mapping itself, which may differ from place to place and over time. Another is the selection criterion by which the efficacy of a technique is tested. Interestingly enough, even when the belief in the humoral theory underlying bleeding practices waned, the technique stubbornly continued to be used until deep into the nineteenth century. The example also illustrates the many difficulties that such a view implies for optimistic scenarios that are based on the hopeful but ahistorical assumption that rational selection processes will eventually yield an outcome that we may recognize as "efficient." That, however, is another story.

Operationalizing Ω is, of course, quite hard, since it skirts questions such as "Who knows that which is known?" and "How easy is the access to this knowledge?" What matters here is that Ω maps into sets of instructions, each of which constitutes a technique which jointly make up λ. Each technique has "traits" that define it. Call these $T_1 \ldots T_n$. We can think of those, say, as "output quality" and "costs of production," although many different product attributes may matter here. Second, each time a technique is "used" it "lives" and a specimen has been "selected." For each technique j, we may then define μ, which is a count of how many times this technique is used, a bit like the size of a population. The fitness equation then defines the basic motion of the system:

$$\mu^*(j) = S(T_1^j, T_2^j \ldots T_n^j; V), \qquad (3.3)$$

where μ^* is some equilibrium level of usage and $\dot{\mu} = f(\mu - \mu^*), f' > 0$ is the change in the frequency of use. For any V (the environmental parameters, assumed exogenous), there are combinations of the Ts which define $\dot{\mu} = 0$.

Assume for simplicity that $\partial \mu / \partial T_1 > 0$ (the trait is favorable) and that $\partial^2 \mu / \partial T_1^2 < 0$ (diminishing returns), and the same holds for T_2. Each technique is defined as a point in this space. We can then define the curve ZZ' in Figure 3.1 which defines the condition of fixed fitness ($\dot{\mu} = 0$). An exogenous deterioration in the environment (possibly due to changes in complementary or rival techniques, changes in preferences, or some other autonomous effect) would be depicted as an outward shift of ZZ'. In addition to the techniques in use, given by the area δ in Figure 3.1, there is a larger set of all feasible techniques λ within which δ is wholly contained. The techniques that are in λ but not in δ are techniques that are feasible but not selected by society.[9]

Selection of techniques thus occurs at two levels: not all techniques in λ are picked to be in δ. In fact only a small minority of all feasible ways of making a pencil, shipping a package from Chicago to New York, or treating a patient suffering from pneumonia are in actual use at any point in time. Second, techniques in use themselves are competing with each other, and in the long run, assuming

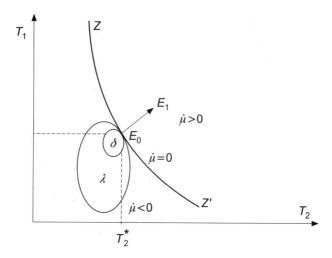

Figure 3.1. Fitness Space for Two Technological Traits and Subsets of Feasible (λ) and Actually Used (δ).

competition is sufficiently stringent, only the ones that are at E_0 actually maintain their numbers. Such an equilibrium, however, is not a prediction of the model. A lot depends on the actual degree of competition, and without more information it is not possible to know if other techniques in λ survive or go extinct. Moreover, points like E_0 are not necessarily unique: the set λ need not be shaped neatly as in Figure 3.1 and could have more than one tangency with ZZ'. Multiple equilibria and path dependence are, of course, standard fare to students of both economic evolution and the history of medical technology.

Beyond the level of the selection of techniques within λ, there is a higher level of selection at the level of knowledge in Ω, precisely of the type that evolutionary epistemology addresses (Hull 1988:ch. 12). It might well be asked why selection is necessary here, since in general knowledge need not displace previous knowledge. If storage costs are sufficiently low, knowledge could just accumulate. The principle of superfecundity (more specimens are born than can be accommodated, which lies at the center of Darwinian thought) does not apply here *strictu sensu*. In practice, however, new knowledge often replaces existing knowledge that must be discarded. Thus, accepting the work of Lavoisier meant that one had to abandon phlogiston theory. Such obsolescence is not the same as extinction in the living world, however, and at times knowledge that was believed obsolete can be resurrected. Knowledge can thus be active or dormant, depending on whether it maps onto λ, regardless of how widely it is accepted.[10]

If the social costs of retaining information are essentially nil, a case could be made for the technological equivalent of biodiversity, that is, preserving seemingly useless old forms of knowledge. Much of this depends on the technology of information storage and the access costs. In effect, that is what postmodernist history of science is trying to do. Rather than reject or accept older forms of knowledge as "correct" or "false," it tries to see them in their social context and has as much interest in innovations that ended up dead ends as in those that led to further progress and more successful forms. Yet unlike some of the more extremist versions of postmodernist history of knowledge, such agnosticism is not *invariably* useful: we need to reject some knowledge in favor of other knowledge as well as the nihilist view that all knowledge about nature is just as interesting as any other knowledge (Kitcher 1993). Even Bruno Latour would not want to be treated by a physician holding on to the humoral theory of disease, or the belief in the spontaneous generation of microorganisms, or the notion that ulcers are caused by stress. Knowledge will be rejected if it is widely believed to be false *and* does not map into any useful technique. It is also often suppressed or delayed by the adherents of an alternative and incompatible set of knowledge who have more political power.[11] Unfortunately, we cannot always tell one form from the other. Unlike genetic information, however, it rarely becomes *irreversibly extinct*. If we wanted to, we could revive medieval herbal treatments or François Broussais's (1772–1838) notorious use of leeches based on his absurd theories that all diseases originated from the digestive tract, just as we can build Roman catapults and Chinese water clocks.[12]

The way subsets of Ω are selected for is largely by persuasion. Selectors who advocate some subset of Ω try to make others see the same. Persuasion contains a large number of rhetorical means such as the proof of mathematical theorems, statistical analysis, experimentation, and other forms of induction. It also contains authoritarian obiter dicta, threats, education, propaganda, and political manipulation. Consequently in many cases useful knowledge came into existence but failed to be "activated," that is, failed to persuade enough people to end up being translated into the set of feasible techniques.[13] Active knowledge is not at all the same as "true" knowledge. Much of the knowledge set of the past may be recognized today as "false" and yet have mapped into a useful technique. One can navigate a ship using stars even under a Ptolemaic geocentric astronomy and improve the techniques of iron production on the basis of phlogiston chemistry. Medical procedures based on Galenian theory could be effective. In short, whether knowledge is "correct" or not seems to matter less than whether it mapped into techniques. Selection thus occurs at both ends of the mapping $\Omega \to \lambda$, although the selection process is different.

Unlike the natural selection mechanism defined by Charles Darwin, in knowledge systems the selection process is not anonymous and decentralized but conscious and deliberate. Techniques are chosen willfully by individuals who are trying to attain certain objectives and tested according to prespecified criteria. However, for a technique to succeed by being chosen is *not* to say that it actually maximizes those objective functions, let alone a social welfare function. In that regard, the historical success of a particular technique may be quite different from what an *ex post* assessment of its fitness would imply. This is particularly true for medical techniques, but there are many other instances of technological choices in which the selection process chooses a technique that seems inferior or—more often—rejects a technique that *ex post* was more efficient. Resistance to innovation is one of the more interesting features of evolutionary systems.[14] A technique's fitness may thus be judged by two criteria: its success in actually being selected and the way it fulfilled the function for which it was intended. The convergence or divergence of these two criteria is still being debated.

3.4   Technological Adaptability and Induced Innovation

Despite the rather confining definitions of information and other simplifications, this setup allows us to make some simple distinctions in an evolutionary framework. For one thing, it allows us to define adaptation and adaptability, which is crucial to the idea of induced innovation. To start with, most techniques have a certain amount of built-in flexibility. The reason is that, like genetic instructions embedded in the DNA, technical instructions often take the form of if-then statements, conditional on a variety of previously experienced or easily predictable contingencies. This allows the technique to adapt at a local level, but such "phenotypical adaptations" are providing only limited flexibility.

What really counts is adaptation to something unexpected requiring a change in technique (consider Figure 3.2). Suppose we are looking at a case, described earlier, in which a society's medical knowledge is very limited, but it has discovered that a certain technique works. Denote this technique by E_0. I will call this a *singleton technique*. Such techniques are based on knowledge that does not extend beyond "such and such a technique works." Under the assumptions stated, the singleton E_0 is the only technique in the feasible set. The accidental discovery of a medicinal herb would be a good example of a singleton. The knowledge set Ω contains the knowledge that such and such an herb is effective against a certain disease. This maps into the feasible techniques set as a single set of instructions: if patients exhibit this symptom, give them the medication. The knowledge does not contain any pharmacological basis of the herb, any information about the disease's etiology, or any hint of the herb's modus operandi against the agent causing the disease. A similar structure is true for most premodern production. In agriculture, fields were fertilized without any underlying knowledge of organic chemistry. Steel was produced for centuries by blacksmiths who had no knowledge about the relation between carbon content and the physical qualities of iron. Much of what we call production was carried out on the basis of "standard operating procedures" passed from generation to generation. The "underlying structure" of knowledge was little more than purely pragmatic: "such and such works." This kind of knowledge was acquired by trial and error and passed on from master to apprentice. When the criteria for efficacy were unclear or the testing procedures flawed, ineffective techniques could be adopted and survive for centuries.

The knowledge base of a controversial technique may be one determinant of its success in the political arena. Joseph Lister, who revived Semmelweiss' discarded antiseptic techniques, could rely on the recent discoveries of Pasteur to defend his insights. Less well-known is the rather stiff resistance against small-pox vaccination by a host of enemies, who exploited the fact that the causation of the disease was unknown and no one had any clue as to the modus operandi of Jenner's successful vaccination technique. It was even disputed whether the disease was contagious, much less understood how it was transmitted. This was particularly true for the United States, where a new epidemic of smallpox flared up in the 1870s after the disease had all but disappeared following the successful vaccination campaigns in the 1810s and 1820s. Many states repealed their manda-tory vaccination programs, and for decades the feared disease returned with often devastating effects (Kaufman 1967).

Another example is the conquest of scurvy. The importance of fresh fruit in the prevention of scurvy had been realized even before James Lind published his *Treatise on Scurvy* in 1746. The Dutch East India Company kept citrus trees on the Cape of Good Hope in the middle of the seventeenth century, yet despite the obvious effectiveness of the remedy, the idea obviously did not catch on and the idea "kept on being rediscovered and lost" (Porter 1995:228). Lind's ideas as-sisted Captain Cook in keeping his crew scurvy-free. Yet modern scholarship has established that Cook's efforts only confused the understanding of the disease and delayed rather than hastened the solution (Carpenter 1986:83). On his voyage, Cook, determined to eradicate the disease, tried a number of different things, and it was difficult to attribute the disappearance of the disease to a specific measure. Even after the curative properties of lemon and lime juice were recognized, it was still thought that the disease itself was caused by breathing foul air in the ship's living quarters. In 1795, Gilbert Blane made the use of oranges and lemons in the British Navy mandatory and scurvy was dramatically reduced. Yet precisely because this was a singleton technique, its persuasive force remained weak. Con-sequently, while controlled on shipboard, scurvy remained a serious problem on land: it survived in jails and poorhouses, and made a serious appearance dur-ing the Irish Famine of 1845–1848. It was still endemic during the Crimean and US Civil Wars and in the Russian army during the First World War. Infantile scurvy was prevalent among wealthier families in which weaning occurred at a relatively early age. The discovery of the germ theory led to decades of futile search for a causative microorganism. Only after the seminal papers by Holst and Fröhlich after 1907 did it become clear that certain diseases were *not* caused by infectious agents but by nutritional deficiencies, and only in 1928–1932 was the crucial ingredient isolated (French 1993).

If the environment changes exogenously in any form as indicated by a change in the slope of ZZ', the primitive production system in which the feasible set is the singleton E_0 cannot adapt and stays at E_0 to its detriment. A singleton tech-nique means that there will be little or no adaptability in the system, so that minor environmental changes can cause very significant losses in fitness. A dramatic example is the appearance of bubonic plague in Europe. While in the very long run adaptations were developed, it took close to three centuries for the disease to disappear.[15] More fortunate were the Europeans of the nineteenth century, who were visited by cholera. While ignorance of contagious disease in 1829 (the date of the first European appearance of the disease) was almost as deep as in 1348, scientific method had progressed and rational investigation based on established procedures was quite different. By the 1850s the mode of transmission of cholera was understood (before its etiology, to be sure), and eventually the disease disappeared again from Europe. The germ theory of disease, arguably one of the most significant increases in Ω in history, mapped into thousands of small and large techniques to avoid infection, both in the domain of public health and in that of household technology, long before antibiotics were developed (Mokyr and Stein 1997). The increase in microbiological and immunological knowledge in the past decades has been pivotal in our ability to deal with human immunodeficiency virus (HIV); it is hard to imagine how any medical adaptation would have been possible had HIV appeared, as cholera did, *ab nihilo*, in the middle of the nineteenth century. Adaptation to a new disease does not necessarily require a cure or a vaccine: in the case of infectious disease the critical piece of knowledge that is required is the understanding of the mode of transmission. Once this is understood, even imperfectly, preventive techniques may be enough to deal with the disease.[16]

One common view of the main change in the process of technological change is that, since the late nineteenth century, engineering and the "knowledge of production" have been far more closely connected to science than previously (Copp and Zanella 1993). What this implies is that the λ "around" modern techniques actually in use is much larger, that the Ω is far more capable of producing new techniques, and that the mapping from Ω to λ is far more flexible and capable of producing novelty "on demand." Yet the "trial and error" and "try every bottle on the shelf" modes of invention have not disappeared, especially in pharmaceutical and biological technology in which many of the underlying processes are very complex and poorly understood.

Flexibility and adaptability, then, have three dimensions. First, the larger the set of feasible techniques λ, the more adaptability the production economy has. If the environment changes from ZZ' to WW' (see Figure 3.2), the system can adapt (as from E_0 to E_1). This is precisely what we normally mean by substitution. The technique E_1 may have "existed" at the time that WW' was in force, in the sense that it was part of the feasible set λ but not "selected." In other words, the society was *capable* of producing E_1 given its knowledge base but chose not to. When circumstances changed, it adjusted by moving along the frontier of λ. Some interpretations of technological progress in economic history are based entirely on this concept: Boserup (1981) has argued that essentially new knowledge is unimportant and what governs the technique in use is population density. When population increases, society will find labor-intensive techniques that will keep living standards from falling.

Second, environmental change may also lead to new searching over Ω and new mappings into λ creating techniques that were previously unknown even if they could be known given that the knowledge base for them existed. In some sense this approach redefines the traditional distinction between substitution and induced technological change.[17] Substitution in the standard microeconomic

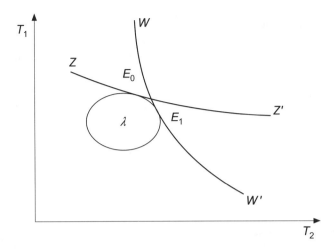

Figure 3.2. Change in Selection Environment (from ZZ' to WW') Affecting Fitness of Known Techniques (λ).

approach concerns a choice between *known* techniques. Induced technological change leads to the emergence of new techniques based on the existing stock of useful knowledge. What really is known is not just a set of blueprints that firms and individuals can pick and choose from freely, but an underlying knowledge set, far more complex and multidimensional. As long as the knowledge set exists, it is possible for society to adapt to a changing environment and innovate as needed by mapping from an existing subset of Ω into a new segment of λ.

The existence of such knowledge is not sufficient for the mapping to occur; nor is an exogenous stimulus due to a change in the environment a necessary condition. The mapping function is one of the more difficult concepts in the history of technological change. How accessible to the searcher is knowledge available to somebody else or stored somewhere? Does existing knowledge translate itself into a new useful technique when the need arises? Historical cases in which that happened, especially in the last century, can be found—one thinks inevitably of the German Fritz Haber's invention of the nitrogen-fixing process in the face of his country's needs to produce fertilizers and explosives during a threatened naval blockade. But exceptions are difficult to explain and are numerous enough to doubt any regularity. There are examples of the mapping of existing knowledge into a technique occurring without any obvious stimulus. One is the invention of spectacles around 1280. The basic elements of the knowledge set—knowledge of glassmaking and the observation that lenses change the refraction of light—had existed in Roman times.[18] It seems hard to believe that a sudden change in demand occurred in the thirteenth century: the physiological changes that cause the need for eye-glasses are more or less constant. The development of printing or, more accurately, of movable and interchangeable type, by Gutenberg runs into a similar dilemma. Famously serendipitous elements in the invention of antibiotics illustrate the precariousness of any easy conclusions about the mapping from Ω into λ. All the same, it is clear that without the concept of pathogenic bacteria, the widespread development and adoption of antibiotic techniques would have been absurd. In areas in which the knowledge base developed more slowly because it was more complex, such as in viral, autoimmune, and psychiatric diseases, progress toward an effective cure was much slower and the techniques used are far more inclined to be singleton techniques based on trial and error or serendipitous finds.

One obvious factor in the mapping function is the *accessibility* of knowledge. Knowledge may exist, either in someone's mind or in some storage device, but a great deal depends on the ability of those who perceive the need for it to find and access it. Another factor is the *technical capability* of society to design and build techniques that its knowledge base suggests might be feasible. In many cases a particular technique is imagined, envisaged, and even designed, but a critical component or complement is missing which makes it impossible to exploit it. A good example is the measurement of longitude, which has recently been popularized thanks to Sobel's (1995) excellent little book. It had been understood how longitude could be measured (by the use of two accurate clocks), but it turned out to be very difficult to construct marine chronometers of sufficient accuracy until the technical difficulties were cracked by John Harrison in the middle of the eighteenth century. Much like longitude, the exploitation of fusion energy in our time seems to elude us despite the knowledge that such energy is possible in principle and probably in practice. The practical application, however, has not materialized. Similarly, President Clinton's recent announcement that AIDS will be either curable or wholly preventable within 10 years is based on a sense that the solution is within reach and that only a few elements are missing before the puzzle is wholly solved.

Third, we can think of *induced knowledge change* as differing from induced technical change in that changes occur in the knowledge set Ω rather than in the set of feasible techniques λ. It might be thought that this is a distinction without a difference. The extension of the knowledge base Ω itself is the underlying force believed to propel human progress. Again, there exists a gray area where the distinctions are blurry. Yet some insights can be gained from it.

Knowledge growth does not develop wholly exogenously. It responds to outside stimuli and search processes and can in that sense be said to be "induced." Scientists do not pick topics at random, they work on problems that they feel other scientists or some patron may be interested in. Knowledge is thus constrained by its own past: the direction of change depends on the state of the world at each moment. In that regard, knowledge can be said to follow an evolutionary path like a Markov chain in which normally innovations are incremental rather than revolutionary. While there is some disagreement among evolutionary theorists as to the likelihood of very rapid, discontinuous evolutionary changes, even "salta-tionists" realize that there are limits to the amount of change that can occur per unit of time. In that sense, all learning is "local." To be sure, some innovations are less localized than others, and at times we observe the birth of something that is radically novel in that it represents knowledge that simply was not available before. It may be argued ad infinitum to what extent Pasteur's famous refutation of spontaneous generation or the appearance of *The Origins of Species* were "local" or "radical" innovations. We can all agree, however, that they were not likely to have been produced in the age of Thomas Aquinas. Yet the "need" or "demand" for them existed as much in 1270 as it did in 1860. People were just as sick and arguably as curious about the development of living beings. The germ theory, of course, had been proposed earlier but lost out in the battle of persuasion.[19] Darwin's insight, while shared with Alfred Russel Wallace, simply had not occurred to anyone before, triggering T.H. Huxley's famous response, "how very foolish not to have thought of this."

This framework allows us to classify all additions to knowledge into three broad classes: new knowledge that will be selected under the current environment; new knowledge that is potentially useful if circumstances change but currently is neutral; and detrimental mutations that will never be useful. In Figure 3.3, which is an elaboration of Figure 3.2, I depict three different increments to knowledge (note that the diagram is drawn for convenience in the space of techniques, but that we really should think of it more in terms of Ω than in terms of λ). The increment α is a classic favorable mutation or invention, in which the traits of the entity "improve," allowing it to become fitter and to augment its numbers. The selection mechanism will choose any technique displaying the traits in this set. The increment β is useless in the current configuration. As long as T_1 and T_2 are positive traits (that is, have positive partials with respect to the objective function), nothing based on the information in it will ever be selected. The mutations in γ are neutral in that they do not affect the phenotype. Yet they could come in handy when the environment changes in such a way as to favor the need for more T_2 as depicted by the curve WW'. Such environmental changes could be changes in factor prices or resource availability, but should be interpreted as including the appearance of a new trait T_3 complementary to T_2 which could "activate" the region γ. A great deal of scientific and mathematical knowledge that seemed useless at the time became useful much later when complementary knowledge became available. Boolean algebra and Hellenistic astronomy, to cite just two examples, eventually became indispensable to technological developments. The "lucky" economy is the one which develops the neutral knowledge γ and is able to access it when circumstances change to move ZZ' to WW'.

Path dependence implies the existence of multiple equilibria, and in the history of medicine this means essentially the development of alternative medical paradigms that emerged more or less along independent trajectories and whose ultimate form was shaped as a result of the history of the system as much as by the objectives that practitioners tried to attain and the constraints on them (David 1997). The importance of "alternative" medicine, from homeopathy to Christian Science, is evidence that multiple approaches to disease are still enormously influential. Tens of millions of Americans resort regularly to one form or another of alternative medicine. The historical development of Chinese and Western medical knowledge along separate historical paths has produced different

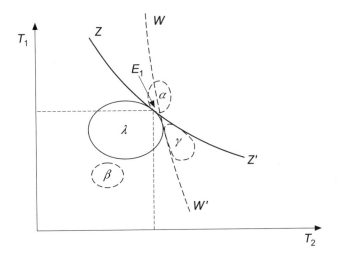

Figure 3.3. Changes in Knowledge (α, β, γ) versus Changes in Selection Environment (ZZ' to WW').

techniques and thus trajectories for change. Some of those trajectories have better options in dealing with changes in the environment than others. For instance, it seems reasonably uncontroversial that, in dealing with the acquired immunodeficiency syndrome (AIDS) infection, modern virology and immunology are better equipped than acupuncture or chiropractic. But in other syndromes that seem to have emerged or worsened in recent years—from skeletal pain caused by exercise to lower back pains to sleeping disorders to carpal tunnel syndrome—this superiority seems less secure. In medical conditions that are psychosomatic in nature, a set of techniques based on the placebo effect, including the hypnotic effects of witch doctors, the touch of holy men, or the couch of the psychoanalyst, maybe better suited to deal with outside shocks than the chemical and surgical interventions of modern medicine. Hence the sheer size of the entire body of medical knowledge, including various forms of nonstandard medicine, provides more adaptability and ability to respond to unexpected shocks.

How do changes in Ω affect what people do and how they produce? Despite dissimilarities with living systems, the process of generating innovations I am proposing here shares some important features with living beings. It may well be the case that a substantial amount of knowledge is "induced," that is, created with a specific purpose in mind. However, the vast majority of all human knowledge, like DNA, is non-coding or "junk" in the sense that it does not apply directly to production. Most scientific (let alone other forms of) knowledge has no applications and does not affect production technology right away, although it may be "stored" and in rare cases called into action when there is a change in the environment or when another complementary invention comes along. Thus most additions to Ω, like mutations, are predominantly "neutral" and are not affected by the selection criterion one way or another, but may become useful when the environment changes and calls for adaptation (Stebbins 1982:76). The activation of such previously inert material may be the evolutionary equivalent of what economists think of as "induced innovation." General knowledge, too, is being created at a much faster rate than technological knowledge, but if it finds no application in production, it is not a part of technological evolution and might be regarded as "neutral." In this important sense, neutral change creates a powerful element of contingency in history: if the knowledge had happened to accumulate for no special reason at some point in the past, it would come in handy when there is an autonomous change in circumstances. In that very sense historians of technology might well adopt the term proposed by Kimura, the proponent of neutral drift in evolutionary biology, who has suggested replacing "survival of the fittest" with "survival of the luckiest" (Kimura 1992:225–230).

Above all, however, there seems to be no obvious social mechanism that predicts endogenous changes in the size of the knowledge set Ω any more than we can predict what direction changes in the gene pool will take. Instead, relative prices and other focusing devices in Rosenberg's phrase may determine *the direction* in which the search for new knowledge takes place, and thus in some cases determine the successful increase in the set of useful knowledge. In that sense they are like a steering mechanism; but it is not the steering mechanism that makes a vehicle move forward and is the primary determinant of its speed. The expansion of the useful knowledge set does not respond neatly to changes in demand because in some sense this demand is always there. This is not just true for medical knowledge even if there the case can be made most persuasively. We can point again and again to societies that "needed" knowledge but did not get it simply because it was out of their reach, or because they looked in the wrong places, or because it did not occur to them to look at all. Classical civilization, an iron-using and seafaring society, made practically no changes in the primitive processes of ironmaking, shipbuilding, and navigation extant around 400 BC for the next half millennium. Ironcasting, which had been known to the Chinese since the second century BC, reached Europe in the fourteenth century. There are obviously many brakes and obstacles to the expansion of knowledge and even more so to their mapping onto the useful techniques set.

3.5   Concluding Remarks

It seems therefore that the ability of the knowledge set to respond to environmental changes depends in large part on the nature of the existing knowledge itself and on the conditions that determine how conducive the knowledge base and society are to its expansion. The more variegated and diverse the knowledge set, the more likely it is to be able to create a response to outside shocks. Useful knowledge is to a large extent cumulative, depending mostly on storage devices and the cost of access. That does not mean that its growth is always harmonious. Thus "modern" medicine regards alternative forms with suspicion bordering on disdain, but so do alternative forms regard each other. New knowledge often encounters stiff resistance from existing forms, when vested interests who have heavily invested in the latter fear the rapid depreciation of their specific human capital. This is equally true of other natural sciences, but there modern science often—if not always—creates the conditions to test different paradigms against each other. All the same, the diversity of the forms of knowledge means a great deal of flexibility and ability to deal with different shocks and needs.

In short, then, past knowledge has developed more or less by its own rules and thus any induced responses to environmental changes at the level of Ω were not very significant. Even in our time, I would insist, it is still true that changes in useful knowledge have remained to a great extent an exogenous variable (though clearly less so than in the past) and that any attempt to endogenize—let alone predict—it is foolhardy. True, if society faces a well-defined and clear-cut problem, it can allocate more resources toward increasing the knowledge set. The "research" part of research and development, unique to the twentieth century, is precisely the conscious attempt to expand Ω in a "needed" direction. It may, therefore, be the case that the modern age will develop a dynamic of knowledge completely different from that of previous ages. Yet the precise form this will take is still quite unclear. The success of "Manhattan project" endeavors is hardly the rule. The relatively modest gains in the war against cancer, declared with pomp by Richard Nixon, attest to the reality of real knowledge constraints. No amount of real resources devoted to medical research would have helped European society in 1348 to solve the riddle of the Black Death. To be sure, the war against HIV is conducted more effectively thanks to the great breakthroughs of our own time, information processing, molecular biology, and genetic engineering. Yet these breakthroughs themselves can hardly be regarded primarily as demand-induced, as most of their uses only became obvious *ex post*.

More seriously, to carry out such research, the problem has to be formulated correctly, and for that prior knowledge is required. Between 1875 and 1890, bacteriologists focused their research on the discovery of pathogenic bacteria and tackled them at the rate of about an organism every two years. But this program required prior knowledge that such a search would indeed have a reasonable probability of yielding results. Consider the question of why in our age we devote so few resources to our search for a substance that would halt or reverse the aging process, the ultimate Faustian dream. It can hardly be argued that no demand exists for such a substance, or that the demand for it has not increased steadily with the rise in the average age of the population. Yet the resources devoted to such research are rather modest simply because few scientists believe it likely that such a substance can be found. In the past such searches, despite being well focused, have often failed: alchemy is perhaps the most striking example. Alchemy was based on the false analogy of the change in the physical properties of compounds to changes in elements. It may well be that the search for an AIDS vaccine is based on the false analogy between a highly stable virus (polio) to one that is genetically unstable (HIV) and that no vaccine is feasible at all—or that no vaccine is feasible with our current knowledge of the molecular processes involved. Rosenberg (1976:51) cites Henry IV, Part I, in which Glendower says that he "can call spirits from the vasty deep" and is met by the deadly response of Hotspur:

"Why, so can I, or so can any man; but will they come when you do call for them?"

Notes

1. Modern empirical studies of technological advance have often claimed that much innovation is "demand-pull." For an effective demolition of much of this work, see Mowery and Rosenberg (1979).

2. A more detailed look would of course nuance this picture; a person's ability to deal with pain and death is subject to social influences as well as a personal hardening of feelings, and there can be little doubt that in societies in which infant mortality rates were, say 350 per 1,000, the pain might be different than in contemporary society where the figures are below 10. All the same, the discomforts of a toothache or an allergic attack are in large part physiological, and the instincts of mammals to protect the lives of their young are genetically determined.

3. The original statement was made in Campbell (1960, 1987). The most powerful statements are made in Hull (1988) and Richards (1987). For a cogent statement defending the use of this framework in the analysis of technology, see especially Vincenti (1990).

4. For an early version, see Mokyr (1991). For more recent reflections, see Mokyr (1996, 2000).

5. The notion of a technique is closely related to and inspired by Nelson and Winter's idea of "routines." For a discussion of the "unit of analysis," see Hodgson (1993:37–51).

6. Note that these instructions contain a first level of interaction with the environment in that they are *conditional* instructions, so that the actual operations carried out can be made contingent on environmental conditions. The need for induced innovation occurs when the environment changes to a point not accounted for in the technique itself.

7. The term is used more or less in this sense by Kuznets (1965:85–87). Kuznets confines his set to "tested" knowledge that is potentially useful in economic production. In what follows below, this definition is far too restrictive. An enormous number of techniques actually in use, from bloodletting to crop rotations to modern slim-down diets, were based on knowledge that was untested and often demonstrably ineffective.

8. In general, increasing Ω will lead to higher Φ, but because adding to the amount of useful knowledge might also make some previously useful knowledge obsolete, the relation between the two is complex. For the purpose here, this ambiguity is not fatal. Note that both in genetics and in technological knowledge only a small fraction of actual existing potentially useful information is actually "switched on." The human gene uses only about 1 percent of the DNA; the rest seems to fulfill no obvious function, but changes in it may at some point in the future become useful.

9. The shape of λ and δ as neat and compact shapes is of course not required: they could well be highly irregularly shaped with multiple tangency points, with ZZ' corresponding to multiple techniques in use with similar features serving similar purposes.

10. An example is the treatment of malaria which, because of the constant mutation of both the mosquitoes and the plasmodium parasite around all medications aimed at them, has become increasingly difficult to treat. It is reported that physicians in their desperation are returning to quinine and even to an extract of wormwood used in China many centuries ago (see Jones 1993:223). The use of artemisinin, the active ingredient in wormwood, was recently reported by Henry Lai at the University of Washington to be successful as a non-toxic treatment of cancer. See http://www.sciencedaily.com/releases/2001/11/011127003905.htm

11. An interesting recent case is the discovery that peptic ulcers are caused by *Helicobacter pylori* (and not by stress), made by the Australian physician Barry Marshall in 1983 and ignored for close to a decade by skeptical opponents and those with vested interests in the status quo. See, for instance, "Why doctors aren't curing ulcers," *Fortune*, June 8, 1997, pp. 100–107.

12. Indeed, the increased demand for leeches in certain surgical purposes (especially in the reattachment of severed limbs) illustrates this point.

13. Thus the germ theory was first proposed by Girolamo Fracastoro in 1546 and proposed repeatedly without having influence on the practice of medicine until late in the nineteenth century. The sad case of Ignaz Semmelweiss, who discovered the need for sterilization of medical tools through the connection between the contamination of physicians and the death rates in maternity wards due to puerperal fever, yet whose work was ridiculed and ignored for 20 years, is another case in point.

14. The literature on the subject has been growing rapidly in recent years. For a recent useful collection, see Bauer (1995). A one-sided and popularized account is Sale (1995); see also Mokyr (1994, 2000).

15. Even more devastating was the appearance of European diseases such as smallpox and measles on the American continent after Columbus, which annihilated most of the native population. The appearance of syphilis in Europe in 1494 (in all probability imported from the New World) had at first devastating effects, but the disease changed its nature in later years and became less fatal.

16. An example is the idea that diseases were transmitted by vectors. For centuries it had been believed that the association of swamps with malaria was caused by the "bad air" that emanated from standing water. The work of Patrick Manson, Ronald Ross, and G.B. Grassi demonstrated the culpability of the *Anopheles* mosquito in the 1890s, and in 1909 Charles Nicholl discovered the louse vector of typhus, five years before the causative germ itself was isolated. These discoveries were decisive in persuading households how such diseases were contracted and thus could be successfully avoided.

17. The evolutionary equivalent of this distinction is, roughly speaking, natural selection from a given set of heterogeneous traits as opposed to the emergence of new phenotypes from a given gene pool, in which fortuitous new combinations—if they emerge—are picked up by selective forces. Whereas the former is a more or less deterministic and predictable process, the latter remains a matter of contingency.

18. Seneca had already observed that letters were enlarged and made more distinct when viewed through a glass globe. Alhazen, who lived around AD 1000 studied the reflection of light from curved mirrors and spheres, yet spectacles were invented in Italy only toward the end of the thirteenth century.

19. Jacob Henle, the main proponent of the germ theory in the 1840s was regarded as "fighting a rearguard action in defense of an obsolete idea." In that regard his student Robert Koch was more successful (cf. Rosen 1993:277).

References

Bauer, M., 1995, *Resistance to New Technology*, Cambridge University Press, Cambridge, UK.

Biraben, J.W., 1975–1976, *Les Hommes et la peste en France et dans les pays Européens et Mediterranéens*, Mouton, Paris, France.

Boserup, E., 1981, *Population and Technological Change*, University of Chicago Press, Chicago, IL, USA.

Campbell, D.T., 1960, 1987, Blind variation and selective retention in creative thought as in other knowledge processes, in G. Radnitzky and W.W. Bartley III, eds, *Evolutionary Epistemology, Rationality, and the Sociology of Knowledge*, Open Court, La Salle, IL, USA (originally published in 1960).

Carpenter, K., 1986, *The History of Scurvy and Vitamin C*, Cambridge University Press, Cambridge, UK.

Cipolla, C., 1981, *Fighting the Plague in Seventeenth Century Italy*, University of Wisconsin Press, Madison, WI, USA.

Copp, N.H., and Zanella, A.W., 1993, *Discovery, Innovation, and Risk*, MIT Press, Cambridge, MA, USA.

David, P.A., 1997, Path Dependence and the Quest for Historical Economics, Discussion Papers in Economic and Social History, No. 20, University of Oxford, Oxford, UK.

French, R., 1993, Scurvy, in K.F. Kiple, ed., *The Cambridge World History of Human Disease*, Cambridge University Press, Cambridge, UK.

Galdston, I., ed., 1958, *The Impact of the Antibiotics on Medicine and Society*, International Universities Press, New York, NY, USA.

Goldstone, J.A., 1991, The causes of long waves in early modem economic history, in J. Mokyr, ed., *The Vital One: Essays in Honor of Jonathan R.T. Hughes*, JAI Press, Greenwich, CT, USA.

Hodgson, G., 1993, *Economics and Evolution*, Polity Press, Cambridge, UK.

Hull, D.L., 1988, *Science as a Process*, University of Chicago Press, Chicago, IL, USA.

Jones, S., 1993, *The Language of Genes*, Anchor Books, New York, NY, USA.

Kaufman, M., 1967, The American anti-vaccinationist and their arguments, *Bulletin of the History of Medicine*, 41(5):463–478.

Kimura, M., 1992, Neutralism, in E.F. Keller and E. Lloyd, eds, *Keywords in Evolutionary Biology*, Harvard University Press, Cambridge, MA, USA.

Kitcher, P., 1993, *The Advancement of Science: Science without Legend, Objectivity without Illusions*, Oxford University Press, New York, NY, USA.

Kuznets, S., 1965, *Economic Growth and Structure*, Norton, New York, NY, USA.

Mokyr, J., 1985, Demand vs. supply in the industrial revolution, in J. Mokyr, ed., *The Economics of the Industrial Revolution*, Rowman and Allanheld, Totowa, NJ, USA (originally published in 1976).

Mokyr, J., 1990, *The Lever of Riches*, Oxford University Press, New York, NY, USA.

Mokyr, J., 1991, Evolutionary biology, technological change and economic history, *Bulletin of Economic Research*, 43(2):127–149

Mokyr, J., 1994, Progress and inertia in technological change, in J. James and M. Thomas, eds, *Capitalism in Context: Essays in Honor of R.M. Hartwell*, University of Chicago Press, Chicago, IL, USA.

Mokyr, J., 1996, Evolution and technological change: A new metaphor for economic history?, in R. Fox, ed., *Technological Change*, Harwood, London, UK.

Mokyr, J., 2000, Innovation and selection in evolutionary models of technology: Some definitional issues, in H. Ziman, ed., *Technological Innovation as an Evolutionary Process*, Cambridge University Press, Cambridge, UK.

Mokyr, J., and Stein, R., 1997, Science, health and household technology: The effect of the Pasteur revolution on consumer demand, in R.J. Gordon and T. Bresnahan, eds, *The Economics of New Goods*, University of Chicago Press and NBER, Chicago, IL, USA.

Mowery, D., and Rosenberg, N., 1979, The influence of market demand upon innovation, *Research Policy*, **8**:102–153

Porter, R., 1995, The eighteenth century, in L. Konrad *et al.*, eds, *The Western Medical Tradition, 800 BC to AD 1800*, Cambridge University Press, Cambridge, UK.

Richards, R.J., 1987, *Darwin and the Emergence of Evolutionary Theories of Mind and Behavior*, The University of Chicago Press, Chicago, IL, USA.

Rosen, G., 1993, *A History of Public Health*, new edn, The Johns Hopkins University Press, Baltimore, MD, USA.

Rosenberg, N., 1976, Science, invention and economic growth, in *Perspectives on Technology*, Cambridge University Press, Cambridge, UK (originally published in 1974).

Sale, K., 1995, *Rebels against the Future: The Luddites and Their War on the Industrial Revolution*, Addison Wesley, Reading, MA, USA.

Sobel, D., 1995, *Longitude*, Penguin Books, Harmondsworth, UK.

Stebbins, G.L., 1982, *Darwin to DNA, Molecules to Humanity*, Freeman, San Francisco, CA, USA.

Vincenti, W., 1990, *What Engineers Know and How They Know it*, The Johns Hopkins University Press, Baltimore, MD, USA.

Chapter 4

Induced Adaptive Invention/Innovation and Productivity Convergence in Developing Countries

Robert E. Evenson

4.1 Introduction

In the 1950s, economists concerned with economic development programs expected the remaining decades of the twentieth century to be characterized by a pattern of "convergence" in per capita income levels. Those countries with low per capita income in 1950 were perceived to enjoy the "advantages of backwardness" and to have the greatest scope for growth in subsequent decades. Development economists also did not generally expect the agricultural sector to be the leading sector in development experience. Many formal models (notably the Latin American structuralist models) were predicated on the expectation that agricultural producers were hampered by the limited visions of the peasant (or minifundia) farmer and that the scope for productivity gains in the sector was limited.[1]

By contrast, industrial development was widely expected to be the vehicle for modernization and convergence. Industrial technology was not regarded as being sensitive to local conditions (as was the case in agriculture) and could presumably be copied or mimicked at relatively low cost. Low-wage countries acquiring this technology would then be highly competitive in both domestic and international markets. Not only was industry expected to drive convergence, it was also expected to be the ultimate source of transformation of the traditional agricultural sector in developing economies.

At the beginning of the twenty-first century, we now have a considerable body of experience to evaluate and contrast with the expectations of the early development modelers. The evidence shows that a general pattern of per capita income convergence has not occurred for developing countries, even though it did occur for the countries of the Organisation for Economic Co-operation and Development (OECD).[2] Many developing economies with low per capita income in the 1950s have made few, if any, gains in income in subsequent decades. By contrast, a small group of developing (newly industrial country, or NIC) economies has achieved very rapid economic growth.

The sectoral evidence shows that in most developing countries, to the degree that productivity convergence has occurred, it has occurred in the agricultural sector, not in the industrial sector. Moreover, while the rapidly growing developing economies did realize rapid industry-led growth, industrial growth did not "transform" the agricultural sectors of the NICs because they had already been transformed by biological invention. Industry-led growth certainly did not transform the agricultural sectors of the majority of developing economies, where most, if not all, productivity gains were realized in agriculture, not industry.

In Section 4.2, two alternative classes of mechanisms of productivity convergence are discussed. The two classes are mimicry mechanisms and induced adaptive invention mechanisms with international invention recharge. Section 4.3 develops the induced adaptive invention model in further detail. Section 4.4 explores evidence for these alternative mechanisms in the agricultural sector. This evidence clearly supports induced adaptive invention mechanisms as the dominant mechanism of convergence. Research organizations for agriculture have been designed to facilitate this form of convergence and have succeeded in doing so.

Section 4.5 explores industrial technology convergence in a setting where mimicry mechanisms have dominated technology policy. Evidence is reported, showing that the induced adaptive invention form of convergence with international invention recharge receives strong support from data on inventions patented. The section concludes that, to a considerable degree, the failure of industrial policy makers to recognize the importance of the adaptive invention/innovation mechanism has retarded industrial development.

4.2   The Mechanics of Convergence

In this section a distinction is made between two classes of convergence mechanisms: mimicry and adaptive invention/innovation.

4.2.1   Mimicry mechanisms

Mimicry mechanisms are appropriate when firms in a developing country can find cost-reducing techniques by simply mimicking current or past techniques used by firms in developed countries. This implies that the cost-reducing character of technological innovation in developed countries is effectively available to developing countries. Mimicking also applies to product change technology.

In this chapter we consider three classes of mimicry mechanisms:

- Simple mimicry—with low tacit knowledge requirements and low transaction costs.

- Transaction cost constrained mimicry—with low tacit knowledge requirements.

- Tacit knowledge constrained mimicry—with low transaction costs.

Simple mimicry is what many development policy makers had in mind in 1950. It is also the model implicitly behind import substitution policies. Had this mechanism actually dominated, convergence in developing country incomes would have occurred.

Transaction cost constrained mimicry could, in principle, explain convergence failure. Countries with inadequate institutions (property rights, bankruptcy laws, etc.) could be prevented from attaining mimicry convergence. However, transaction cost constrained mimicry is not consistent with super-convergence. It could, however, explain the division between converging and non-converging economies.

Tacit knowledge constrained mimicry is perhaps the favored explanation for convergence failure. The basic idea is that the mastery of technology requires more than experience and normal skills. Some degree of engineering competence is required for mimicry. This mechanism is consistent with the technical assistance provisions in many foreign direct investment contracts. It is, at least in principle, consistent with convergence failure (where the engineering skills are insufficient) and super-convergence (where sufficient skills are available).[3]

4.2.2   Adaptive invention/innovation mechanisms

Adaptive invention/innovation mechanisms are appropriate when a developing country finds that the developed country technology available for mimicking has impaired cost-reduction effectiveness because of differences in production conditions between developed and developing countries. These differences may be differences in prices, institutions, or natural environments such as soil or climate conditions.

This impairment of cost reduction effectiveness opens up scope for adaptive invention/innovation activities to modify the developed country technology so as to reduce the impairment factor. Thus, in order to exploit this mechanism, the recipient developing country must have an invention/innovation capacity in place.[4]

In a globalized trading world, price differences between produced goods will not differ very much between countries. With cost-efficiency differences between developed and developing countries, the chief price difference will be the price of labor. This has two effects on convergence mechanisms. First, some developed economy technological improvements in machine processes may effectively be unavailable at all to developing countries with low wages. For example, rice harvesting equipment improvements may be of no value to economies where rice is harvested by hand because of low wages. Second, even where machine improvements are of value in developing countries, their value may be reduced because the improvements were induced by high wage conditions.

Production differences associated with natural conditions—soil, temperature, rainfall, day length, etc.—most clearly affect production requiring biological activity. Plants and animals perform differently in different environments.

Darwinian natural selection produced plant and animal species reflecting comparative natural advantages. Farmer-produced improvements in cultivated crop species and in domesticated animals were also governed by Darwinian sensitivity. Similarly, today's modern plant breeding crop genetic improvements remain conditioned by these forces. Thus, the development of improved crop rice varieties in Japan had no direct value in India. But the genetic resources and the methodology behind the Japanese development were very important to the adaptive invention of crop varieties in India.

Models of adaptive invention incorporate international invention recharge as a central feature. The next section develops this feature further.[5]

4.3   The Induced Invention/Innovation Convergence Model with an Application to Plant Breeding

The traditional induced innovation model postulates an "innovation-possibility frontier" (IPF) in economic dimensions or traits. The key insight of this model is that the economic traits have different values that should guide inventive effort. The model developed in this section is presented in terms of plant traits sought by plant breeders. These plant traits can be given economic values. An important part of this model, however, is the invention recharge specification, particularly the international recharge mechanism that is subjected to test in Sections 4.4 and 4.5.

Plant breeders have two alternative search strategies in their research programs. The first of these is the search for "quantitative" plant traits governing yields. Quantitative traits are controlled by multiple genes (or genetic alleles) and require complex strategies for crossing parental materials and selecting improved cultivars. The second strategy is the search for "qualitative" traits such as host plant resistance (HPR) to the tungro virus in rice. Qualitative traits typically are controlled by a single gene.

Both breeding strategies rely on searching for genetically controlled traits in collections of crop genetic resources (CGRs), which include landraces of the cultivated species (distinct types selected by farmers over centuries from the earliest dates of cultivation and diffused across different ecosystems), "wild" (related) species, and related plants that might be combined. CGR collections also include "combined" landraces, including varieties (officially recognized uniform populations of combined landraces, often with many generations of combinations). The systematic combining of landraces and evaluation is termed "pre-breeding."[6]

4.3.1   The simple one-trait, one-period model

Here, we consider the following representation of the single-trait, one-period model.

In period 1, the existing breeders' techniques and breeders' CGR collections determine a distribution of potential varieties indexed by their economic value, x.

Following Evenson and Kislev (1976), suppose this distribution to be an exponential distribution:

$$f(x) = \lambda e^{-\lambda(x-\theta)}, \theta \leq x \,. \tag{4.1}$$

The cumulative distribution is

$$F(x) = 1 - e^{-\lambda(x-\theta)} \,, \tag{4.2}$$

with mean and variance

$$E(x) = \theta + 1/\lambda \,, \tag{4.3}$$

$$Var(x) = 1/\lambda^2 \,. \tag{4.4}$$

The cumulative distribution of the largest value of $x(z)$ from a sample of size (n) is the "order statistic" (Evenson and Kislev 1976),

$$H_n(z) = \left[1 - e^{-\lambda(z-\theta)}\right]^n \,, \tag{4.5}$$

and the probability density function for (z) is

$$h_{n(z)} = \lambda n \left[1 - e^{-\lambda(z-\theta)}\right]^{n-1} e^{-\lambda(z-\theta)} \,. \tag{4.6}$$

The expected value and variance of $h_n(z)$ are

$$E_n(z) = \theta + \frac{1}{\lambda} \sum_{i=1}^{n} \frac{1}{i} \approx \theta + \lambda \ln(n) \,, \tag{4.7}$$

$$Var_n(z) = \frac{1}{\lambda^2} \sum_{i=1}^{n} \frac{1}{i^2} \,. \tag{4.8}$$

Evenson and Kislev discuss the applicability of expressions (4.7) and (5.8) to plant breeding research. Basically, expression (4.7) can be thought of as the breeding production function. The approximation $\ln(n)$ is a reasonable approximation for any symmetric distribution $f(x)$ including the uniform distribution and the normal distribution. The marginal product of breeding effort is simply

$$\partial E_n(z)/\partial n = \lambda/n \,. \tag{4.9}$$

Given a measure of the units over which (z) applies (e.g., the areas in a specific ecosystem), the value of the marginal product V can be computed and set equal to the marginal cost of search to solve for optimal n:

$$\lambda V/n = MC(n) \,. \tag{4.10}$$

For two or more traits, each can be characterized by expression (4.7) with different parameters:

$$
\begin{aligned}
E_n(Z_1) &= t_1 = \theta_1 + \lambda_1 \ln(N_1)\,, \\
E_n(Z_2) &= t_2 = \theta_2 + \lambda_2 \ln(N_2)\,, \\
E_n(Z_n) &= t_n = \theta_n + \lambda n \ln(N_n)\,.
\end{aligned}
\tag{4.11}
$$

When these traits are qualitative traits, breeders typically search for them independently because there are techniques that enable the breeders to incorporate only the single trait in a cultivar (i.e., by back crossing the other methods, unwanted traits can be discarded). Thus, even if traits are highly correlated, the breeder will search independently for them.[7]

Figure 4.1 depicts distributions for two traits: x_1 and x_2. If we set $N = N_1 + N_2$ at some level (say, the optimizing level) where

$$
MC(N_1) = MC(N_2) = \alpha_1 V_1 N_1 = \alpha_2 V_2 N_2\,,
\tag{4.12}
$$

we have the IPF depicted in Figure 4.1. (Note that the depiction is in terms of traits, but that these can be translated into economic units through values.) The point TD in Figure 4.1 is the technology determination point determined by the maximum value of the traits in each distribution (or an arbitrary stopping point in each distribution).

4.3.2   Multiple periods without recharge

Now consider periodicity. In practice, we do not observe the single-period optimal search implied by expressions (4.7), (4.11), and (4.12). Typically, we observe multiple-year research and development (R&D) programs even for narrowly defined objectives. Could we treat this multiple-year sequence as simply a long period instead of a sequence of periods? Certainly not in plant breeding. Plant crosses (genetic combinations) must be evaluated and selected over several generations. Plant "types" (quantitative) are built with multiple-generation crosses, where the crossing decisions for the second generation can effectively be made only after the first generation has been observed.[8]

In addition to this periodicity, two related types of shifts in the invention distributions are relevant. The first is periodicity associated with search field narrowing (elimination of unpromising search avenues). The second is recharge shift, discussed below.

The search field narrowing case is depicted in Figure 4.2, where a rightward shift in the mean of the distribution, but not the right-hand tail of the distribution, is depicted from period 1 to period 2. This shift can be thought of as the systematic elimination of unpromising search avenues. It may be possible to classify material into n groups. In period 1 sufficient sampling is undertaken to enable the estimation of the mean and variance for each of the n groups. On the basis of these estimates, several groups may be eliminated, with the resultant distributional shift depicted in Figure 4.2.[9]

The shift in the mean for both trait distributions is proportional to the period 1 optimal discovery as depicted in the IPF diagram for the two traits.

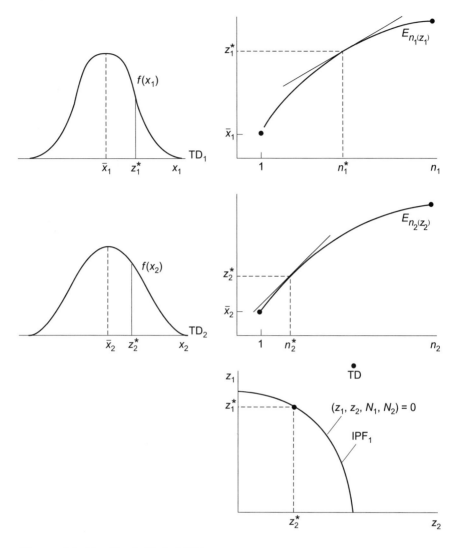

Figure 4.1. The Single-Period IPF.

The shape of IPF2 is affected by the period 1 search. Because the period 1 search was induced by prices to produce more n_{11} search for t_1 than n_{21} search for t_2, there is more exploitable search scope for t_2 in period 2. Thus the resultant optimal point on IPF_2 is not on the same ray from the origin as was the optimal point of IPF_1. The search exhaustion phenomenon has moved the optimal point in the direction of the TD point.

Implications of this search exhaustion case are that multiple-period searches can take place but that they will eventually stop. During the multiple-period

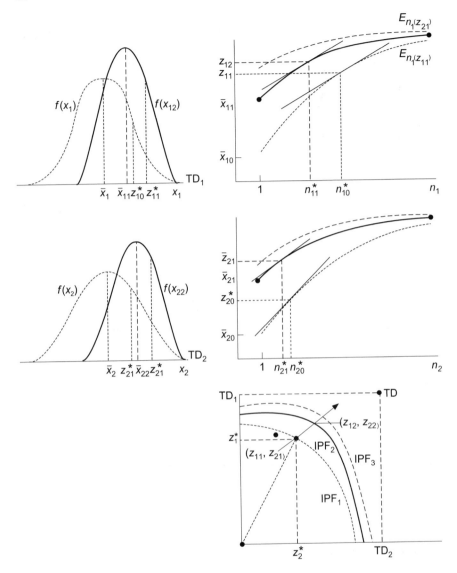

Figure 4.2. Multiple-Period Search with Search Field Narrowing.

search, the ratio of inventions (proportional to $t_{11} - t_{10}, t_{12} - t_{11}$, etc.) to search resources (R&D) n will decline.

4.3.3 Multiple periods with recharge

The second type of invention distribution shift is associated with several types of recharge mechanisms.[10] These include the following:

- Genetic resource collection and evaluation programs. These programs are designed to discover uncollected materials and make them available to breeders.

- Pre-breeding programs where landrace materials are systematically combined into potential breeding lines by specialized research programs. These programs do not seek to develop "final products," that is, new cultivars. Instead they seek to evaluate and produce "advanced lines" that are then used by final product inventors.

- Wide-crossing programs where techniques for inter-specific combinations of genetic resources (between related species) are made possible. This expands the size and scope of the original materials that can be utilized in breeding programs.

- Transgenic breeding programs where DNA insertion techniques allow traits associated with alien genes (i.e., from unrelated species) to be incorporated into cultivated plants.

These programs are "pre-invention" science or applied science programs. They provide recharge to the invention distributions by shifting both the mean and the right-hand tail of the invention search distribution. The actual mechanism of recharge, however, is often in the form of biological invention or varieties that serve as parents in the recharged invention distribution.

Figure 4.3 depicts the nature of these shifts for search distributions and IPFs with recharge. Note that the technological determinism point, TD, moves with recharge. The reader can readily see that one could have cases of "super-recharge" for a number of periods where inventions per inventor might increase over time (e.g., in sugarcane breeding; see Evenson and Kislev 1975). But recharge science itself is likely to be subject to diminishing returns, unless it is also recharged by the more basic sciences.

These ideas can be clarified with a little algebra. Describe the breeding (invention) process as

$$
\begin{aligned}
T_1 &= T_1(G_1, B_1) = \theta_1 + \lambda_1 G_1 \ln(B_1) + \alpha_1 G_1 \,, \\
T_2 &= T_2(G_2, B_2) = \theta_2 + \lambda_2 G_2 \ln(B_2) + \alpha_2 G_2 \,, \\
T_3 &= T_3(G_3, B_3) = \theta_3 + \lambda_3 G_3 \ln(B_3) + \alpha_3 G_3 \,.
\end{aligned}
\tag{4.13}
$$

This system of equations describes the incorporation of traits as a function of germplasm, G_i, and breeding activity, B_i. The functional form is based on the search model. Note that the germplasm term enters linearly in this model. The implications of this are used in the empirical estimates reported below.

The first-order condition for allocating breeding research between any two traits when the marginal cost of T_i is equal to the marginal cost of T_j is

$$
\frac{B_i}{B_j} = \frac{\lambda_i G_i V_i}{\lambda_j G_j V_j} \,,
\tag{4.14}
$$

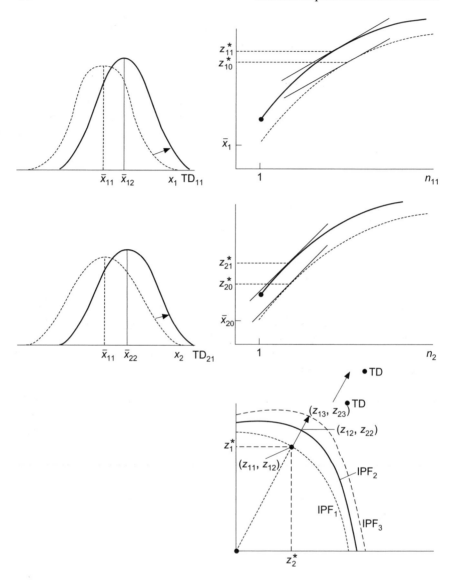

Figure 4.3. Multiple Search with Recharge.

where V_i and V_j are measures of the marginal contribution to crop value of traits i and j, respectively. (Note that each trait may appear in several varieties and that each variety may be planted in different areas.)

Now consider the production of germplasm G_i. This is characterized as being produced in a pre-breeding process:

$$G_1 = G_1(G_c, PB_1) = \delta_1 + \phi_1 G_c \ln(PB_1) \,,$$
$$G_2 = G_2(G_c, PB_2) = \delta_2 + \phi_2 G_c \ln(PB_2) \,,$$
$$\vdots$$
$$G_n = G_n(G_c, PB_n) = \delta_n + \phi_n G_c \ln(PB) \,. \qquad (4.15)$$

In this pre-breeding process, pre-breeding activity converts evaluated genetic resources G_c into germplasmic breeding materials. This process is also a search process:

$$G_c = G_c(G_n, E, C) \,. \qquad (4.16)$$

Evaluated germplasm is produced by the natural stock of genetic resources G_n and collection (C) and evaluation (E) activities.

The following features of this model can be noted:

- If the marginal search coefficients are equal ($\lambda_i = \lambda_j$), plant breeding activity obeys "the congruence rule," where inventive activity is proportional to the value of the units affected [see expression (4.13)].

- Departures from congruence (a strong form of induced innovation) are justified when search parameters differ.

It can be further noted that the optimal conditions for pre-breeding or germplasmic recharge science [expression (4.14)] also imply that, if the germplasmic search coefficients are equal, then at least partial congruence occurs for both pre-breeding and breeding (partial, because of the common G_c term). This is a strong form of multiple-period induced innovation. The multiple-period invention path is a ray from the origin (if prices do not change) that is parallel to the TD expansion path. A change in prices (values) will result in a change in both the invention path and the TD path.

Agricultural experiment stations were developed to produce inventions for farmers. Extension systems were developed to diffuse these inventions. Over the years, these institutions have been continuously in tension over the relative weights to place on extension, invention, and pre-invention or recharge science. Figure 4.4 (Huffman and Evenson 1993) reflects the institutionalization of the level II recharge sciences in modern agricultural research systems.

Farmers express a demand for inventions (level III) and invention products (level IV). The agricultural experiment stations, however, have been able to convince state legislatures in the United States that level II pre-invention sciences are necessary to recharge the invention distributions. This represents a type of public sector competition to provide services to farmers at the state level. Comparable systems exist for medical technology but are weak for many other fields of technology.

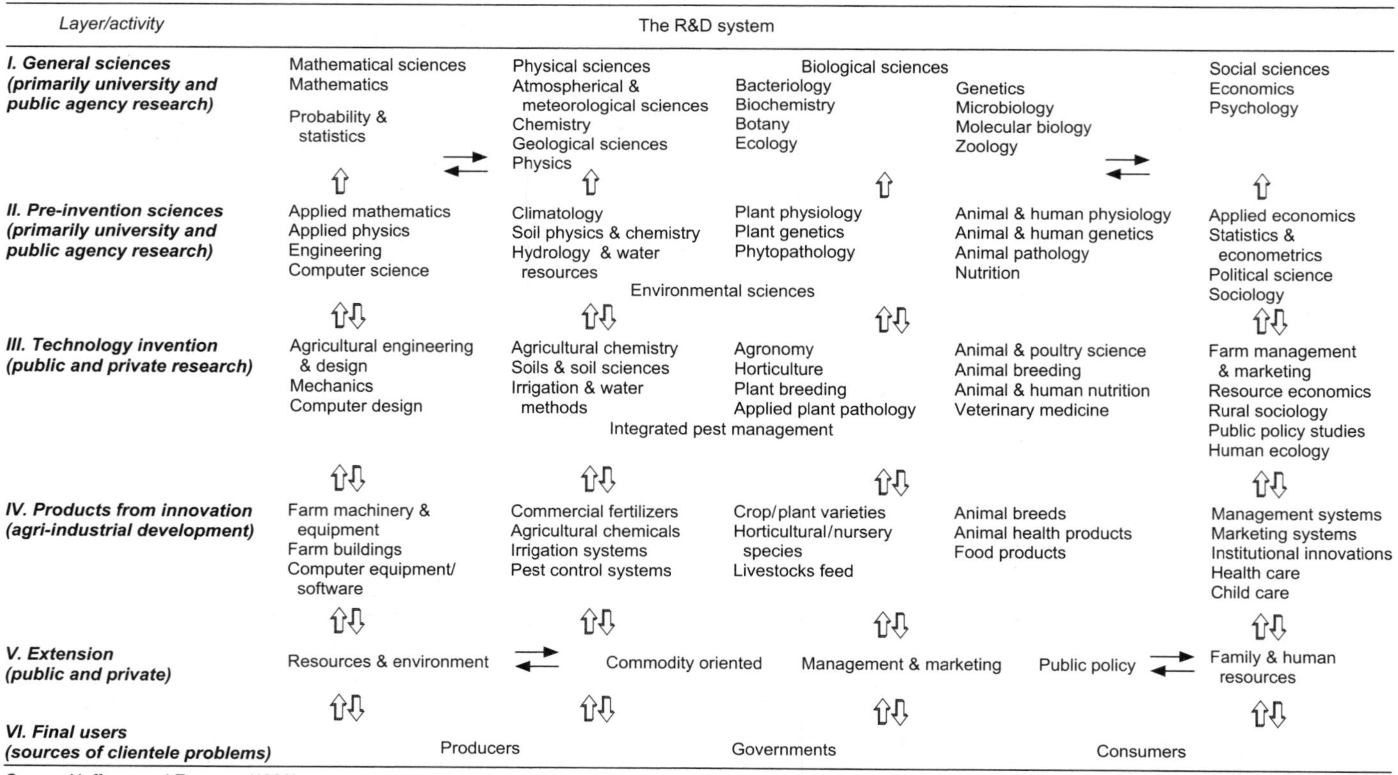

Layer/activity	The R&D system			
	Mathematical sciences	Physical sciences	Biological sciences	Social sciences
I. General sciences (primarily university and public agency research)	Mathematics Probability & statistics	Atmospherical & meteorological sciences Chemistry Geological sciences Physics	Bacteriology Biochemistry Botany Ecology	Genetics Microbiology Molecular biology Zoology	Economics Psychology
II. Pre-invention sciences (primarily university and public agency research)	Applied mathematics Applied physics Engineering Computer science	Climatology Soil physics & chemistry Hydrology & water resources *Environmental sciences*	Plant physiology Plant genetics Phytopathology	Animal & human physiology Animal & human genetics Animal pathology Nutrition	Applied economics Statistics & econometrics Political science Sociology
III. Technology invention (public and private research)	Agricultural engineering & design Mechanics Computer design	Agricultural chemistry Soils & soil sciences Irrigation & water methods *Integrated pest management*	Agronomy Horticulture Plant breeding Applied plant pathology	Animal & poultry science Animal breeding Animal & human nutrition Veterinary medicine	Farm management & marketing Resource economics Rural sociology Public policy studies Human ecology
IV. Products from innovation (agri-industrial development)	Farm machinery & equipment Farm buildings Computer equipment/software	Commercial fertilizers Agricultural chemicals Irrigation systems Pest control systems	Crop/plant varieties Horticultural/nursery species Livestocks feed	Animal breeds Animal health products Food products	Management systems Marketing systems Institutional innovations Health care Child care
V. Extension (public and private)	Resources & environment	Commodity oriented	Management & marketing	Public policy	Family & human resources
VI. Final users (sources of clientele problems)	Producers		Governments		Consumers

Source: Huffman and Evenson (1993).

Figure 4.4. Institutional Specialization in R&D Systems for Agriculture.

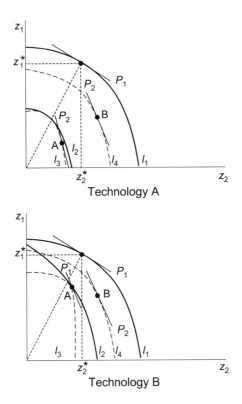

Figure 4.5. Spatial Spillovers.

4.3.4 Technology spillovers

The problem of spillovers has been recognized in agricultural research systems and is reflected in numerous locations of agricultural experiment stations around the world (see discussion of rice below) with a high degree of germplasmic spillovers and adaptive invention for targeted ecosystems.[11] It is well known that plants and animals perform differently in different ecosystems and that modern plant-breeding has only partially overcome the "Darwinian" adaptation to ecosystem niches in nature. It is also well known that relative prices affect the real value of an invention (an improved rice harvesting machine is valuable in Texas but has no real value in Bangladesh, where wages are low and rice is harvested by hand).

Figure 4.5 illustrates these issues for two types of technology, A and B.

- I_2 shows how non-price factors (ecosystem institutions, etc.) remediate the real performance of z_1 and z_2 in location 2 and lead to an interior IPF.

- I_3 is the real-value IPF in location 2 given that location 1 produced (z_1^*, z_2^*). This lies below I_2. Location 2 will then have direct spill-in, shown as point A.

- I_4 is the IPF now available to location 2 should it choose to undertake research. Location 2 has a choice between no research (point A on I_3) and conducting its own research (point B on I_4).

- Technology B has the same I_1 as technology A but lower non-price remediation.

It is generally thought that agricultural technology is characterized by technology A, where non-price remediation is high and adaptive research potential is good. For biological traits, relative prices also may not differ between location 1 and location 2. This will lead to strong incentives to locate research capacity in both locations. These research programs "feed" off each other and sometimes on international recharge programs (see below).

Mechanical technology is thought to be more like technology B, where non-price remediation is low, price differences are great, and adaptive and germplasmic potential are low (at least at location 2's prices).

4.4   Evidence from Agriculture in Developing Countries

Public sector agricultural research systems have been built in most countries of the world. These systems were among the earliest cases where governments recognized that incentive systems, chiefly intellectual property rights (IPRs) systems, were not sufficient to bring forth adequate invention from the private sector. In response, public sector colleges of agriculture and mechanics (A&Ms) were designed to train agricultural and engineering practitioners, and a system of public experiment stations was established to undertake biological invention. (See Figure 4.4 for further details.)

The seminal study of hybrid corn by Griliches (1957) provides insight concerning spillovers and recharge mechanisms. Figure 4.6, from the Griliches study, illustrates these factors. Griliches noted that the invention of hybrid corn was an invention of a technique or method of invention. It represented "recharge science." In this case, the recharge science was undertaken in public sector agricultural experiment stations (notably the Connecticut Agricultural Experiment Station in New Haven). Most actual invention was undertaken by private firms, although many state experiment stations also produced hybrid corn varieties.

Griliches also noted that corn plants have a high degree of genotype–environment interactions, that is, their performance is sensitive to soil type, day length, etc. (non-price remediation in Figure 4.5). Accordingly, the rapid adoption of hybrid corn varieties produced for Iowa and Illinois did not transfer to rapid adoption in Alabama, because the varieties suited to Iowa were not suited to Alabama. It was only after breeding programs designed for Alabama conditions were developed that Alabama farmers had access to the hybrid technology (and it was only after similar programs were developed in the Philippines in the 1980s

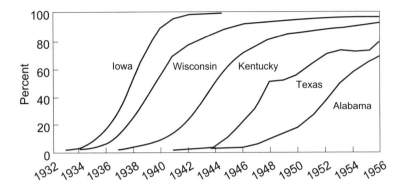

Figure 4.6. Adoption Patterns: Hybrid Corn in the United States, Percent of Total Corn Acreage Planted with Hybrid Seed.
Sources: USDA (various years); Griliches (1957).

that the technology was available there). Sub-Saharan Africa did not have access to the hybrid technology until the 1990s.

4.4.1   Investment strategy in developing countries

During the first half of the twentieth century, plant breeding programs were established in most developed countries and the specializations depicted in Figure 4.4 were established. Agricultural experiment stations were located in US states, and public competition to serve farmers emerged as the motivating force for investment. The lessons implicit in the hybrid corn experience regarding adaptive innovation and pre-breeding were incorporated into system design.

For the developing countries, agricultural research programs were not as well developed. Most developing country systems were guided by colonial politics. This did produce effective research programs for crops destined for export to the mother country (tea, coffee, sugar, spices), but for most food crops little real research capacity was in place.

During the 1950s, considerable emphasis was placed on extension programs providing farmers with technical advice. These programs were motivated by the "easy mimicry" model that also guided industrial policy. By the end of the 1950s, however, the broad outlines of the population expansion in developing countries were becoming clear. Most countries were faced with the prospect of a doubling, and in some cases a tripling, of population in the next four decades. This called for an unprecedented expansion of food production.

The response of development agencies, both bilateral and multilateral, was to support National Agricultural Research System (NARS) development and to support NARS programs by developing a system of commodity-focused International Agricultural Research Centers (IARCs). The design of the IARC system (supported by a consortium of donors—the Consultative Group for International

Agricultural Research) was based on two predecessor projects, each of which facilitated early "green revolution" achievements.

The first of these was a special project of the Rockefeller Foundation in Mexico. Beginning in the early 1940s, Norman Borlaug instituted a wheat breeding program designed to incorporate genetic improvements in temperate zone wheats into wheat varieties suited to the subtropical wheat regions in developing countries. By the early 1960s this program had achieved considerable success, and by the late 1960s wheat varieties from this program were credited with creating a green revolution in Asia.

The second program was the program for Japonica–Indica rice development supported by the Food and Agriculture Organization of the United Nations (FAO). This program was dedicated to incorporating features of temperate zone Japonica rice improvements into tropical zone Indica rice cultivars. While the program was discontinued at the end of the 1960s, it did lead to the development of several important rice varieties and established the foundations for the early development of semi-dwarf plant type rice varieties in the 1960s at the first IARC, the International Rice Research Institute (IRRI). This development was also credited with creating a green revolution.

The design of the IARCs was guided by past experience with agricultural experiment stations. The features of recharge science and the provision of germplasm to NARS in the crop-focused IARCs were guided by induced innovation perspectives. In a sense, the IARCs were built to remedy the lack of level II pre-invention science in developing country NARSs.

The IARCs were also designed to be governed by independent boards instead of member countries. Again, this organization was based on experience gained from regional and United Nations (UN) programs, where the member country model of control was clearly shown to be inconsistent with the effective conduct of science. This enabled the IARC system to recruit highly qualified scientists into the system.

4.4.2   Varietal production and germplasm

Early accounts of the green revolution in developing countries are incomplete in two dimensions. First, they are based primarily on crop genetic improvement (CGI) in the form of "modern" varietal (MV) development in wheat and rice. Second, they are based on varietal development for the period from 1965 to 1980. A more complete picture of the green revolution is reported in Table 4.1, where a summary of the production of modern improved crop varieties for 10 food crops is presented. These 10 crops account for 85 to 90 percent of food crop production in the developing world. (Rice is the most important crop, followed by wheat and maize.) Table 4.1 also presents data for five-year periods extending through the 1980s and most of the 1990s. Summaries of varietal releases by region are also reported.[12]

Table 4.1. Average Annual Varietal Releases by Crop and Region, 1965–1998.

Crop	Average annual releases							1965–1998 IARC content			
	1965–1970	1971–1975	1976–1980	1981–1985	1986–1990	1991–1995	1996–1998*	IX	IP	IA	IN
Wheat	40.8	54.2	58.0	75.6	81.2	79.3	(79.3)	.49	.29	.08	.14
Rice	19.2	35.2	43.8	50.8	57.8	54.8	58.5	.20	.25	.07	.48
Maize	13.4	16.6	21.6	43.4	52.7	108.3	71.3	.28	.15	.04	.53
Sorghum	6.9	7.2	9.6	10.6	12.2	17.6	14.3	.16	.07	.06	.71
Millet	0.8	0.4	1.8	5.0	4.8	6.0	9.7	.15	.41	.09	.35
Barley	0.0	0.0	0.0	2.8	8.2	5.6	7.3	.49	.20	.01	.30
Lentils	0.0	0.0	0.0	1.8	1.8	3.9	(3.9)	.54	.05	.01	.40
Beans	4.0	7.0	12.0	18.5	18.0	43.0	(43.0)	.72	.05	.01	.19
Cassava	0.0	1.0	2.0	15.8	9.8	13.6	(13.6)	.53	.15	.01	.31
Potatoes	2.0	10.4	13.0	15.9	18.9	19.6	(19.6)	.17	.06	.02	.75
All crops											
Latin America	37.8	55.9	65.9	92.5	116.2	177.3	139.2	.39	.14	.04	.43
Asia	27.2	59.6	66.8	86.3	76.7	81.2	79.9	.18	.29	.10	.43
Middle East/North Africa	4.4	8.0	10.2	12.2	28.4	30.5	82.2	.62	.22	.04	.12
Sub-Saharan Africa	17.7	18.0	23.0	43.2	46.2	50.1	55.2	.45	.21	.07	.27
All regions	87.1	132.0	161.8	240.2	265.8	351.7	320.5	.36	.17	.05	.42

*Numbers in parentheses are simple repetitions of 1991–1995 rates because of insufficient data.

Note: IX = variety based on IARC cross; IP = variety based on NARS cross with at least one IARC parent; IA = variety based on NARS cross with at least one non-parent IARC ancestor; IN = variety based on NARS cross with no IARC ancestors.

IARC content in released varieties is summarized by crop and region as well. For all crop varieties, 36 percent were produced in an IARC breeding program where the cross leading to the varietal release was made in the IARC program (IX). The remaining varieties were based on crosses made in a NARS program.

NARS-crossed releases can be further classified according to whether one or both parents in the cross was an IARC release. For all NARS-crossed varieties, roughly 17 percent had at least one IARC-crossed parent (IP). (These varieties accounted for 25 percent of MV acreage in 1998.) This attests to a strong germplasmic recharge effect, since NARS breeders found success in using IARC parental material. When grandparents and other ancestors of NARS-crossed varieties are considered (IA), IARC-crossed germplasm appears in 23 percent of all NARS-crossed varieties. This IARC germplasm proportion is also rising over time.

There are significant differences in these patterns among crops. There are also differences in what might be termed the "maturity" of the breeding programs by crop. In wheat, the total number of releases has stayed relatively constant since 1985, but with a high proportion of varieties based on IARC crosses or parents. The varieties/breeder ratio has been constant since 1985. Wheat is produced over a narrower range of climate and biotic diversity than is rice and has a relatively high level of multiple releases. This might be termed a mature pattern.

Rice also exhibits a mature pattern, but of a different type. Total releases have also been roughly constant since the mid-1980s, but the IARC cross proportion has declined from the "green revolution" levels of the 1970s. This appears to be a case of maturing and strengthened NARS programs. Previous work by Gollin and Evenson (1997) supports this interpretation.

Maize, the third most important cereal grain in developing countries, offers another pattern. Public sector releases appear to be rising, with a relatively low IARC germplasm component. Private sector varietal production, primarily of hybrids, is clearly increasing. It is also clear that IARC germplasm has been useful to private breeders, along with NARS germplasm. This is a case of public sector research creating a "platform" on which the private sector can be productive. (Note that this also provides a platform for modern biotechnology products.)

The pattern for sorghum is roughly similar to that for maize, again with a growing proportion of hybrid sorghum varieties being produced by private sector breeders. The pearl millet pattern indicates relatively weak NARS production in early years. This is a case where the IARC program not only provided germplasm to NARS but initiated expanded CGI work generally. Until ICRISAT began its CGI work, there was little useful raw material for NARS programs to work with.[13]

The pattern where the IARC programs effectively initiated breeding work on a crop holds for barley, lentils, and cassava, as well as for millet. In each of these crops the IARC cross proportion is high and total varietal production is generally rising. The NARS CGI programs for these crops are not very mature at this point.

For beans, IARC programs have also stimulated increased varietal production, with IARC crosses accounting for high proportions of released varieties. In sub-Saharan Africa, the IARC programs are dominant in beans, cassava, and potatoes.

The conditions for potato CGI differ from those of other crops because of different taste and management factors. IARC cross shares are low but have been rising.

Table 4.1 also reports release data for all crops by region. These data show that the highest rate of increase in varietal production in the 1980s and 1990s occurred in the Middle East/North Africa and sub-Saharan Africa regions. These regions were also the most dependent on IARC crosses and germplasm.

A specific study of rice varieties (Gollin and Evenson 1997) reported that, of the rice varieties released over the 1965–1991 period, only 6 percent were crosses made in one NARS system that were released in another NARS system. However, for parents of releases, 18 percent were crossed in one NARS program and served as a parent in another NARS program. IRRI was responsible for 17 percent of varietal crosses and 24 percent of parental crosses. Only 8 percent of all modern rice varieties released since 1965 did not have international landrace content in their genealogy.

Total varietal releases for all crops show a steady increase over time. Annual varietal releases in the 1990s were more than double the releases during the 1970s.

The induced innovation model outlined earlier provides a basis for testing for a germplasm impact on NARS breeding programs. The "breeding with recharge" function actually imposes a specific functional form for the germplasm impact, and this form can be tested against a more general form. This test is carried out using NARS data for three periods, 1965–1975, 1976–1985, and 1986–1996, for varietal releases in wheat, rice, maize, beans, and potatoes.

The specific functional form implied by the induced innovation model is

$$V_N = a + b \ln(B_N) + c\, G_I + d\, G_I \ln(B_N)\,, \tag{4.17}$$

where V_N is the number of varieties produced by a NARS CGI program in a given period, B_N measures the NARS plant breeding resources employed during the period, and G_I is a measure of IARC germplasm available to the program.

The principle of diminishing returns dictates the logarithmic specification. The G_I variable is not in logarithmic form because it is not part of the NARS search per se. That is, IARC germplasm affects NARS productivity, but (except indirectly) NARSs do not produce IARC germplasm. G_I is thus a linear shifter of the search distribution for NARS and has the specific form noted in expressions (4.17) and (4.13).

A specification that does not rely on search theory would use a general production function. Perhaps the most widely used production functional form is the Cobb-Douglas form:

$$V_N = A\, B_N^{\alpha}\, G_I^{\beta}\,. \tag{4.18}$$

The variables are defined as follows:

- V_N is the number of varietal releases based on NARS crosses over the period, where N indexes countries.

- B_N is the number of scientists engaged in CGI research on the crop at the beginning of the period. B_N was estimated in two stages. First, the

Table 4.2. Estimates: Indirect Impacts of IARC Germplasm on NARS CGI Programs.

Dependent variable = NARS-bred varietal releases by period: 1965–1975, 1976–1985, 1986–1997 (t ratios in parentheses).

	Wheat	Rice	Maize	Beans	Potatoes	Pooled
Observations	84	81	51	66	33	315
R^2 (adj.) [specification (4.17)]	0.403	0.383	0.568	0.505	0.740	0.459
ALT R^2 (adj.) [specification (4.18)]	0.311	0.215	0.098	0.174	0.630	(0.289)
Constant	4.616	−2.473	−4.241	6.905	−14.690	7.387
	(0.27)	(0.30)	(0.40)	(1.91)	(1.53)	(1.20)
ln(BN)	4.247	2.665	−0.777	−0.771	4.635	2.915
	(1.15)	(1.55)	(0.35)	(0.88)	(2.84)	(2.72)
ln(BN)GI	0.552	0.534	0.491	0.748	1.458	0.568
	(3.35)	(2.33)	(3.44)	(5.82)	(3.69)	(7.82)
GI	−2.020	−1.433	−0.718	−1.925	−2.457	−1.904
	(2.52)	(0.99)	(0.95)	(3.56)	(1.47)	(5.05)
D 65–75	−0.917	3.313	9.030	−5.480	6.751	−1.492
	(0.08)	(0.46)	(1.19)	(1.72)	(1.19)	(0.35)
D 76–85	0.890	−0.416	9.500	−2.220	8.460	−0.360
	(0.09)	(0.06)	(1.56)	(0.82)	(1.72)	(0.10)
D Beans						−9.070
						(2.04)
D Maize						−7.820
						(1.81)
D Potatoes						−5.760
						(1.09)
D Rice						−7.570
						(1.85)

total number of senior agricultural scientists for the period and country was computed from the International Service to National Agricultural Research (ISNAR) database (Pardey and Roseboom 1989). Then a search of the FAO Agrostat database was conducted for publications on plant breeding and related activities by crop, as well as on social science research, animal and pasture research, and other fields of agricultural science. Publication shares for plant breeding on the crops in question were then formed for each country. These shares were then multiplied by the ISNAR scientist data to obtain our measure of B_N.

- G_I is measured as the cumulated number of IARC crosses released as varieties in the country. This definition of germplasmic input attempts to correct for the fact that only a subset of IARC-crossed material is relevant in a given country. If the country actually released an IARC cross as a variety, this is taken to be an indication of relevance.

Table 4.2 reports a goodness of fit test (adjusted R^2) for the two specifications. In all cases, specification (4.17) fit the data significantly better than specification (4.18). This supports the interpretation of the coefficients as real research effects. This is a case where model-guided specification is possible and supported by the data.

Coefficients for specification (4.17) are reported in Table 4.2. The key germplasm impact variable is $\ln(B_n)G_I$. The coefficients for this variable are positive and highly significant in all specifications. The coefficients are similar for crops separately and pooled. The net effect of the variable G_I depends on both the coefficient of G_I and on the coefficient of $\ln(B_n)G_I$. Since G_I is growing over time, the negative coefficient on G_I is adjusting for this to some degree. However, the net effect of G_I is positive and large. The production elasticity of G_I is approximately .3, evaluated at the mean of the data in the pooled estimates.

The varietal production elasticity of the NARS breeding effect depends on the coefficient of $\ln(B_n)$ as well as on the coefficient of $\ln(B_n)G_I$. For the pooled estimate, the production elasticity of B_n is approximately .7 (note the dependent variable is not in logarithms, so this is not a constant elasticity).

Thus, NARS breeding effects are subject to diminishing returns. NARS programs have approximately doubled over the periods studied. This would have produced a 70 percent increase in varietal production in the absence of G_I effects. The G_I effects were quite large and contributed roughly 30 percent more varietal production than would have occurred in their absence. Thus, IARC germplasm impacts on NARS CGI programs were sufficient to offset the diminishing returns to NARS breeding effects over the periods covered.

Evenson (2001) also estimated the net effect of IARC germplasm on NARS plant breeding investments. Since IARC programs both compete with and complement NARS programs, IARC germplasm could either crowd out or stimulate NARS investments. The study estimated NARS investment specifications and concluded that IARC germplasm stimulated NARS investments.

4.4.3 Modern variety adoption and location specificity

As noted in the previous section, roughly 35 or 36 percent of varieties released were crossed in an IARC program, and many of these crossed international borders in the form of releases. An additional 8 to 10 percent of these varieties were crossed in one NARS program and released in another. (Data on these international flows will be compared with data on patented inventions in the next section.)

Farmers actually have to adopt modern varieties if they are to have a production impact. Data on MV adoption are presented by crop and region in Table 4.3. These data are characterized by two features. First, adoption rates differ by region for the same crop, attesting to high degrees of location specificity. Second, adoption rates are correlated with varietal production data, but the correlation is far from perfect. In particular, both the Middle East/North Africa and sub-Saharan

Table 4.3. Modern Variety Diffusion: Percent area Planted with Modern Varieties, 1970, 1980, 1990, and 1998.

	Latin America				Asia (including China)				Middle East/North Africa				Sub-Saharan Africa			
	1970	1980	1990	1998	1970	1980	1990	1998	1970	1980	1990	1998	1970	1980	1990	1998
Wheat	11	46	82	90	19	49	74	86	5	18	38	66	5	22	32	52
Rice	2	22	52	65	10	35	55	65					0	2	15	40
Maize	10	20	30	46	10	25	45	70					1	4	15	17
Sorghum					4	20	54	70					0	8	15	26
Millet					5	30	50	78					0	0	5	14
Barley									2	7	17	49				
Lentils									0	0	5	23				
Beans	1	2	15	20	0	0	2	12					0	0	2	15
Cassava	0	1	2	7	30	50	70	90					0	0	2	18
Potatoes	25	54	69	84									0	25	50	78
All crops	8	23	39	52	13	42	62	80	4	13	29	58	1	4	12	26

Africa regions have low levels of MV adoption in 1970 and 1980, even though there were significant levels of varietal release in preceding periods. Many earlier releases were not widely adopted because of susceptibility to plant diseases, insect pests, and abiotic stresses. It is when varieties are developed in response to these problems that high levels of MV adoption are observed.

MV adoption rates can be converted into growth contributions with estimates of the production gain associated with MV adoption. The CGI study from which these data were drawn included three country studies (India, China, and Brazil) where these production gains were estimated. Estimates were based on cross-section variability in adoption. Adoption was treated as an endogenous variable in these studies. The resultant growth contributions are reported by decade and region in Tables 4.4 and 4.5.

Table 4.4 reports annual CGI contributions for the crops in the study. These rates differ by crop because of differences in MV adoption rates. For all crops combined, 35 percent of the area under MVs in 1998 was based on an IARC cross (IX). Another 30 percent was based on a NARS cross with an IARC-crossed ancestor (IA).

Table 4.5 reports CGI production growth components for all crops by region. This table essentially describes the "transformation" of developing country agriculture. Comparisons are often made between countries in the sub-Saharan Africa region and to a lesser extent in the Middle East/North Africa region showing low rates of productivity growth relative to Asian economies. Tests of these CGI component estimates with actual yield changes indicate that, for countries with low levels of research investment, the CGI component makes up more than half the total factor productivity (TFP) growth measured for the region.

Given this, Table 4.5 explains the poor growth performance of the sub-Saharan Africa and Middle East/North Africa regions. In the 1960s and 1970s, sub-Saharan Africa experienced virtually no CGI growth. Even in the 1980s, the region experienced only one-third the growth seen in Asian economies. The Middle East/North Africa region also lagged in CGI growth until the 1990s. Note, however, that the IARC content in both released and adopted varieties was highest in these regions, indicating a strategy of IARC compensation for weak NARS programs.[14]

4.5 Industry Evidence

As noted in the introduction to this chapter, most economists concerned with industrial development have adopted a "mimicry" policy philosophy and hence have not stressed the development of an industrial R&D capacity in developing countries except in advanced stages of development. This is in sharp contrast to agriculturalists, who, by and large, conclude that technology–ecosystem interactions produce a high degree of location specificity calling for a research (especially CGI) capacity in all important regions at all stages of development.

Table 4.4. Crop Genetic Improvement Contributions to Yield Growth, by Crop.

Crop	1960s	1970s	1980s	1990s	1960–1998	Contribution shares			
						Adoption (1998)		Varieties (1965–2000)	
						IX	IA	IX	IA
Wheat	0.514	0.981	1.125	0.975	0.960	0.32	0.32	0.49	0.37
Rice	0.342	0.940	0.959	0.747	0.794	0.29	0.29	0.20	0.32
Maize	0.311	0.481	0.733	0.906	0.665	0.23	0.32	0.28	0.19
Sorghum	0.055	0.391	0.716	0.676	0.504	0.22	0.16	0.16	0.11
Millet	0.228	0.428	0.537	0.854	0.565	0.27	0.38	0.15	0.50
Barley	0.073	0.199	0.424	1.010	0.490	0.50	0.30	0.49	0.20
Lentils	0.000	0.000	0.193	0.750	0.283	0.70	0.20	0.54	0.65
Beans	0.022	0.027	0.367	0.331	0.208	0.80	0.20	0.72	0.05
Cassava	0.000	0.006	0.087	0.636	0.222	0.74	0.19	0.53	0.16
Potatoes	0.708	0.711	0.749	0.846	0.739	0.08	0.09	0.17	0.08
All crops	0.321	0.676	0.832	0.823	0.718	0.35	0.30	0.36	0.22

Note: IX = varietal cross made in IARC program; IA = varietal cross in NARS program with IARC ancestor.

Table 4.5. Crop Genetic Improvement Contributions to Yield Growth, by Region.

| | | | | | | Contribution shares | | | |
| | | | | | | Adoption (1998) | | Varieties (1965–2000) | |
Region	1960s	1970s	1980s	1990s	1960–1998	IX	IA	IX	IA
Latin America	0.312	0.600	0.781	0.751	0.658	0.28	0.30	0.39	0.18
Asia (including China)	0.452	0.932	1.030	0.890	0.884	0.30	0.37	0.18	0.39
Middle East/North Africa	0.141	0.270	0.681	1.228	0.688	0.51	0.31	0.62	0.28
Sub-Saharan Africa	0.017	0.142	0.358	0.497	0.280	0.44	0.27	0.45	0.28
All regions	0.321	0.676	0.832	0.823	0.718	0.35	0.34	0.36	0.19

Note: IX = varietal cross made in IARC program; IA = varietal cross in NARS program with IARC ancestor.

Because industrial technology policy is dominated by the mimicry perspective, it has stressed direct foreign investment as a vehicle for acquiring tacit knowledge and has downplayed the building of domestic R&D capacity. Experience with industrial R&D and invention in the chemical, electrical, and mechanical fields has been quite different regarding public research organizations than has been the case for biological invention. Public sector research programs have been effective in biological inventions. They have not been as effective in other fields of invention. Nor have public sector extension programs for industry been as effective as they have been for agriculture.

As a consequence of the mimicry-based technology policy for industry, only 25 or so developing countries have a bona fide R&D capacity in producing industrial firms. Of these, only 15 to 20 have effective intellectual property rights systems (Evenson and Westphal 1994).

4.5.1   Indirect evidence on location specificity: International patent data by industry

International patent data afford a related measure of location specificity and enable a comparison between agricultural inventions and industry inventions. International patent classifications (IPCs) are given to patents for virtually all countries maintaining patent protection systems. The International Patent Documentation (INPADOC) database includes IPCs. The Yale Technology Concordance (YTC), based on Canadian Patent Office industry assignments, enables the assignment of patents to industries of manufacture (IOMs) and sectors of use (SOUs). This database also records the country of origin (the priority country) and the granting country for each patent.

Table 4.6 illustrates the priority country versus granting country for seven countries for the 1975–1988 period for six IPC-defined fields of agriculturally related inventions. According to Paris Convention rules, a priority inventor has a relatively short period (one year) to obtain patent protection in another member country while maintaining the priority date. The decision to obtain protection abroad depends on several factors, including the degree of location specificity of the invention. These factors also include the perceived importance of the invention and the size of the granting country market.[15]

Table 4.6 illustrates the differences in these decisions by field of technology. For non-fertilizer agricultural chemicals, the off-diagonal elements are higher relative to the diagonal elements than is the case for harvesting machinery. This reflects differences in location specificity as well as differences in the distribution of invention values. It is arguable that the patent field comparisons control for value distribution differences, so that the ratio of off-diagonal elements to diagonal elements in these matrices is an indicator of location specificity.

Table 4.7 reports ratios of off-diagonal to diagonal elements for patents granted in all fields in eight OECD countries over the 1969–1987 period. The

Table 4.6. International Spill-In Comparison, by Number of Patents, 1975–1988.

Priority country	Granting country						
	USA	UK	France	Brazil	Japan	Germany	Canada
Non-fertilizer agricultural chemicals							
USA	3,316	670	287	552	1,152	293	485
UK	249	726	147	203	625	153	113
France	136	120	595	116	247	124	77
Brazil	26	18	15	83	26	15	19
Japan	421	298	236	215	6,156	276	154
Germany	540	122	122	599	1,295	2,389	263
Canada	68	24	21	21	46	28	132
Harvesting machinery							
USA	623	36	40	26	0	39	121
UK	26	156	21	5	1	14	10
France	20	16	268	0	0	23	2
Brazil	0	1	0	87	0	0	0
Japan	7	15	10	0	13	12	1
Germany	23	28	59	9	2	365	4
Canada	31	9	3	3	0	3	58
Fertilizer							
USA	366	31	36	42	56	39	64
UK	14	75	9	15	13	15	10
France	21	10	147	26	20	23	10
Brazil	5	5	2	81	7	4	4
Japan	34	26	30	24	761	37	21
Germany	50	24	39	31	69	313	19
Canada	4	5	4	4	5	6	31
Post-harvest technology fruit							
USA	110	17	22	17	3	18	14
UK	5	19	4	0	0	3	1
France	10	6	102	3	1	9	2
Brazil	1	1	2	37	1	2	0
Japan	12	5	5	2	129	5	4
Germany	14	11	18	3	3	61	6
Canada	8	3	2	1	1	3	10
Biotechnology: Mutation/genetic engineering							
USA	115	65	33	24	183	33	12
UK	6	35	7	12	39	7	2
France	4	6	36	0	12	3	2
Brazil	0	0	0	1	0	0	0
Japan	27	40	35	4	322	21	1
Germany	3	11	10	0	29	56	0
Canada	0	1	0	1	0	0	0
Biotechnology: Micro-organism and tissue culture							
USA	194	47	27	14	97	0	44
UK	18	36	9	7	38	16	4
France	13	12	38	2	16	13	11
Brazil	0	0	1	6	2	0	0
Japan	58	54	46	5	512	64	24
Germany	20	13	10	11	47	98	7
Canada	6	1	2	0	5	3	10

Source: Evenson and Westphal (1994).

Table 4.7. Invention Spill-In Indexes by Industrial Class for Eight OECD Countries, 1969–1987.

	Spill-in indexes		Number of inventions	
	IOM	SOU	IOM	SOU
Electrical machinery	2.122	2.185	193,017	123,780
Electronic equipment	2.199	2.201	27,453	170,053
Instruments	2.015	2.239	15,253	52,841
Office machinery	2.071	4.345	46,416	25,225
Chemicals	2.854	2.788	251,203	154,047
Drugs	2.696	3.039	25,473	47,384
Petroleum refineries	2.179	2.264	5,449	19,998
Aerospace	1.876	1.929	10,415	15,116
Motor vehicles	2.009	2.044	89,415	123,828
Ships	1.664	1.779	7,509	9,287
Other transport	1.642	1.961	13,296	17,892
Ferrous metals	2.195	2.217	12,902	25,854
Nonferrous metals	2.548	2.483	5,182	16,081
Fabricated metals	1.806	1.887	194,292	92,863
Food, drink, and tobacco	2.271	2.106	15,652	53,146
Textiles and clothing	2.019	2.488	32,622	60,642
Rubber and plastics	1.952	2.381	72,086	58,526
Stone, clay, and glass	2.093	2.260	27,366	28,162
Paper and printing	1.900	2.470	24,955	44,367
Wood and furniture	1.620	1.705	32,076	28,738
Other machinery	2.060	2.084	530,158	197,678
Other manufacturing	1.719	1.814	14,998	39,033
Agriculture		1.966		62,920
Mining	1.842	1.903	1,016	48,060
Construction		1.735		152,438
Transportation service	1.767		41,084	
Communication, utilities		2.002		79,992
Retail, wholesale trade	1.827		38,290	
Finance business		1.687		42,200
Government and education	1.694		27,679	
Health services		2.031		93,040
Other services		1.866		50,739

Source: Evenson and Westphal (1994).
Note: IOM = industry of manufacture; SOU = sector of use.

patents in this case are classified by IOM and SOU using the YTC. The agricultural sector is not an important IOM, but it is an important SOU of inventions. While these indices are not entirely comparable with crop varietal data, they are instructive for comparison purposes. They suggest that inventions intended for use in transport equipment, fabricated metals, wood and furniture, other machinery, mining, construction, finance, and other services may be as location specific as inventions intended for use in agriculture.

4.5.2 Indirect evidence for location specificity: International patterns

Table 4.8 reports a more complete pattern of international choices regarding obtaining patent protection abroad. The table includes inventions in all fields; almost all inventions, however, are produced in the industrial sector. Table 4.8 essentially covers three blocks of countries: the developed market economy block, the recently industrialized countries (RICs) block, and the newly industrialized countries (NICs) block. The table does not include the countries of the former Soviet Union and the substantial majority of developing countries that do not have functioning patent systems.[16]

Table 4.8 thus has several "blocks." The upper portion of Table 4.8a shows invention flows between developed market economies. These flows are extensive. The upper portion of Table 4.8b shows flows from developed countries to RICs and to NICs. These flows are also substantial. Developed market economies protect their inventions in these "downstream" countries.

The lower portion of Table 4.8a shows upstream invention flows from RICs and NICs to developed countries. These upstream flows are minor for RICs and negligible for NICs.

The lower portion of Table 4.8b shows minor flows from RICs to NICs and negligible flows from NICs to RICs. This portion of the table also shows negligible flows from RICs to other RICs and from NICs to other NICs. These data then do not support the proposition of technology "cascading" for industrial invention. In other words, it does not appear that RICs are adapting developed country inventions and selling them further downstream in the NICs.

The diagonal elements in both the RIC and NIC blocks, however, are of special interest to the analysis of convergence mechanisms. Are these adaptive inventions? If so, they will be related to the invention flows from developing countries (upper portion of Table 4.8b) as well as to domestic R&D in much the same way as the IARC germplasmic inventions are.

4.5.3 International recharge in industrial invention

An international recharge test for industrial invention that is roughly comparable to the test for agricultural inventions (in the form of varieties produced by NARS) is reported in Table 4.9. This test requires variables measuring the following:

- Domestic invention

- Domestic R&D or scientist and engineering resources

- International recharge

Domestic invention (DOMINV) was measured by numbers of domestic patents obtained in 1980, 1985, 1990, and 1995.[17]

Table 4.8a. International Patent Flows, All Sectors, 1990.

Applicant country	Austria (AT)	Switzerland (CH)	Germany (DE)	Denmark (DK)	Europe (EP)	France (FR)	Britain (GB)	Italy (IT)	Japan (JP)	Netherlands (NL)	Norway (NO)	Sweden (SE)	United States (US)
AT	2,782	64	0	38	325	41	31	115	262	11	28	14	304
CH	137	1,892	15	149	777	105	84	176	1,067	53	79	40	1,074
DE	944	671	2,889	937	6,045	1,253	965	1,757	6,8767	280	364	240	7,339
DK	5	3	0	766	89	16	27	17	69	11	43	18	116
EP	4	1	0	47	1,171	2	1	0	661	0	23	0	530
FR	66	0	11	353	2,486	7,653	149	409	2,515	124	210	45	2,625
GB	98	99	1	535	1,833	240	4,248	309	2,960	112	231	69	3,050
IT	71	88	4	146	730	207	181	16,591	901	49	62	41	1,127
JP	129	130	2	298	4,129	934	1,738	662	281,027	226	88	123	18,470
NL	41	5	0	135	688	48	21	19	745	180	51	11	769
NO	34	1	0	63	55	14	29	12	32	8	458	32	86
SE	40	9	0	375	529	60	51	28	488	39	276	1,506	653
US	222	202	1	928	5,643	905	1,507	1,161	14,917	411	650	320	59,692
ES	4	10	1	1	40	75	41	135	73	2	5	0	84
HU	92	35	1	53	58	52	50	77	85	6	18	30	80
IE	1	1	0	0	18	1	44	2	17	0	1	1	37
KR	1	1	1	0	6	6	26	61	340	0	1	1	194
PL	0	0	0	0	4	4	2	1	5	0	0	1	13
PT	3	0	0	0	2	1	1	3	0	0	1	0	6
TR	0	0	0	0	0	0	0	0	1	0	0	0	0
AR	0	0	0	0	1	0	0	6	3	0	0	0	13
BR	2	0	0	0	6	12	7	27	28	0	1	1	33
CN	0	0	0	0	8	1	3	2	6	0	0	0	36
EG	0	0	0	0	0	0	0	0	0	0	0	0	1
IN	0	0	0	0	1	0	3	3	1	0	0	0	1
MX	0	0	0	0	2	0	0	0	2	0	0	0	10
MY	0	0	0	0	0	0	0	0	1	0	0	0	0
PH	0	0	0	0	1	0	2	0	0	0	0	0	1
ZM	0	0	0	0	0	0	0	0	0	0	0	0	1

Granting country

Developed country

Table 4.8b. International Patent Flows, All Sectors, 1990.

| Applicant country | Granting country | | | | | | | | | | | | | | | |
| | RICs | | | | | | | NICs | | | | | | | | |
	Spain (ES)	Hungary (HU)	Ireland (IE)	Korea (KR)	Poland (PL)	Portugal (PT)	Turkey (TR)	Argentina (AR)	Brazil (BR)	China (CN)	Egypt (EG)	India (IN)	Mexico (MX)	Malaysia (MY)	Philippines (PH)	Zambia (ZM)
AT	14	35	0	7	10	3	2	3	32	0	3	10	5	4	5	0
CH	38	52	27	46	21	43	18	13	131	184	8	66	27	46	40	0
DE	323	286	78	152	79	154	67	36	522	768	20	211	122	47	93	3
DK	27	15	8	3	3	8	0	0	19	45	2	6	3	2	12	0
EP	5	10	6	15	7	9	5	3	57	114	3	43	9	10	5	0
FR	137	61	92	122	21	119	20	33	288	434	36	138	62	12	21	0
GB	137	179	223	112	70	143	39	22	432	495	33	226	74	184	118	11
IT	144	28	22	63	18	33	16	21	161	158	12	49	29	4	11	0
JP	93	120	56	3,883	27	28	9	7	258	1,712	6	126	67	301	92	0
NL	5	27	14	54	6	8	9	2	40	225	3	19	7	7	12	0
NO	11	1	6	4	3	1	1	1	16	22	3	6	6	1	0	0
SE	63	20	0	26	7	14	2	2	89	88	6	20	10	18	8	0
US	543	307	378	1,410	90	320	83	138	1,783	2,486	90	602	872	266	500	11
ES	1,805	0	0	5	0	5	0	7	15	11	1	4	1	0	3	0
HU	20	1,496	3	4	35	10	1	9	14	44	5	9	2	2	12	0
IE	1	0	121	0	0	1	0	0	4	4	0	1	0	0	5	0
KR	2	0	0	3,294	0	0	1	1	3	40	1	4	0	0	0	0
PL	1	15	0	0	5,591	0	0	0	0	6	0	3	0	0	1	0
PT	1	1	0	3	0	16	0	0	2	0	0	0	0	0	0	0
TR	0	0	0	0	0	0	107	0	0	0	0	0	0	0	0	0
AR	3	0	0	0	0	0	0	201	15	73	0	0	0	1	0	0
BR	9	1	3	1	0	1	0	2	2,555	5	2	2	7	0	0	0
CN	1	1	0	0	0	0	0	0	3	7,339	0	0	0	0	0	0
EG	0	0	0	0	0	0	0	0	0	0	62	0	0	0	0	0
IN	0	0	0	0	0	0	0	0	0	0	0	616	0	0	0	0
MX	3	0	0	0	0	0	0	0	2	2	0	0	268	0	0	0
MY	0	0	0	0	0	0	0	0	0	0	0	0	0	17	0	0
PH	0	0	0	0	0	0	0	0	0	0	0	0	0	0	120	0
ZM	0	0	0	0	0	0	0	0	0	0	0	0	0	0	0	11

Note: RIC = recently industrialized country; NIC = newly industrialized country.

Table 4.9. Estimates: Industrial Adaptive Invention Specifications.

Dependent variable = domestic patents.

Independent variable	Specification				
	(1)	(2)	(3)	(4)	(5)
L_N (SC)	3119 (2.67)	1636 (1.40)	2121 (2.13)		
L_N (SCBE)				1706 (2.08)	1745 (1.78)
FPATSTK		−4.238 (2.35)			
FPATSTK × ln (SC)		0.436 (2.56)	0.178 (2.98)		
CFPATSTK			−1.788 (2.07)	−2.944 (4.30)	
CFPATSTK × ln (SC)			0.178 (2.98)	0.182 (4.74)	
ln (CFPATSTK)	11.16 (1.51)				
FPAYSTK					−9.953 (3.11)
FPAYSTK × ln (SC)					0.740 (2.49)
R^2	0.9204	0.9421	0.9489	0.9661	0.9601
Root mean square error	1095.8	960.3	901.8	734.7	797.3

Note: All specifications include time and country dummy variables.

Domestic R&D was measured by two alternative variables:

- SC is the total number of scientists and engineers in all economic sectors according to United Nations Educational, Scientific, and Cultural Organization (UNESCO) data.[18]

- SCBE is the reported proportion of R&D financed by the business enterprise sector reported by UNESCO.[19]

International recharge is measured by three alternative variables:

- FPATSTK is the current period number of foreign origin patents granted in the country.

- CFPATSTK is the cumulative number of foreign origin patents granted in the country.[20]

- FPAYSTK is the cumulative royalty payments (in US dollars) to foreigners for technology rights.[21]

Table 4.9 reports five specifications incorporating the alternative variables. Observations are for four periods: 1980, 1985, 1990, and 1995. All specifications include country and year fixed effects.

- Specifications (1) and (2) provide a comparison between a general specification (1) and the search-based recharge specification (2). The data clearly prefer the search-based recharge specification.

- Specifications (2) and (3) provide a comparison of current foreign patents and cumulated foreign patents as the recharge measure. The data prefer specification (3).

- Specifications (3) and (4) provide a comparison between *SC* as a measure of scientists and *SCBE* as a measure of scientists. The data prefer *SCBE* as a measure of scientists.

- Specifications (4) and (5) provide a comparison between *FPATSTK* and *FPAYSTK* as measures of foreign recharge. Interestingly, both measures appear to be relevant indices of foreign technology.

The negative coefficients for *FPATSTK* (*FPAYSTK*) combined with the positive coefficients for ln(*SCBE*) × *FPATSTK* (*FPAYSTK*) can be interpreted as both a competition effect (the negative effect) and a germplasm or recharge effect (the positive coefficient). The partial elasticities of domestic invention, with respect to *SCBE* or *SC*, are .8 to .85 in the absence of foreign recharge and rise to 1.15 to 1.25 when mean foreign recharge is present. The partial elasticities of domestic patents to foreign patenting or technology payments are actually negative for countries with low *SC* or *SCBE* levels (below the 40th percentile) and rise as *SC* or *SCBE* rises.

Thus, it appears that the competitive effect of foreign technology payments discourages small countries from investing in R&D even though the complementary effect encourages it.

4.6 Conclusion

Empirical tests of the induced adaptive invention model with international recharge are reported for both agricultural biological invention in developing countries and for industrial invention for a subset of developing countries. Both sets of tests support the proposition that induced adaptive invention/innovation is a vehicle for convergence. Technology policy for agriculture is based on this proposition. Technology policy for industry is generally based on mimicry as the chief mechanism for convergence. In recent decades, convergence failures in industrial development have moved policy in the direction of mimicry with high tacit knowledge requirements as the basis for policy design.

Does the evidence for the adaptive invention/innovation model imply a glaring inconsistency in industrial development policy? Or is the evidence also consistent with the mimicry with high tacit knowledge mechanism? More work is required to settle this issue, but we do know that virtually all cases of rapid industrial productivity growth have taken place in economies with an R&D capacity in industry (Evenson and Westphal 1994). We also know that all cases of modest industrial productivity growth have taken place in economies purchasing foreign technology. The evidence in this paper indicates a strong complementarity between domestic R&D and foreign technology purchase.

While agricultural research institutions were built in response to food security issues associated with the population expansion of the post–World War II period, the longer-term investment strategy for development requires more explicit attention to convergence mechanisms. Regardless of whether the adaptive invention/innovation mechanism or the mimicry with high tacit knowledge mechanism dominates, it is relatively clear that investment in research and engineering capacity is required to achieve convergence.

Notes

1. Many of these models also stressed institutional constraints that were perceived to be more important for agriculture than for industry. Gerschenkron (1952) provides a discussion of economic backwardness. Landes (1990) provides a more recent historical perspective.

2. Jones (1998) provides evidence for this. Barro and Sala-i-Martin (1996) summarize this evidence. Mankiw *et al.* (1992) also report estimates.

3. Most endogenous growth models are predicated on the tacit knowledge constrained mechanism. Lucas (1988), for example, postulates human capital as the key factor. Jones (1995) has a learning-by-doing model. Mokyr (1996) provides a broad historical perspective on technological change.

4. The models of Romer (1986, 1994) and to some extent the Jones model (1995) recognize the importance of research and development and inventions but do not model international recharge mechanisms. Arrow (1962) provided the original model of "learning by doing."

5. Invention recharge is implied by the term adaptive inventions.

6. The pre-breeding activities of the International Agricultural Research Centers are an important part of the recharge mechanism discussed in Section 4.4.

7. This independent search for traits is dictated by the fact that traits have different genetic sources.

8. The selection process following crossing requires multiple-period evaluation.

9. The elimination of unpromising search avenues is one of the products of research programs.

10. Note that, for inventors receiving recharge, the recharge elements shift the invention distribution linearly. The diminishing returns to pre-invention or recharge activities are not incorporated into the recharge recipient's invention functions. This implication is tested in the application in Sections 4.4 and 4.5.

11. Spillovers can be of two types, direct and indirect. Direct spillovers occur when an invention made in one location or industry is directly used in another. Indirect spillovers occur through the invention recharge mechanism.

12. These data are based on a CGI study commissioned by the International Agricultural Evaluation Group—a body of the CGIAR. The author was the principal investigator of that study. The three country studies were commissioned to provide consistent estimates of the impact of MVs on production.

13. ICRISAT is the International Center for Research in the Semi-Arid Tropics, one of the 16 IARCs.

14. Related work on technology capital (or infrastructure) evaluates other sources of TFP growth in acquisitions and shows that direct technology spillovers from industry are an important component of TFP growth (Evenson 2000, 2001).

15. In many cases inventors consider political factors in decisions to obtain patent protection abroad. Most developing countries exhibit hostility toward IPRs generally. The perception of widespread "piracy" of IPRs by developing countries was a factor in the inclusion of IPRs in the Uruguay round of the General Agreement on Trade and Tariffs.

16. The classification of developing countries in the RIC and NIC categories is somewhat arbitrary. It should be noted that Zambia is not generally regarded as an NIC and that Indonesia, Thailand, and Chile are regarded as NICs (or near NICs) but do not have functioning patent systems.

17. These data are tables for the INPADOC database and are reported in Johnson and Evenson (2000).

18. These data are from the UNESCO database: http://unescostat.unesco.org/stat sen/st

19. These data are also from UNESCO.

20. These data are reported in Johnson and Evenson (2000).

21. This variable is from the 2000 World Development Indicators CD-ROM, World Bank.

References

Arrow, K.J., 1962, The economic implications of learning by doing, *Review of Economic Studies*, **29**(June):153–173.

Barro, R.J., and Sala-i-Martin, X., 1996, Convergence Across States and Regions, Brookings Papers on Economic Activity, Brookings Institution, Washington, DC, USA, pp. 107–158.

Evenson, R.E., 2000, Agricultural production and productivity in developing countries, in *The State of Food and Agriculture*, Food and Agriculture Organization of the United Nations, Rome, Italy.

Evenson, R.E., 2001, Economic impacts of agricultural research and extension, in B. Gardner and G. Rausser, eds, *Handbook of Agricultural Economics*, Vol. 1, Elsevier Science B.V., Amsterdam, Netherlands.

Evenson, R.E., and Kislev, Y., 1975, *Agricultural Research and Productivity*, Yale University Press, New Haven, CT, USA.

Evenson, R.E., and Kislev, Y., 1976, A stochastic model of applied research, *Journal of Political Economy*, **84**(2):265–282.

Evenson, R.E., and Westphal, L., 1994, Technological change and technology strategy, in T.N. Srinivasan and J. Behrman, eds, *Handbook of Development Economics*, Vol. 3, North Holland Publishing Company, Amsterdam, Netherlands.

Gerschenkron, A., 1952, Economic backwardness in historical perspective, in B.F. Hoselitz, ed., *The Progress of Underdeveloped Areas*, University of Chicago Press, Chicago, IL, USA.

Gollin, D., and Evenson, R.E., 1997, Genetic resources, international organization, and rice varietal improvement, *Economic Development and Cultural Change*, **45**(3):471–500.

Griliches, Z., 1957, Hybrid corn: An exploration in the economics of technological change, *Econometrica*, **25**:501–522.

Huffman, W., and Evenson, R.E., 1993, *Science for Agriculture*, Iowa State University Press, Ames, IA, USA.

Johnson, D.K.N., and Evenson, R.E., 2000, R&D spillovers to agriculture, measurement and application, *Contemporary Economic Policy*, **17**(4):432–456.

Jones, C.I., 1995, R&D-based models of economic growth, *Journal of Political Economy*, **103**(August):759–784.

Jones, C.I., 1998, *Introduction to Economic Growth*, W.W. Norton & Company, New York, NY, USA.

Landes, D.S., 1990, Why are we so rich and they so poor?, *American Economic Association Papers and Proceedings*, **80**(May):1–13.

Lucas, R.E., Jr., 1988, On the mechanics of economic development, *Journal of Monetary Economics*, **32**(July):3–42.

Mankiw, N.G., Romer, D., and Weil, D., 1992, A contribution to the empirics of economic growth, *Quarterly Journal of Economics*, **107**(May):407–38.

Mokyr, J. (1996), *The Lever of Riches*, Oxford University Press. New York, NY, USA.

Pardey, P.G., and Roseboom, J., 1989, *A Global Data Base on National Agricultural Research Systems*, ISNAR Agricultural Research Indicator Series, Cambridge University Press, Cambridge, UK.

Romer, P., 1986, Increasing returns and long-run growth, *Journal of Political Economy*, **94**(October):1002–1037.

Romer, P., 1994, The origins of endogenous growth, *Journal of Economic Perspectives*, **8**(Winter):3–22.

USDA, various years, Agricultural Statistics, United States Department of Agriculture, Washington, DC, USA.

Chapter 5

The Induced Innovation Hypothesis and Energy-Saving Technological Change

Richard G. Newell, Adam B. Jaffe, and Robert N. Stavins

5.1 Introduction

There is currently much interest in the potential for public policies to reduce energy consumption because of concerns about global climate change linked with the combustion of fossil fuels. Basic economic theory suggests that if the price of energy relative to other goods rises, the energy intensity of the economy will fall as a result of a series of behavioral changes: people will turn down their thermostats and drive more slowly; they will replace their furnaces and cars with more efficient models available on the market; and, over the long run, the pace and direction of technological change will be affected, so that the menu of capital goods available for purchase will contain more energy-efficient choices.

This last conjecture—that increasing energy prices will lead to technological change that facilitates the commercialization of capital goods that are less energy intensive in use—is a modern manifestation of the "induced innovation" hypothesis of Sir John Hicks: "a change in the relative prices of the factors of production is itself a spur to invention, and to invention of a particular kind—directed to economizing the use of a factor which has become relatively expensive" (1932:124–125).

There is a considerable theoretical and empirical literature on the induced innovation hypothesis, often formulated as the principle that increases in real wages will induce labor-saving innovation. That literature typically analyzes the inducement effect in the framework of an aggregate production function.[1] Technological change, however, is inherently a microeconomic, product-level phenomenon. If the inducement mechanism operates with respect to energy, it does so largely by leading firms to develop and introduce new models of cars, appliances, and

This chapter was originally published as Richard G. Newell, Adam B. Jaffe, and Robert N. Stavins, The induced innovation hypothesis and energy-saving technological change, *The Quarterly Journal of Economics*, **114**:3 (August 1999), pp. 941–975.

industrial equipment that deliver greater services per unit of energy consumed. From this perspective, it seems natural to formulate the inducement hypothesis in terms of a product-characteristics framework, summarizing the technological possibilities for the production of a good as a menu of feasible vectors. Each vector represents the characteristics of technically feasible models, including the resource cost of producing such models. Innovation is the introduction into the relevant menu of a vector that was previously not available.

In this we follow Schumpeter (1939), who used "invention" for the act of creating a new technological possibility and "innovation" for the commercial introduction of a new technical idea. Both are to be distinguished from the third stage of Schumpeter's trichotomy, diffusion, which is the gradual adoption by firms or individuals of commercially available products.[2] Thus, the induced innovation hypothesis implies that when energy prices rise, the characteristic "energy efficiency" of items on the capital goods menu should improve faster than it otherwise would.

In this chapter we formalize the inducement hypothesis in this framework and test it empirically. We also generalize the Hicksian notion of inducement to investigate whether government regulations have affected energy-efficiency innovation. We find evidence that both energy prices and government regulations have affected the energy efficiency of the models of room air conditioners, central air conditioners, and gas water heaters available on the market over the past four decades, although there have also been substantial improvements in energy efficiency that do not appear to be induced by price changes or regulations.

In Section 5.2 we describe technological change in terms of product characteristics and lay out our econometric approach for estimating induced innovation using "characteristics transformation surfaces." In Section 5.3 we describe our data and present empirical estimates of such transformation surfaces for three products over the past several decades, including the extent to which technological change in these products has been induced by prices and regulations. In Section 5.4 we develop the distinction between improvements in efficiency due to changes in technological possibilities and improvements due to the "substitution" of models along a given set of technological possibilities. We econometrically assess the importance of these factors in generating changes in the composition of models actually offered along the frontier. In Section 5.5 we offer some concluding observations.

5.2   The Characteristics Transformation Surface

5.2.1   Innovation in the product characteristics framework

Theoretical analysis within the product characteristics framework has been discussed by numerous authors.[3] Innovation in this framework can be thought of as the introduction of a product model with a bundle of characteristics that was not previously available, or the production of a previously available bundle of characteristics at a cost that is lower than was previously feasible. To incorporate both of

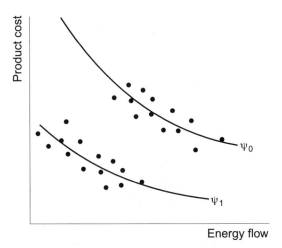

Figure 5.1. Innovation in the Characteristics Transformation Surface.

these possibilities, we characterize a product model by a vector of dimensionality $n + 1$, where n is the number of product attributes or characteristics that consumers care about, and the additional vector element is an index of the quantity of real inputs used to produce that model. In effect, we treat the real cost of producing a model as an additional characteristic of that model. At any point in time, the frontier of the technologically feasible production set can then be described in terms of a functional relationship between the bundle of characteristics and the input quantity necessary to produce that bundle.

To be concrete, consider an air conditioner with two characteristics: energy flow per unit of time f and cooling capacity c. Let k represent the real cost of producing a model i with a particular bundle of characteristics. We approximate the transformation surface as a simple log-linear function with k as the dependent variable.[4] Thus, at a particular point in time we have

$$\ln k_i = \alpha + \beta_1 \ln f_i + \beta_2 \ln c_i . \tag{5.1}$$

The β parameters are interpretable as elasticities of product cost with respect to each characteristic. Figure 5.1 illustrates a projection of the transformation surface onto the k–f plane for a fixed level of capacity, at two points in time. For the specific example at hand, because energy use is a bad, the curve is downward sloping, and we would expect β_1 to be negative.[5]

Referring to Figure 5.1, suppose that the curves Ψ_0 and Ψ_1 represent econometrically estimated functions based on the set of models offered for sale at time t_0 and a later time t_1, where individual models are represented by the two sets of points in the figure. Suppose further that the price of energy increased between time t_0 and t_1. As drawn, three things have occurred. The frontier has moved toward the origin, making it possible to produce models that are simultaneously cheaper and more energy efficient than was previously possible. Second, the slope of the frontier has decreased, meaning that the elasticity of product cost with respect to energy flow is lower, or, equivalently, that the trade-off at a point in time between production cost and energy efficiency has shifted so that energy efficiency is less expensive on the margin. Finally, the subset of feasible models that are actually offered for sale has shifted noticeably toward less energy-intensive models. We refer to these three kinds of shifts as overall technological change, directional technological change, and model substitution. We take the term "innovation" to encompass the combined effect of all these changes in the product menu.

Figure 5.1 is representative of what occurred (to varying degrees) between the early 1970s and the early 1990s in the technologies that we examined. In terms of the overall energy efficiency of the menu of models offered for sale, we have observed significant improvement. The Hicksian hypothesis is that this improvement is related to the rise in energy prices. The goal of this chapter is to develop an empirical framework for measuring the extent to which that improvement can be associated with changes in energy prices; we also generalize the notion of "inducement" to include the possibility that government-mandated efficiency standards may have induced energy-efficiency innovation. In order to do this in a sensible way, we decompose the overall change in the energy efficiency of the menu into the parts due to overall technological change, directional technological change, and model substitution. The first two components are related to changes in the parameters of the transformation surface, represented by a functional relationship as in Equation (5.1). Model substitution corresponds to "movements along" this surface. In Section 5.4 we show that this decomposition can be carried out in a straightforward way, once the parameters of the transformation surface and their changes over time have been estimated. We now turn to that estimation.

5.2.2   Econometric specification

We investigate technological change by estimating the parameters of the transformation surfaces and simultaneously estimating how these parameters change over time. For room air conditioners, central air conditioners, and gas water heaters, respectively, we separately estimate the following versions of the transformation surface:

$$\ln k_{it} = \alpha + \beta_1 \ln f_{it} + \beta_2 \ln c_{it} + \beta_2 2\,speed + \beta_4 3\,speed + \epsilon_{it} \qquad (5.2)$$

$$\ln k_{it} = \alpha + \beta_1 \ln f_{it} + \beta_2 \ln c_{it} + \epsilon_{it} \qquad (5.3)$$

$$\ln k_{it} = \alpha + \beta_1 \ln f_{it} + \beta_2 \ln c_{it} + \beta_5 \ln g_{it} + \epsilon_{it} \,, \qquad (5.4)$$

where k is product cost, f is energy flow, c is cooling or heating capacity, $2\,speed$ and $3\,speed$ are dummy variables indicating the number of fan speed settings in room air conditioners, g is storage capability in gas water heaters, i indexes product models, t indexes time, and ϵ is an independently distributed error term with zero mean. Note that we have simplified notation by omitting product-specific subscripts on the α, β, and γ parameters; they are not restricted to be equal across products.

To allow for autonomous technological change, we allow the parameters of the surfaces to vary flexibly as second-order functions of time.[6] We introduce induced technological change by allowing the relevant parameters to vary as functions of the relative price of energy p and the level of energy-efficiency standards s. We show in Section 5.4 that "overall" improvements in technology are associated primarily with changes in α, while "directional" technological change relative to energy efficiency is associated with changes in β_1. It is not clear whether Hicks should be interpreted as saying that rising energy prices stimulate overall technological change, directional technological change favoring energy, or both. Newell (1997) shows that an effect of changing energy prices on the *direction* of technological change can be derived from a model of the firm's optimal investment in research. An effect on the overall rate of technological change could perhaps be motivated by a satisficing or evolutionary model in which any "shock" to the economic environment stimulates innovation. We will estimate versions in which inducement is permitted in both the α and β_1 terms, and also versions in which it is limited to affecting β_1.

Thus, the varying coefficients of the estimated surfaces take on the following form:

$$\alpha = \alpha_0 + \alpha_1 t + \alpha_2 t^2 + \alpha_3 \ln q_t + \alpha_4 \ln p_{t-j} + \alpha_5 s \tag{5.5}$$

$$\beta_1 = \beta_{10} + \beta_{11} t + \beta_{12} t^2 + \beta_{13} \ln p_{t-j} + \beta_{14} s \tag{5.6}$$

$$\beta_2 = \beta_{20} + \beta_{21} t + \beta_{22} t^2 \,, \tag{5.7}$$

where t is time, p is the relative price of energy, s is the level of energy-efficiency standards, and q is aggregate product shipments.[7] To control for any effects of aggregate production levels on product cost, we allow the constant term to vary as a function of product shipments, q.[8] The subscript j indicates that the associated price occurred j years prior to year t. Based on assessments in the literature of the tooling and redesign time required to bring energy-saving product innovations to market,[9] we estimate equations using the three-year lag in the relative price of energy (i.e., $j = 3$). Note again that we have simplified notation on the α and β parameters, which are not restricted to be equal across products.

5.3 Estimation on the Transformation Surface over Time

5.3.1 Data

Using the Sears catalog (Sears, Roebuck and Co. 1958–1993) and other publicly available data sources, we compiled a database of information on 735 room air conditioner models offered for sale from 1958 through 1993; 275 central air conditioner models from 1967 through 1988; and 415 gas water heater models from

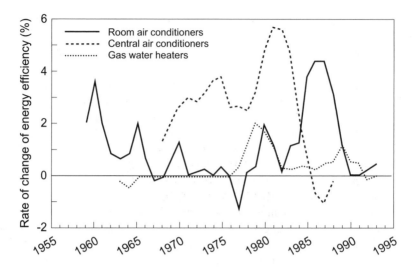

Figure 5.2. Changes in Energy Efficiency.

The figure shows a three-year moving average of the annual rate of change of mean energy efficiency of models offered for sale.

1962 through 1993. The catalogs contain a wide variety of product models over many decades, a comprehensive set of descriptive data on product characteristics, and importantly, transaction prices, as opposed to list prices (which may be subject to discount).[10] Below, we describe the variables used in the analysis and summarize their construction. Table 5.1 provides descriptive statistics and units of measurement; see Newell (1997) for additional detail on data sources and methods of variable construction.[11] Note that all references to "mean" or "average" values for characteristics refer to the mean value of models *offered for sale*; the means are not weighted by the number of units sold, for example.

Model Characteristics

We assembled data on the cooling/heating capacity, energy flow, energy efficiency, nominal price, and other characteristics of all models of room and central air conditioners and gas water heaters from the Sears catalog over the 36-year period from 1958 to 1993. We also included storage capability for gas water heaters and dummy variables indicating whether a room air conditioner had multiple fan speeds (i.e., two or three rather than one). The mean capacity of available models of all three products changed very little over the sample period, falling on average by less than 0.3 percent per year. Energy flows fell faster, leading to a net rise in the mean energy efficiency of the three products (Figure 5.2). Central air conditioners experienced the greatest annualized rates of change in energy efficiency (2.6 percent), followed by room air conditioners (1.2 percent), and gas water heaters (0.3 percent).

Table 5.1. Summary Statistics for Variables.

Variable	Symbol	Overall Mean	Overall Std dev	Initial year Mean	Initial year Std dev	Final year Mean	Final year Std dev	Mean growth rate (%)
Room air conditioners (1958–1993; N = 735)								
Energy flow (1,000 watt)	f	1.5	1.0	1.9	0.6	1.2	0.8	–1.3
Energy efficiency (Btu/hr/ watt)	e	7.6	1.4	5.9	1.0	9.0	0.6	1.2
Cooling capacity (1,000 Btu/hr)	c	11.4	6.7	10.8	3.1	10.6	6.2	–0.1
Nominal price (US$)		376	166	248	64	548	166	2.3
Product cost (overall normalized mean = 1)	k	1.00	0.46	1.85	0.47	0.86	0.26	–2.2
Shipments (millions/year)	q	3.39	1.21	1.67	–	3.08	–	1.8
Relative price of electricity (1975 = 1)	p_f	1.08	0.10	1.25	–	1.14	–	–0.3
Central air conditioners (1967–1988; N = 275)								
Energy flow (1,000 watt)	f	4.4	1.5	6.1	1.8	3.5	1.4	–2.6
Energy efficiency (Btu/hr/watt)	e	8.3	1.7	6.4	0.1	10.8	0.4	2.5
Cooling capacity (1,000 Btu/hr)	c	35.1	10.0	39.3	12.1	37.2	14.0	–0.3
Nominal price (US$)		911	404	531	158	1,299	313	4.4
Product cost (overall normalized mean = 1)	k	1.00	0.26	1.23	0.37	0.85	0.21	–1.8
Shipments (millions/year)	q	2.66	0.91	1.01	–	4.35	–	7.2
Relative price of electricity (1975 = 1)	p_f	1.04	0.10	1.02	–	1.11	–	0.3
Gas water heaters (1962–1993; N = 415)								
Energy flow (1,000 Btu)	f	44.1	12.2	47.0	12.0	40.0	7.7	–0.5
Energy efficiency (90° gal/1,000 Btu)	e	0.98	0.05	0.94	0.03	1.05	0.05	0.3
Heating capacity (90° gal/hr)	c	43.0	11.0	44.4	11.6	42.0	8.5	–0.2
Storage capability (gallons)	g	41.8	11.1	36.3	7.4	46.8	14.0	0.8
Nominal price (US$)		173	96	79	21	284	104	4.2
Product cost (overall normalized mean = 1)	k	1.00	0.30	1.20	0.31	0.90	0.33	–0.9
Shipments (millions/year)	q	3.56	0.45	3.22	–	4.54	–	1.1
Relative price of natural gas (1975 = 1)	p_f	1.29	0.33	1.13	–	1.54	–	0.9
Other variables (1958–1993)								
Housing starts (millions/year)		1.51	0.32	1.34	–	1.29	–	–0.1
Median household income (US[1994]$)		36,196	4,149	26,055	–	37,905	–	1.1

Note: Means are for models offered for sale; they are not weighted by units sold, for example. Growth rates are geometric means over the period available for each technology. Product cost is equal to the nominal price divided by an input price index.

The transformation surface has as a dependent variable an index of the quantity of real inputs necessary to produce the particular product model. We do not observe model-specific input quantities. We do, however, observe model-specific prices. The nominal price of a given product model can be thought of as its production cost multiplied by a price/cost markup. Further, the production cost can be thought of as the quantities of physical inputs needed to produce that model, multiplied by the prices of those inputs. Thus, to use the model price as a proxy for the model's product cost, we must assume that the price/cost markup is constant across models and time for a particular product.[12] To the extent that this is untrue, our computation of the rate of product cost reductions could be biased. We address this issue econometrically by including national annual product shipments[13] for each technology in the transformation surface estimation. This provides a control for possible markup changes due to economies of scale or demand fluctuations, although it does not control for possible changes in industrial structure or economic regulation.

Assuming constant markups converts the model's price into a proxy for its nominal product cost. To convert nominal product cost into an index of input quantity, we deflate each model's nominal unit price (from the Sears catalog) by an index of input prices. We developed separate input price indices for air conditioners and gas water heaters, based primarily on census data on capital, labor, and materials for the corresponding four-digit SIC (Standard Industrial Classification) industries.[14]

To summarize, the (real) product cost for each product model is taken to be its catalog price, divided by an index of input prices that varies across time, and is calculated separately for air conditioners and water heaters, but does not vary across models. The mean nominal prices of the three products rose at lower rates than their respective input price indices over the sample periods, implying that their mean real product costs fell by an annualized 2.2 percent, 1.8 percent, and 0.9 percent, respectively, for room air conditioners, central air conditioners, and gas water heaters (Table 5.1).

Relative Price of Energy

We assume that the inducement mechanism is driven by the price of energy *relative* to the price of product inputs. We estimate this relative price by dividing the Consumer Price Index (Bureau of the Census 1975, 1996) for electricity (for room and central air conditioners) and natural gas (for water heaters) by the input price indices described above. The relative price indices for electricity and natural gas have varied substantially over the past four decades, falling during much of the period, but rising sharply in the mid-1970s and early 1980s, coinciding with the Arab oil embargo of 1973–1974, several major domestic natural gas shortages, and other energy shocks (Figure 5.3). Both fuels experienced their lowest price levels in the early 1970s, with the peak electricity and natural gas price levels of the mid-1980s being about 35 percent and 85 percent higher, respectively, than their lowest levels.

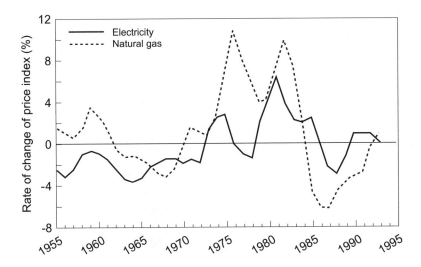

Figure 5.3. Changes in Electricity and Natural Gas Prices.
The figure shows a three-year moving average of the rate of change of the relative price index. See text for details on data construction.

Our use of past energy prices as a measure of consumers' expectations about the future path of energy prices raises the possibility of a conservative bias in our estimates. This could arise from an "errors in variables" problem associated with using actual energy prices that exhibit greater variation than the true price expectations for which they act as a proxy, thereby imputing a coefficient bias toward zero relative to the true effect of expected energy price changes. In addition to the expected path of energy prices, the relevant "price" of energy flow in the consumer's choice of optimal energy efficiency will be influenced by the consumer's discount rate, their expected utilization of the product, and how long they expect to have the good in service. Omission of these additional variables from our analysis will only bias our results, however, if they have *changed* over time in a manner that is systematically correlated with the variables we include. Although this possibility exists, we believe that any bias is likely to be small and in the direction of making our results *understate* the true inducement effects.[15]

Energy Efficiency Standards

The National Appliance Energy Conservation Act of 1987 (NAECA) mandated that minimum energy-efficiency standards be met by room air conditioners and gas water heaters after 1 January 1990, and central air conditioners after 1 January 1992, or 1993 (*United States Code of Federal Regulations* 1995b). Since manufacturers did not wait until the deadline to meet the standards, we model the effect of efficiency standards as cumulative over the period of time between passage and enforcement; that is, we let $s = 0,1,2,3,4$ for $t < 1987$, $t = 1987$,

$t = 1988$, $t = 1989$, and $t \geq 1990$, respectively. We do not analyze the effect of efficiency standards on central air conditioners because the compliance deadlines were beyond the time frame of the data available to us.

5.3.2 Estimation issues

Although we do not focus on the equilibrium that determines which product models are offered in a given year, the existence of this equilibrium process raises the possibility of an endogeneity problem in using the product characteristics as regressors. We interpret the regression function as tracing out the transformation surface, which could be thought of as a "supply function" for characteristics. There is, of course, a large literature on hedonic price regression, in which model prices are regressed on characteristics, and the regression function is sometimes used as the basis for estimating a demand function for characteristics.[16] It is well known that a regression of prices on characteristics embodies a variant of the classic supply/demand simultaneity problem, in which the regression coefficients are not, in general, identified as parameters of either the supply or demand function [Rosen (1974); see also Epple (1987) and Triplett (1987)]. The ordinary hedonics interpretation of a regression of model prices on characteristics as a demand function requires an assumption that the data are generated by heterogeneous suppliers' distinct supply functions tracing out the demand curve of homogeneous consumers. Conversely, our interpretation of the deflated price (i.e., product cost) regression as a supply function requires an assumption of heterogeneous consumers' distinct demand functions tracing out the supply curve of a homogeneous production sector.

This latter interpretation is plausible in the current context. Indeed, all models in the sample are supplied by the same firm, although the models in the Sears catalog span the space of available models. Different consumers purchase different models with different efficiencies, presumably because they have different discount rates or different anticipated intensities of usage, or face different energy prices. Further, energy efficiency is different from other characteristics that are typically the focus of hedonic analysis. In general, we expect a given consumer to have a downward-sloping demand curve for a given characteristic; for example, at some point additional computer memory has little value, particularly holding constant other attributes. In contrast, the value of energy savings is essentially constant for any given discount rate and expected usage. Hence, each consumers' demand curve is infinitely elastic, so that there is unlikely to be simultaneity bias in interpreting our estimates as parameters of the supply structure.

Finally, to compensate for heteroskedasticity, we compute robust standard errors using White's (1980) method. There was no evidence of autocorrelation of residuals along the time series dimension of the estimated equations. To avoid problems associated with the potential endogeneity of aggregate shipments, we estimate the surfaces using two-stage least squares, instrumenting for the log of the quantity of shipments using levels and changes in the log of housing starts (Bureau of the Census 1975, 1996) and real household income (Bureau of the Census

1975; Council of Economic Advisers 1997), in addition to the other explanatory variables.[17]

5.3.3 Results of estimation of characteristics transformation surfaces

The estimation results for the three technologies are presented in Tables 5.2, 5.3, and 5.4. We present results for a "pure" autonomous technological change model and the induced innovation model. Since it is not obvious theoretically that the inducement mechanism should affect the intercept term, and these effects are indeed generally insignificant, we also present the results when these terms are suppressed. Overall, the results are consistent with the economic interpretation of the parameters. The estimated elasticities for the various characteristics all have the expected signs and reasonable magnitudes; and the coefficient on time is negative in all cases, indicating positive autonomous technological change. The results confirm that the cost of durable goods increases with increasing energy efficiency, capacity, and other desirable characteristics, and that the cost of producing a given bundle of characteristics tends to fall over time as a result of technological change in production techniques and product design.

The coefficient on $\ln f$ (i.e., β_{10}) in each table measures the elasticity of product cost with respect to energy flow in 1975; the elasticity for other years depends on the year and the magnitudes of β_{11}, β_{12}, β_{13}, and β_{14}.[18] The elasticity is negative for all three products, as expected, indicating that reductions in energy flow come at the expense of higher product cost. However, the magnitude of this trade-off varies significantly among the technologies; a 10 percent decrease in energy flow (or increase in energy efficiency) was associated with a 4 percent increase in product cost for room air conditioners, a 12 percent increase for central air conditioners, and a 40 percent increase for gas water heaters (in 1975). The estimates also indicate that multiple fan speeds and increases in capacity are costly, although, not surprisingly, there are "economies of product scale" as cost increases less than proportionately with capacity.[19]

The coefficients α_1 in each table indicate the rate of change of the intercept in 1975; the rate of change in other years depends on α_2. The quantitative significance of these changes is discussed further in the following section; we note here only that all three technologies have α_1 significantly negative, meaning that there is autonomous overall technological change.

There is also evidence of autonomous "directional" change, that is, changes over time in the slope of the transformation surface. For example, the results indicate that in 1975 ($t = 0$) the absolute magnitude of the elasticity of product cost with respect to energy flow was decreasing autonomously by 0.10 and 0.06 annually for central air conditioners and gas water heaters, respectively, indicating an autonomous "bias" of technological change *against* energy efficiency. In all three cases, however, this component has shifted over the course of time toward favoring energy efficiency.

Table 5.2. Transformation Surface Estimates: Room Air Conditioners.

Param-eter	Explanatory variable	Autonomous innovation	Induced innovation	
			Specification 1	Specification 2
β_{10}	$\ln f$	−0.387	−0.362	−0.383
		(0.027)	(0.026)	(0.026)
β_{11}	$t \ln f$	0.80e-3	1.17e-3	1.51e-3
		(2.68e-3)	(2.88e-3)	(2.94e-3)
β_{12}	$t^2 \ln f$	8.33e-4	0.70e-4	3.28e-4
		(2.42e-4)	(3.14e-4)	(2.98e-4)
β_{13}	$\ln p \ln f$	−	0.410	0.361
			(0.125)	(0.127)
β_{14}	$s \ln f$	−	0.028	0.034
			(0.011)	(0.012)
β_{20}	$\ln c$	0.919	0.914	0.937
		(0.028)	(0.027)	(0.027)
β_{21}	$t \ln c$	−2.73e-3	−1.04e-3	−1.16e-3
		(2.95e-3)	(3.05e-3)	(3.10e-3)
β_{22}	$t^2 \ln c$	−6.78e-4	−5.90e-4	−8.69e-4
		(2.68e-4)	(2.93e-4)	(2.75e-4)
β_3	$2\,speed$	0.197	0.202	0.201
		(0.016)	(0.016)	(0.016)
β_4	$3\,speed$	0.300	0.299	0.298
		(0.016)	(0.016)	(0.016)
α_0	$Constant$	−0.215	−0.234	−0.220
		(0.017)	(0.019)	(0.016)
α_1	t	−0.026	−0.026	−0.027
		(0.001)	(0.001)	(0.001)
α_2	t^2	1.05e-3	1.05e-3	0.93e-3
		(0.19e-3)	(0.19e-3)	(0.06e-3)
α_3	$\ln q$	−0.083	−0.083	−0.102
		(0.024)	(0.024)	(0.016)
α_4	$\ln p$	−	0.043	−
			(0.088)	
α_5	s	−	−0.016	−
			(0.010)	
	# of observations	735	735	735
	R^2	0.96	0.96	0.96

Note: Dependent variable is the log of product cost ($\ln k$). Variables are described in more detail in Table 5.1 and in the text. Estimation method is two-stage least squares, with instrumentation for shipments ($\ln q$) using levels and changes in the log of housing starts and real household income in addition to the other explanatory variables. Robust standard errors are reported in parentheses.

Table 5.3. Transformation Surface Estimates: Central Air Conditioners.

Param- eter	Explanatory variable	Autonomous innovation	Induced innovation	
			Specification 1	Specification 2
β_{10}	$\ln f$	−1.247	−1.205	−1.177
		(0.077)	(0.087)	(0.082)
β_{11}	$t \ln f$	−0.103	−0.107	−0.103
		(0.014)	(0.016)	(0.014)
β_{12}	$t^2 \ln f$	4.87e-3	4.04e-3	2.81e-3
		(1.41e-3)	(2.14e-3)	(1.67e-3)
β_{13}	$\ln p \ln f$	–	0.968	1.291
			(0.566)	(0.558)
β_{20}	$\ln c$	1.978	1.991	1.978
		(0.079)	(0.083)	(0.078)
β_{21}	$t \ln c$	0.101	0.107	0.105
		(0.013)	(0.015)	(0.014)
β_{22}	$t^2 \ln c$	−4.43e-3	−5.26e-3	−4.60e-3
		(1.41e-3)	(1.80e-3)	(1.42e-3)
α_0	$Constant$	0.086	0.064	0.086
		(0.010)	(0.018)	(0.010)
α_1	t	−0.051	−0.055	−0.052
		(0.004)	(0.005)	(0.004)
α_2	t^2	−1.48e-3	−0.64e-3	−1.4ge-3
		(0.32e-3)	(0.65e-3)	(0.31e-3)
α_3	$\ln q$	0.320	0.385	0.339
		(0.055)	(0.082)	(0.055)
α_4	$\ln p$	–	−0.421	–
			(0.286)	
	# of observations	275	275	275
	R^2	0.90	0.90	0.90

Note: Dependent variable is the log of product cost ($\ln k$). Variables are described in more detail in Table 5.1 and in the text. Estimation method is two-stage least squares, with instrumentation for shipments ($\ln q$) using levels and changes in the log of housing starts and real household income in addition to the other explanatory variables. Robust standard errors are reported in parentheses.

Turning to the induced innovation specifications, there is little evidence of significant inducement effects on overall technological change. Although four of the five coefficients (α_4 in Tables 5.3 and 5.4; α_5 in Tables 5.2 and 5.4) are negative, none is statistically significant at conventional levels. In terms of energy-price-induced changes in the slope, indicated by β_{13}, there is a statistically significant and robust effect in the predicted direction for room air conditioners.[20] For central air conditioners the energy price effect is in the predicted direction; it is statistically significant at only the 0.10 level in the full specification, but is significant at the 0.05 level after the α_4 term is deleted.[21] For gas water heaters the inducement effects are always insignificant. Since many of the specific technological changes that foster efficiency are common between room and central air conditioners, while water heaters are quite different, these results suggest that changes

Table 5.4. Transformation Surface Estimates: Gas Water Heaters.

Param-eter	Explanatory variable	Autonomous innovation	Induced innovation Specification 1	Specification 2
β_{10}	$\ln f$	−3.918	−3.829	−3.925
		(0.235)	(0.267)	(0.221)
β_{11}	$t\ln f$	−0.055	−0.074	−0.061
		(0.032)	(0.023)	(0.028)
β_{12}	$t^2\ln f$	0.012	0.013	0.013
		(0.002)	(0.002)	(0.002)
β_{13}	$\ln p\ln f$	–	−0.056	−0.088
			(0.263)	(0.227)
β_{14}	$s\ln f$	–	−0.079	−0.032
			(0.058)	(0.051)
β_{20}	$\ln c$	4.670	4.557	4.659
		(0.238)	(0.271)	(0.226)
β_{21}	$t\ln c$	0.071	0.094	0.077
		(0.032)	(0.023)	(0.028)
β_{22}	$t^2\ln c$	−0.011	−0.012	−0.011
		(0.002)	(0.002)	(0.002)
β_5	$\ln g$	0.381	0.383	0.383
		(0.024)	(0.025)	(0.025)
α_0	$Constant$	−0.006	−0.010	−0.004
		(0.012)	(0.012)	(0.012)
α_1	t	−0.018	−0.014	−0.018
		(0.002)	(0.003)	(0.002)
α_2	t^2	1.156e-4	4.02e-4	0.74e-4
		(1.05e-4)	(2.37e-4)	(1.01e-4)
α_3	$\ln q$	0.640	0.594	0.646
		(0.092)	(0.103)	(0.092)
α_4	$\ln p$	–	−0.073	–
			(0.065)	
α_5	s	–	−0.025	–
			(0.015)	
	# of observations	415	415	415
	R^2	0.92	0.92	0.92

Note: Dependent variable is the log of product cost ($\ln k$). Variables are described in more detail in Table 5.1 and in the text. Estimation method is two-stage least squares, with instrumentation for shipments ($\ln q$) using levels and changes in the log of housing starts and real household income in addition to the other explanatory variables. Robust standard errors are reported in parentheses.

in energy prices probably did affect technological change in cooling technology, but not in water heating technology.

Results regarding the effect of energy-efficiency standards on the direction of technological change (β_{14}) are qualitatively similar. The direction of technological change shifted substantially in favor of energy efficiency during the period that federal energy efficiency standards were being implemented for room air conditioners, but not gas water heaters.[22] Of course, the concern remains that the representation of changing regulatory standards by a constant time dummy is a blunt instrument that could potentially act as a proxy for factors other than energy-efficiency standards that were influential during the same time period.

5.4 Overall Changes in the Menu of Models Offered

The results of the previous section indicate that, for all three technologies, there were changes in the position and slope of the transformation surfaces over time. We now turn to the question of whether the changes in the menu of products actually offered can be related to these estimated changes in the positions of the curves, as well as to "movements along the curves." The question is illustrated by Figure 5.4, which shows the position and slope of the estimated surface for room air conditioners at five-year intervals, based on the estimates of the previous section. The heavy dot on each line is the mean of the characteristics of all models offered in that year. Consistent with the previous section, the surfaces move toward the origin and become flatter over time. The figure also shows that the mean model offered for sale got much cheaper and slightly more energy efficient from 1960 to 1980, and then got much more energy efficient and slightly cheaper from 1980 to 1990. We would like to explain the movements in this mean model as being driven by the inward shift of the curve (overall innovation), the change in the slope of the curve (directional innovation), and movements along the curve at a point in time (model substitution).

5.4.1 Decomposition of characteristics innovation

Because the transformation surface for any given product is continually shifting, tilting, and changing its composition, there is no unique way to measure these separate effects. We adapt to the current context a standard approach to decomposition from the aggregate technical change literature. Figure 5.5 presents another projection of the transformation surface onto the product cost/energy flow plane, holding constant capacity (and other characteristics, if relevant). For the time being, assume that there is a single consumer whose optimal product cost/energy flow combination at time t_0, given technical possibilities represented by Ψ_0, is at point a. The line p_f^0 represents the relative "price" of energy relevant for the choice of optimal energy efficiency; it is determined by the expected path of energy prices, the discount rate, and the expected utilization and service life of the capital good. Assume that, at some time t_1, technical possibilities have improved, as represented by the shift of the transformation surface from Ψ_0 to Ψ_1, and that energy prices have also changed so that the price line now has the slope of p_f^1 instead of p_f^0. Accordingly, optimal energy flow and product cost are now located at point d.

Measured in terms of energy flow, we can decompose the improvement between points a and d into the distances labeled R, D, and P on the horizontal axis, which correspond, respectively, to the movements from a to b, from b to c, and from c to d. Point b is the point on the new transformation surface that lies on a ray to the origin from the initial point a. Hence, the movement from a to b

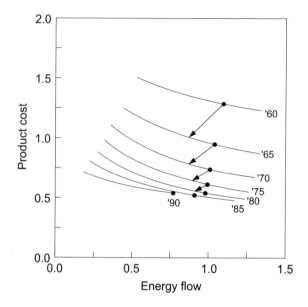

Figure 5.4. Innovation in Characteristics Transformation Surface: Room Air Conditioners.

The figure illustrates the estimated transformation surfaces over five-year intervals. Variables are normalized to equal one at their grand means. The dots represent the mean characteristics bundle of models offered for sale at each point in time, and the arrows represent the overall rate of innovation, measured radially from the origin.

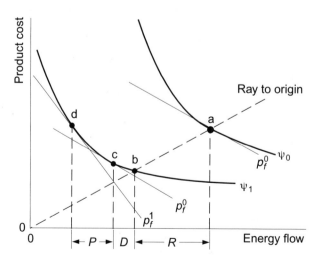

Figure 5.5. Decomposition of Characteristics Innovation.

represents equiproportionate improvement in up-front product costs and energy operating costs, and the distance R corresponding to this movement is a measure of the rate of overall technological change. In percentage terms, it is the rate of decrease in the total cost of the good to its user (i.e., product cost plus energy cost). Point c lies at the tangency between the new transformation surface and the old price line. Thus, the horizontal distance D between b and c measures the effect on energy use of the tilt in the transformation surface between time t_0 and t_1; that is, what we call "directional technological change." Finally, point d is the new optimum and the movement from c to d, which we label P, is the "model substitution" effect (given the new technology) brought about by the change in prices from p_f^0 to p_f^1.[23]

Expressions for R, D, and P can be derived in continuous time by taking the first-order condition that corresponds to the tangency shown in Figure 5.5 and differentiating it with respect to time. If the transformation surface is expressed as a log-linear function like those estimated in the previous section, then it is shown in Appendix 5.A that the rate of change in energy flow for the optimal product model will be given by

$$
\begin{aligned}
\frac{\dot{f}^*}{f^*} &= \frac{1}{1-\beta_1} \times (\dot{\alpha} + \dot{\beta}_1 \ln f^* + \dot{\beta}_2 \ln c^*) \\
&+ \frac{1}{1-\beta_1} \times \frac{\dot{\beta}_1}{\beta_1} - \frac{1}{1-\beta_1} \times \frac{\dot{p}_f}{p_f},
\end{aligned}
\tag{5.8}
$$

where optimal values are indicated by an asterisk and p_f is the discounted present value of the expected stream of relative energy prices, as in Figure 5.5. The last term in Equation (5.8) corresponds to P in Figure 5.5; it indicates that the optimal rate of change of energy flow has a term that is proportional to the rate of change of the effective energy price. The middle term captures optimal responses to changes in the slope of the surface; it corresponds to D. The first term includes changes in the intercept, as well as a weighted average of changes in β_1 and β_2. (Recall that β_2 relates to the capacity characteristic, which is held constant in Figure 5.5) It is shown in Appendix 5.A that this term is, in fact, the rate of change of f^* (and of k^*) if these rates of change are set equal to each other. Thus, it is equal to what is labeled R in Figure 5.5. The factor $1/(1-\beta_1)$ appearing in each term indicates that the optimal responsiveness of energy efficiency to any of these changes depends on the "costliness" of energy efficiency. β_1 is negative; when it is large in absolute magnitude, energy efficiency is expensive, so the energy-efficiency response to any of these changes is muted.

5.4.2   Decomposition of annual changes in energy efficiency

Using the values of the parameters estimated in Section 5.3 and data on changes in energy prices, it is possible to calculate each of the terms on the right-hand side of Equation (5.8) for any time period.[24] There is, of course, nothing that ensures that *actual* changes in mean efficiency of models offered for sale will correspond to the calculated changes in *optimal* efficiency. We explore this issue by approximating Equation (5.8) in discrete time and treating it as a regression equation. In this way, we allow explicitly for error between the computed "optimal" changes and the actual changes in energy efficiency. We can also generalize to allow for effects of the minimum efficiency standards described above, as well as labeling standards that require prominent display of energy-efficiency information for all product models.[25] We conjecture that these labeling standards may have affected consumers' actual responsiveness to energy prices, and thereby affected the extent to which firms faced incentives to offer more energy-efficient models as energy prices rose.

To allow for all these effects, we estimate for each product a time-series regression of the form

$$
\Delta \ln \bar{e}_t = \sigma + \mu R_t + \xi D_t + l_{0t} \sum_{j=0}^{\hat{j}} \tau_{0j} \frac{1}{1 - \beta_{1t}} \Delta \ln p_{t-j}
$$

$$
+ \; l_{1t} \sum_{j=0}^{\hat{j}} \tau_{1j} \frac{1}{1 - \beta_{1t}} \Delta \ln p_{t-j} + \theta \Delta s_t + \epsilon_t , \tag{5.9}
$$

where the dependent variable is the rate of change in mean energy efficiency of models offered for sale, t indicates values in year t, Δ indicates annual changes from year $t - 1$ to year t, ln is the natural logarithm, R and D are the overall and directional changes in the transformation surface based on the estimated parameters in that year, p is the relative price of electricity or natural gas to production inputs, l_0 is a dummy variable indicating that energy-efficiency labeling *was not* yet in effect, l_1 is a dummy variable indicating that labeling *was* in effect, Δs is a dummy variable indicating that energy-efficiency standards had been legislated but not yet achieved (i.e., $s = 1$ for $1987 \leq t \geq 1990$), and ϵ is an independently distributed error term with zero mean. The subscript j indicates that the associated price change occurred j years prior to year t, where $j = 0$ is the contemporaneous price and $j = \hat{j}$ is the most distant price lag included in the lag structure.[26] We use this distributed lag because we do not have a theoretical basis on which to determine how quickly the menu of product offerings should change in response to energy price changes.

To summarize, if the mean model is optimal for the typical consumer, the theory behind Equation (5.8) says that σ should be zero, that μ and ξ should each be unity, and the τ_1 terms should sum to unity. Our specification allows for this but does not impose it, and also allows for the possibility that labeling regulation increases the elasticity of average efficiency with respect to energy price, and that minimum efficiency regulations have an independent effect on average efficiency. The estimation results are summarized in Table 5.5, while the results for the full distributed lag version are provided in Appendix 5.B.

Because of the dynamic nature of the equations, as well as the existence of inducement effects on movements both *of* and *along* the surfaces, it is difficult to assess fully the empirical significance of the price and standards effects on the

Table 5.5. Factors Affecting Changes in Energy Efficiency.

Parameter	Explanatory variable	Description	Room air conditioners	Central air conditioners	Gas water heaters
η_0	$\frac{1}{1-\beta_1}l_0\Delta\ln p$	Prelabeling price effect	0.001 (0.630)	1.394 (0.423)	0.326 (0.529)
η_1	$\frac{1}{1-\beta_1}l_1\Delta\ln p$	Postlabeling price effect	1.175 (0.391)	–	0.577 (0.277)
θ	Δs	Standards	0.024 (0.025)	–	0.017 (0.007)
μ	R_t	Rate of innovation	0.055 (0.417)	0.844 (0.882)	–2.045 (2.872)
ξ	D_t	Direction of innovation	–0.053 (0.145)	0.047 (0.059)	0.479 (0.761)
σ	$Constant$		0.007 (0.007)	0.001 (0.026)	0.007 (0.009)
		# of observations	35	21	31
	$1 - U$	Goodness of fit	0.67	0.66	0.61

Note: Dependent variable is the rate of change of mean energy efficiency of models offered for sale ($\Delta\ln\bar{e}$). Estimation method is ordinary least squares. Robust standard errors are reported in parentheses. U is Theil's U statistic, where $1 - U$ is a measure of dynamic goodness of fit. See Appendix 5.B for parameter estimates of the individual distributed lag price effects.

basis of the parameter estimates themselves. As an overall estimate of those effects, we therefore carry out dynamic simulations using the estimated parameters of Equation (5.8) in which total changes in energy efficiency are compared, both with and without the historical changes in prices, and with and without minimum efficiency standards. The results are presented in Table 5.6, including standard errors based on the underlying parameter estimates.[27] We first carry out a "baseline" simulation that fits the model to the actual data,[28] and then compare the baseline with counterfactual simulations that isolate the effects of energy price changes and energy efficiency standards, as well as autonomous influences.

Effect of Changes in the Relative Price of Energy

In general, the results suggest that there is a substantial positive relationship between changes in the price of energy relative to production inputs and the rate of energy-efficiency improvements. We also find that there was a marked *change* in the responsiveness of energy-efficiency innovation to relative prices from the period before to the period after energy-efficiency labeling of room air conditioners and gas water heaters took effect. In the cases of both room air conditioners and water heaters, these parameters are statistically and economically significant for the period after energy-efficiency labeling took effect, but the effects are smaller and not significantly different from zero during the prelabeling period. Indeed, the postlabeling effect for both products is not statistically distinguishable from the theoretically predicted value of unity.[29] Labeling for central air conditioners

Table 5.6. Historical Effects of Price Changes and Efficiency Standards on Energy Efficiency, 1973–1993.

Historical simulations of cumulative percent changes in energy efficiency.

	Room air conditioners		Central air conditioners		Water heaters	
	Relative to 1973	Share of total change	Relative to 1973	Share of total change	Relative to 1973	Share of total change
Total change (%) (baseline)	29.7	–	58.9	–	11.2	–
	(4.5)		(3.5)		(2.4)	
Price-induced portion (%)	8.2	28	16.1	27	5.1	46
	(5.0)		(5.0)		(2.4)	
Standards-induced portion (%)	7.1	24	–	–	7.6	68
	(3.1)				(1.8)	
Autonomous portion (%)	12.7	43	36.8	62	–1.1	–10
	(2.7)		(3.7)		(1.9)	

Note: The baseline simulation uses the coefficient estimates from Table 5.5 to estimate the cumulative change in energy efficiency of models offered for sale from 1973 to 1993, assuming that energy prices and efficiency standards took their historical values (simulations for central air conditioners extend only to 1988). We decompose the total change into price-induced, standards-induced, and autonomous portions using the parts of the estimated model corresponding to these effects, and including any induced influences on the movement of the transformation surfaces (based on parameter estimates for Specification 2 from Tables 5.2, 5.3, and 5.4). The portions do not sum to the total change owing to the nonlinear nature of the model. Standard errors are reported in parentheses.

did not take effect until after the period covered by our data, so the table presents a single price elasticity estimate; it approximates the theoretical value of unity.[30]

To assess the cumulative effect of energy price changes, we compare the baseline simulations with counterfactual simulations that hold real energy prices at their 1973 levels, approximately their minimum for our study period.[31] The relative prices of electricity and natural gas rose 24 percent and 69 percent, respectively, over the simulation period (1973–1993). If the relative prices of electricity and natural gas had remained at their low 1973 levels, the model says that about one-quarter to one-half of the increase in the energy efficiency of the available menu experienced since then would not have occurred. Energy efficiency would have been about 8 percent lower for room air conditioners, 16 percent lower for central air conditioners, and about 5 percent lower for gas water heaters.

Effect of Energy-Efficiency Standards

Direct energy-efficiency standards appear to have had a modest positive effect on energy-efficiency changes during the compliance period from enactment of legislation to the time of the compliance deadline (1987–1990). For both room air conditioners and gas water heaters, the point estimate implies that energy efficiency improved about 2 percent per year faster during this implementation period than would otherwise have occurred, implying a cumulative effect of 7 percent for room air conditioners and 8 percent for water heaters based on the simulations. Note that while the "direct" effect of standards on average efficiency levels in room air conditioners is statistically insignificant in the regression, the over-all effect of standards *is* significant when combined with the "indirect" effect of standards through D and R, as shown through the simulations. This analysis does not reveal the extent to which this occurred because inefficient models were dropped versus new efficient models added, although new, more efficient models were being added during the compliance period.[32]

Effects of the Rate and Direction of Technological Change

The parameter estimates for the influence of R and D on changes in mean energy efficiency do not provide any consistent support for the theoretical prediction of unitary coefficients. It is possible that this failure is due to the arbitrariness of the mean model as representative of the preferences of the typical consumer. Another problem is that the predicted changes due to the direction of technological change are extremely sensitive to the estimated curvature of the surface, particularly given the relatively slight curvature that we find in some cases. More generally, it is em-pirically difficult to distinguish the effects of R and D, because of a high degree of correlation between the rate and direction of technological change, as well as relatively constant rates of overall technological change, which makes R difficult to distinguish from the constant term.

On the other hand, the simulations indicate that a substantial portion of the overall change in energy efficiency for all three products cannot be associated with either price changes or government regulations. The autonomous drivers of energy-efficiency changes (including the constant term) explain up to 62 per-cent of the total change in energy efficiency. Thus, we can view the induce-ment hypothesis as either half-full or half-empty; a substantial portion of energy-efficiency changes appear to be induced, but a large portion cannot be explained in this way. Of course, as with the "residual" in standard analyses of technologi-cal change, the association of these noninduced changes with "time" means only that we cannot explain exactly where they come from. To the extent that they are driven by forces such as government-funded research, some portion of what we label autonomous is probably endogenous in a broader sense.

5.5 Conclusion

The reemergence of energy efficiency as a policy concern has drawn new atten-tion to an old question: To what extent does the innovation process respond to economic incentives, making it systematically easier over time to economize on inputs that become more expensive? A natural way to approach this question is to think of capital goods as products, and their resource-consuming properties as product characteristics. In this way, we place microeconomic structure on the induced innovation hypothesis. In principle, this structure permits econometric identification of the extent to which the pace of technological advance, the di-rection of technological advance, and changes in the "menu" of offered product models each respond to changes in resource prices.

Our application of this analytical framework to the evolution of three energy-using household durable goods yields several important findings. First, the substantial observed increases in the energy efficiency of two of the three products over the past several decades appear to have been associated with overall technological advance. In the early part of the period, autonomous improvement in these products appears to have been biased away from energy efficiency. That is, the up-front costs of the products were decreasing faster than their operating costs. But the significant increase in energy prices that occurred in the 1970s and 1980s had noticeable effects, slowing or reversing this process. Second, increasing energy prices had an observable effect on which technically feasible models were offered for sale. Third, this effect of energy-price increases on "model substitution" was particularly strong after product-labeling requirements went into effect. Indeed, our simulations suggest that the post-1973 energy price increases account for one-quarter to one-half of the observed improvements in the mean energy efficiency of models offered for sale over the past two decades. Fourth and finally, government energy-efficiency standards also had a significant impact on the average energy efficiency of the product menu.

The recent resurgence in interest in endogenous technological change has focused attention on the mechanisms by which economic agents' optimizing decisions affect the overall pace of technological change. But the endogeneity of the direction or composition of technological change is surely at least as significant. Further, whereas empirical implementation of endogenous growth models has been hampered by the difficulty of measuring the underlying exogenous factors that drive the system, variations in relative prices provide interesting "natural experiments" that permit empirical investigation of induced changes in the direction of technological change. We suggest that the product characteristics approach provides a useful framework in which to look at these natural experiments.

Appendix 5.A
Theory

A straightforward model of consumer optimization over the product cost k and energy flow f of a durable good shows that the consumer will desire a model for which the marginal cost of f is equal to the consumer's willingness to pay for f (i.e., p_f). This first-order condition can be represented in elasticity form, as follows:

$$\frac{\partial k^*/k^*}{\partial f^*/f^*} = \epsilon_{kf} = \frac{p_f f^*}{k^*} , \qquad (5.10)$$

where ϵ_{kf} is the elasticity of product cost with respect to energy flow. This condition shows that the consumer would like to purchase a model such that the ratio of operating costs to product costs (in present value terms) implied by the purchase is equal to the elasticity of the transformation surface, that is, a point of tangency such as is shown in Figure 5.5.

Solving Equation (5.10) for f^* and taking natural logs,

$$\ln f^* = \ln k^* + \ln \epsilon_{kf} - \ln p_f . \qquad (5.11)$$

The transformation surface [Equation (5.1)] gives $\ln k^*$ as a function of f, so we can use that to solve out both $\ln k^*$ and ϵ, and, assuming that $\beta_1 < 0$, we get

$$\ln f^* = \alpha + \beta_1 \ln f^* + \beta_2 \ln c^* + \ln(-\beta_1) - \ln p_f . \qquad (5.12)$$

By taking the derivative of Equation (5.12) with respect to time, holding capacity constant, we get Equation (5.8) in the text, which shows how consumers' desired levels for f, and therefore k, might change over time:

$$\begin{aligned}
\frac{\dot{f}}{f^*} &= \frac{1}{1 - \beta_1} \times \dot{\alpha} + \dot{\beta}_1 \ln f^* + \dot{\beta}_2 \ln c^* \\
&+ \frac{1}{1 - \beta} \times \frac{\dot{\beta}_1}{\beta_1} - \frac{1}{1 - \beta_1} \times \frac{\dot{p}_f}{p_f} ,
\end{aligned} \qquad (5.13)$$

where time subscripts on all variables and parameters are suppressed for notational convenience.

As described further in the text, the three terms of Equation (5.8) correspond to R, D, and P in Figure 5.5. To see this, return to the basic transformation surface of Equation (5.1) and differentiate with respect to time, considering both the characteristics and the parameters as functions of time:

$$\frac{\dot{k}}{k} = \beta_1 \times \frac{\dot{f}}{f} + \beta_2 \times \frac{\dot{c}}{c} + \dot{\alpha} + \dot{\beta}_1 \ln f + \dot{\beta}_2 \ln c . \qquad (5.14)$$

Now recall that R is the overall rate of technological change, defined as the percent reduction in k and f that is implied by movements of the transformation surface, assuming that these reductions occur proportionately, and holding other characteristics constant. If $\dot{k}/k = \dot{f}/f = R$, then this equation can be solved for R to yield

$$R = -1/(1 - \beta_1) \times (\dot{\alpha} + \dot{\beta}_1 \ln f^* + \dot{\beta}_2 \ln c^*) , \qquad (5.15)$$

where the minus sign indicates that cost reductions represent positive innovation.[33]

Appendix 5.B
Statistics

Factors Affecting Energy-Efficiency Innovation (Full Distributed Lag).

Parameter	Variable	Description	Room air conditioners	Central air conditioners	Gas water heaters
η_{00}	$\frac{1}{1-\beta_{1t}}$ $\times(l_0\Delta\ln p_t)$	Prelabeling percent change in relative price of energy	−0.271 (0.194)	0.670 (0.270)	0.059 (0.414)
η_{01}		1-year lag	0.127 (0.189)	0.724 (0.341)	−0.113 (0.458)
η_{02}		2-year lag	0.309 (0.236)	−	−0.230 (0.466)
η_{03}		3-year lag	−0.573 (0.323)	−	0.611 (0.356)
η_{04}		4-year lag	−0.116 (0.215)	−	−
η_{05}		5-year lag	0.526 (0.268)	−	−
η_{0}		Total prelabeling price effect	0.001 (0.630)	1.394 (0.423)	0.327 (0.529)
η_{10}	$\frac{1}{1-\beta_{1t}}$ $\times(l_1\Delta\ln p_t)$	Postlabeling percent change in relative price of energy	−0.062 (0.191)	−	0.317 (0.183)
η_{11}		1-year lag	0.098 (0.194)	−	−0.140 (0.233)
η_{12}		2-year lag	0.235 (0.314)	−	0.159 (0.234)
η_{13}		3-year lag	0.743 (0.277)	−	0.242 (0.214)
η_{14}		4-year lag	−0.429 (0.224)	−	−
η_{15}		5-year lag	0.590 (0.194)	−	−
η_{1}		Total postlabeling price effect	1.175 (0.391)	−	0.577 (0.277)
θ	Δs	Standards	0.024 (0.025)	−	0.017 (0.007)
μ	R_t	Rate of innovation	0.055 (0.417)	0.844 (0.882)	−2.045 (2.872)
ξ	D_t	Direction of innovation	−0.053 (0.145)	0.047 (0.059)	0.479 (0.761)
σ	$Constant$		0.007 (0.007)	0.001 (0.026)	0.007 (0.009)
		# of observations	35	21	31
	$1-U$	Goodness of fit	0.67	0.66	0.61

Note: Dependent variable is the rate of change of mean energy efficiency of models offered for sale ($\Delta\ln\bar{e}$). Estimation method is ordinary least squares. Robust standard errors are reported in parentheses.

Acknowledgments

This chapter is based on Newell's Ph.D. dissertation at Harvard University. We thank, without implicating, Olivier Blanchard, Robert Deacon, William Hogan, Lawrence Katz, Raymond Kopp, Albert Nichols, William Pizer, Martin Weitzman, seminar participants at several universities, and anonymous referees for useful comments. We also thank Suzanne Kim, Sandip Madhavareddy, and Karthik Muralidharan for excellent research assistance. The research was supported by US Department of Energy award No. DE-FGO2-95ERG2106, a Resources for the Future Joseph L. Fisher Dissertation Award, and a John F. Kennedy School of Government Joseph Grump Fellowship. Such support does not constitute an endorsement by those institutions of the views expressed in this chapter.

Notes

1. See, for example, Binswanger and Ruttan (1978). See Thirtle and Ruttan (1987) for a summary of this literature.

2. For an empirical analysis of the diffusion of energy-efficient technologies, see Jaffe and Stavins (1995).

3. See, for example, Rosen (1974), Triplett (1985, 1987), Trajtenberg (1990), and Berry *et al.* (1995) .

4. We also estimated translog versions. The decomposition of innovation discussed below is considerably more complex in the translog world, and we found that the translog estimation yielded similar results (Newell 1997). For other products there are additional characteristics, which are accommodated by simply adding more logarithmic terms. For notational convenience we generally suppress the model subscript i in subsequent equations.

5. An alternative would be to redefine all characteristics so that they were desirable, using energy efficiency rather than energy flow, for example. Given our interest in energy-saving technological change, however, we found it useful to formulate the problem in terms of capital inputs k and energy inputs f in a manner analogous to a standard production function when viewed from the perspective of the *user* of the product, be it a consumer or a firm.

6. It is of course possible that what we call "autonomous" and measure with t is at some deeper level itself endogenous, being associated with research investments, the state of knowledge and technical experience, or other factors for which we do not have measures.

7. Traditional estimation of "neutral" technical change would allow only α to vary, and only as a linear function of time. We add quadratic terms to allow for acceleration or deceleration in technological change. We allow the β_1 and β_2 parameters to vary over time to permit autonomous changes in the slope of the function with respect to the important characteristics. We allow energy prices and standards to affect α and β_1 to test the inducement hypothesis; since there is no theoretical reason why energy prices or standards should affect the shape of the surface in the other characteristic dimensions, we do not enter these variables into β_2. Time interaction terms for the fan speed dummies for room air conditioners and for storage capability for gas water heaters were eliminated because we found them to be both very small and statistically insignificant.

8. It is not clear, *a priori*, whether we should expect the coefficient on $\ln q$ to be positive, owing to an association with demand shocks or the business cycle, or negative, owing

to economies of scale in production. Because shipments for the products we investigated have been generally increasing over time, we also face the difficult issue of distinguishing scale effects from technological change. We therefore do not place too much interpretative emphasis on the effect of product shipments, viewing it rather as a useful control variable along the "price-quantity" dimension.

9. An Energy Information Administration (1980) study found that major tooling and redesign changes to incorporate energy-saving design options require lead times of about 1.5 to 2 years for a single model and longer for an entire product line. Thus, a typical cycle for introducing new appliance models can be 3 or more years. This time frame is consistent with other assessments in the literature of the time required to develop and bring new product innovations to market (Levin *et al.* 1987).

10. There are, however, occasional sales on products in the Sears catalog. See Gordon (1990) for a thorough review of the advantages and disadvantages of using various data sources for such analysis.

11. To facilitate interpretation of the parameter estimates of the characteristics transformation surfaces, we normalized the time variable to equal zero in 1975 and we normalized all other purely time-series variables (i.e., energy prices and product shipments) to equal unity in 1975, or zero after taking natural logarithms. We normalized all other variables so their normalized means equal unity, or zero after taking natural logarithms.

12. Given data on nominal product prices, we construct a measure of real product cost, and over time a measure of technological change, by controlling for changes in input prices. Thus, our measure of product cost k is calculated as $k_{it} = p_{it}/p_{xt}$, where p_{it} is the nominal price of product model i at time t and p_{xt} is an input price index for that product at time t (subscripting for product models and time is suppressed in what follows). To clarify the procedure we used, including its possible limitations, consider the relationship $p = \mu c'$, where c' is the product's nominal marginal cost and μ is the markup of price above marginal cost. We thank an anonymous referee for pointing out that the markup can be decomposed into components representing pure profit π and the degree of returns to scale γ, leading to $p = \pi\gamma c'$, where $\pi = p/\bar{c}$, $\gamma = \bar{c}/c'$, and $\bar{c}$ is average cost. Our input cost deflation procedure can be represented explicitly within this framework by expressing marginal cost as the product of the input price index and an input index $x : p = \pi\gamma p_x x$ or $p/p_x = k = \pi\gamma x$. Over time, this implies that $\dot{k}/k = \dot{\pi}/\pi + \dot{\gamma}/\gamma + \dot{x}/x$, confirming that our measure of real product cost can provide an unbiased measure of technological change (i.e., reductions in the amount of inputs used to produce a given output) as long as the components of markup are together relatively constant over time.

13. We could not obtain model-specific data on product shipments; aggregate data are from Bureau of the Census (1947–1958, 1959–1994).

14. We constructed Divisia price indices having rates of growth equal to a weighted average of the rates of growth of labor, capital, and material prices; the weights are the relative shares of each component in total value. The data are primarily from the census (Bureau of the Census 1954–1992) using Standard Industrial Classification (SIC) Code 3585, Refrigeration and Heating Equipment for room and central air conditioners, and SIC Code 3639, Household Appliances—Not Elsewhere Classified for gas water heaters.

15. The "price" of energy flow to an optimizing consumer, p_f, can be written as $p_f = pu\rho$, where p is the price of energy per unit time, u is the utilization level (e.g., hours), and ρ is a present value factor (which takes into account the discount rate and product service life) (Newell 1997). Given this relationship, if expected utilization is negatively correlated with expected energy prices (as we might expect), an estimated coefficient involving the price of energy alone could be biased *downward* relative to the true value. In

practice, Hausman (1979) estimated the elasticity of air conditioner utilization with respect to the price of electricity to be −0.04, suggesting that any such bias would likely be quite small. In any event, the utilization data that would be necessary for our analysis do not exist. Regarding discount factors, one could use market interest rates as a proxy, but it is not clear what interest rate to use. If the nominal credit card rate is the relevant discount rate, this suggests that the discount rate has changed very little because nominal credit card rates have been remarkably stable over the past three decades. *Real* interest rates, on the other hand, have historically been positively correlated with real energy prices, suggesting that any potential bias from their omission would again *underestimate* the effect of energy prices (recall that the discount rate is inversely related to the present value factor). Finally, we found no evidence for significant trends in the service iifetimes for these appliances based on communications with industry experts and other information sources at our disposal. In practice, we found that our results were generally robust to a wide variety of alternative specifications for the energy "price" variable, including specifications that included market interest rates.

16. Gordon (1990), for example, uses hedonics to generate quality-adjusted price indices for a large number of durable goods, including room air conditioners and gas water heaters.

17. Shipments of many durable goods, including air conditioners and water heaters, are correlated with both housing starts and real household income. Housing starts tend to be correlated with demand for most appliances for two primary reasons: almost every new house requires a new set of appliances; and housing starts are a good indicator of a healthy economy, which would encourage replacement and discretionary purchase of appliances.

18. Recall that, to facilitate interpretation of the parameter estimates of the transformation surfaces, we normalized the variables; see endnote [11]. Coefficients on those variables that do not involve interactions with t, t^2, $\ln p$, or s have the interpretation of being the elasticity for that variable in 1975.

19. The response of product cost to changes in capacity is best measured by adding β_1 and β_3, which gives the elasticity of product cost with respect to capacity holding constant energy efficiency, rather than energy flow. Thus, a 10 percent increase in capacity was associated with a 5 percent increase in product cost for room air conditioners, an 8 percent increase for central air conditioners, and a 7 percent increase in product cost for gas water heaters (in 1975). Multiple-fan-speed room air conditioners were estimated to cost 20 percent and 30 percent more, respectively, for a two- or three-speed than for a one-speed model (see β_3 and β_4). To some extent, the multiple-fan-speed dummies may be picking up characteristics not included in the analysis that are also associated with higher-quality room air conditioners (e.g., adjustable thermostats, rotating louvers, and filter monitors).

20. A positive coefficient for the inducement effect means that an increase in the energy price or standard makes β_1 less negative, thereby reducing the elasticity of product cost with respect to energy flow.

21. These effects are also of modestly significant magnitude. For example, the estimate for β_{13} in Table 5.2 of 0.41 means that a 10 percent increase in energy prices is associated with a change in the elasticity of product cost with respect to energy flow of about 0.04. The base elasticity (β_{10}) is about −0.4, so this represents a reduction of about 10 percent in absolute magnitude. The relative magnitude is similar for central air conditioners.

22. Recall that we cannot estimate a standards-inducement effect for central air conditioners because there were no standards during the period of our data.

23.   We call P model "substitution" because of its similarity to the concept of input substitution along a production isoquant or output substitution along a production possibility frontier, recognizing that this term is not strictly appropriate because the range of model offerings can be expanded or enriched without a new model always displacing an old one.

24.   Since energy use per unit of capacity has been falling, and it becomes awkward to constantly talk about the absolute value of negative numbers, at this point we will switch from looking at changes in energy flow (f), to changes in "energy efficiency," which we denote e. Since all of our analyses are carried out holding other characteristics (including capacity) constant, the rate of change of energy efficiency is simply the negative of the rate of change of energy flow.

25.   Title V of the Energy Policy and Conservation Act of 1975 (EPCA) requires product labels providing information on the energy efficiency, estimated annual energy costs, and operating cost ranges for similar products for 13 categories of appliances and equipment (Office of Technology Assessment 1992; *United States Code of Federal Regulations* 1995a). The compliance deadline was May 1980 for room air conditioners and water heaters, but the deadline was delayed until February 1988 for central air conditioners. The Sears catalog includes label information in tabular form. We model the potential effect of product labeling on energy-efficiency changes by allowing the coefficient on the relative price of energy to change from the prelabeling to the postlabeling period, where the change occurs in 1981 for room air conditioners and in 1977 for water heaters.

26.   Three conventional methods for selecting the number of lags (i.e., the Akaike Information Criterion, the Schwarz Bayesian Information Criterion, and the adjusted R^2) recommended the same distributed lag structure for our estimated equations: five years for room air conditioners, one year for central air conditioners, and three years for gas water heaters.

27.   Recall that we have allowed for inducement effects in the overall movement of the surface, the change in the slope of the surface, and the movement along the frontier. The first of these was insignificant, and the second was significant in some specifications for some products. Hence the inducement effects we simulate and present in Table 5.6 are based on the parameter estimates from Table 5.5 in conjunction with any induced influences of energy prices and standards on the measures of R and D, which are in turn based on parameter estimates for Specification 2 from Tables 5.2, 5.3., and 5.4.

28.   The baseline simulation replicates actual experience quite well, including capturing turning points in the innovation trajectory. This is supported by a conventional quantitative measure for dynamic goodness of fit, $1 - U$, where U is Theil's U statistic (Theil 1961). In addition, decomposing U into its component sources revealed that the vast majority of the simulation error for the three products was due to unsystematic error—a desirable property for simulation models (Newell 1997).

29.   The $1/(1 - \beta_1)$ elasticity adjustment is quite important here; the unadjusted price elasticity for gas water heaters is much lower than for air conditioners ($\eta_1 = 0.87, 0.73$, and 0.14 for room air conditioners, central air conditioners, and gas water heaters, respectively) (Newell 1997). This is as expected given the less favorable product cost trade-off inherent in central air and gas water heater technology.

30.   The elasticity estimate for central air without labeling suggests that the labeling effect for the other two technologies may be due to some sectoral shift other than the labeling itself. To test this, we estimated an equation for central air conditioners with dummy variables for pre- and post-1981 observations as for the other products. Because there were no labeling requirements for central air, these periods should not differ if labeling is the true explanation for the shifts found for the other technologies. The result is that

the postlabeling coefficient is higher than the prelabeling coefficient, but the difference is small and insignificant. This suggests that while the difference we ascribe to labeling in the room air conditioner and water heater equations may be partly due to other (unobserved) factors (e.g., other information sources), labeling does seem to have some effect. Possibly, the fact that central air is a "bigger" purchase makes consumers (and hence manufacturers) sensitive to price trade-offs even without mandatory labeling, but this is only speculation.

31. The simulations assume that product labeling occurred as it did historically, as measured in our regressions by a shift in the coefficient on energy prices from a pre- to a postlabeling level.

32. An important limitation of our approach that bears consideration, especially in the context of the effect of standards, is our reliance on changes in the mean energy efficiency of the menu as our measure of energy-efficiency improvements. Obviously, the mean efficiency can rise because of the disappearance of inefficient models, without any introduction of new models. The elimination of inefficient models was, in fact, the primary intention of these regulations. Inspection of the distribution of the efficiencies of room air conditioners and gas water heaters over the time period when these standards were taking effect suggests that the primary effect was the elimination of the distribution's lower tail.

33. As is clear from the definition of R, the rate of technological change will depend on where measurements take place along the transformation surface. The usual approach to addressing this measurement issue is to create a productivity index using the mean of the distances from the old to the new surface that are found using rays from the origin through the realized point on the old and on the new surfaces (which will not generally lie on the same ray). In the typical aggregate or firm-level context, the points on the old and new surfaces will correspond to points of tangency between the surface and a price hyperplane. In the empirical application we calculate the rate of innovation as the mean of the rates found using the mean characteristic levels at the initial and subsequent points in time.

References

Berry, S., Levinsohn, J., and Pakes, A., 1995, Automobile prices in market equilibrium, *Econometrica*, **LXIII**:841–890.

Binswanger, H.P., and Ruttan, V.W., 1978, *Induced Innovation: Technology, Institutions, and Development*, The Johns Hopkins University Press, Baltimore, MD, USA.

Bureau of the Census, 1954, 1958, 1963, 1967, 1972, 1977, 1982, 1987, 1992, *Census of Manufactures*, US Department of Commerce, Washington, DC, USA.

Bureau of the Census, 1959–1994, *Current Industrial Reports, Issues for Air Conditioning and Refrigeration Equipment, Heating and Cooking Equipment, and Major Household Appliances*, US Department of Commerce, Washington, DC, USA.

Bureau of the Census, 1947–1958, *Facts for Industry, Issues for Air Conditioning and Refrigeration Equipment and Heating and Cooking Equipment*, US Department of Commerce, Washington, DC, USA.

Bureau of the Census, 1975, *Historical Statistics of the United States: Colonial Times to 1970*, US Department of Commerce, Washington, DC, USA.

Bureau of the Census, 1996, *Statistical Abstract of the United States, 1996*, US Department of Commerce, Washington, DC, USA.

Council of Economic Advisers, 1997, *Economic Report of the President*, US Government Printing Office, Washington, DC, USA.

Energy Information Administration, 1980, *National Interim Energy Consumption Survey (NEICS)*, US Department of Energy, Washington, DC, USA.

Epple, D., 1987, Hedonic prices and implicit markets: Estimating demand and supply functions for differentiated products, *Journal of Political Economy*, **XCV**:59–80.

Gordon, R.J., 1990, *The Measurement of Durable Goods Prices*, University of Chicago Press, Chicago, IL, USA.

Hausman, J.A., 1979, Individual discount rates and the purchase and utilization of energy-using durables, *Bell Journal of Economics*, **X**:33–54.

Hicks, J., 1932, *The Theory of Wages*, Macmillan, London, UK.

Jaffe, A.B., and Stavins, R.N., 1995, Dynamic incentives of environmental regulations: The effects of alternative policy instruments on technology diffusion, *Journal of Environmental Economics and Management*, **XXIX**:S43–S63.

Levin, R.C., Klevorick, A.K., Nelson, R.R., and Winter, S.G., 1987, Appropriating the returns from industrial research and development, *Brookings Papers on Economic Activity*, **3**:783–820.

Newell, R.G., 1997, Environmental Policy and Technological Change: The Effects of Economic Incentives and Direct Regulation on Energy-Saving Innovation, Ph.D. dissertation, Harvard University, Cambridge, MA, USA.

Office of Technology Assessment, 1992, *Building Energy Efficiency*, US Congress, Washington, DC, USA.

Rosen, S., 1974, Hedonic prices and implicit markets: Product differentiation in pure competition, *Journal of Political Economy*, **LXXXII**:34–55.

Schumpeter, J.A., 1939, *Business Cycles*, Vols I and II, McGraw-Hill, New York, NY, USA.

Sears, Roebuck and Co., 1958–1993, *Spring/Summer Catalog*, Sears, Roebuck and Co, Chicago, IL, USA.

Theil, H., 1961, *Economic Forecasts and Policy*, North-Holland Publishing Co, Amsterdam, Netherlands.

Thirtle, C.G., and Ruttan, V.W., 1987, The role of demand and supply in the generation and diffusion of technical change, in F.M. Scherer, ed., *Fundamentals of Pure and Applied Economics*, Vol. 21 in the Economics and Technological Change Section, Harwood Academic Publishers, New York, NY, USA.

Trajtenberg, M., 1990, *Economic Analysis of Product Innovation*, Harvard University Press, Cambridge, MA, USA.

Triplett, J.E., 1985, Measuring technical change with characteristic-space techniques, *Technological Forecasting and Social Change*, **XXVII**:283–307.

Triplett, J.E., 1987, Hedonic functions and hedonic indexes, in J. Eatwell, M. Milgate, and P. Newman, eds, *The New Palgrave: A Dictionary of Economics*, Vol. 2, Macmillan Press, New York, NY, USA.

United States Code of Federal Regulations, 1995a, 16 C.F.R., Chapter 1, Federal Trade Commission, Part 305-Appliance Labeling Rule, US Government Printing Office, Washington, DC, USA.

United States Code of Federal Regulations, 1995b, 10 C.F.R., Chapter 11, Department of Energy, Part 430, Subpart C-Energy Conservation Standards, US Government Printing Office, Washington, DC, USA.

White, H., 1980, A heteroskedasticity-consistent covariance matrix estimator and a direct test for heteroskedasticity, *Econometrica*, **XLVIII**:817–830.

Chapter 6

Inter-Firm Technology Spillover and the "Virtuous Cycle" of Photovoltaic Development in Japan

Chihiro Watanabe, Charla Griffy-Brown, Bing Zhu, and Akira Nagamatsu

6.1 Introduction

Despite a locational disadvantage as a mid-latitude country, Japan has taken a leading role in world photovoltaic (PV) development. Two main factors have been critical to the successful development of PV technology in Japan. First, like semiconductors, PV technology is central to a complex web of related technologies and can therefore benefit from both learning effects and economies of scale. Second, because of the interdisciplinary nature of PV development, technology spillover benefits are high, in turn further stimulating learning effects (Watanabe 1997).

The above-mentioned interdependent success factors in PV development highlight the importance of both an endogenous technological innovation perspective and the inducement mechanisms of technological change. In the case of PVs, Japan's Ministry of International Trade and Industry (MITI) initiated development under its Sunshine Program, a research and development (R&D) program on new energy (MITI 1970–1990). In particular, the Sunshine Program aimed to encourage broad cross-sector industry participation; to stimulate cross-sector technology development; and, finally, to induce substantial industry investments in PV R&D, leading to a rapid increase in the industry's knowledge stock on PV technology (Watanabe 1995c; Watanabe and Clark 1991).

Supported by a PV R&D acceleration strategy adopted in 1979 (Industrial Technology Council 1979) and the creation of a research association for PV development (established in 1990), the technology knowledge stock arising from proprietary PV R&D and via spillover effects increased dramatically. The increase in this technology knowledge stock contributed to a significant decrease in the cost of solar cell production, which induced a further increase in demand for solar cells (and hence production). In turn, this increase in demand/production induced further PV R&D, thus creating a "virtuous cycle" (a positive feedback loop)

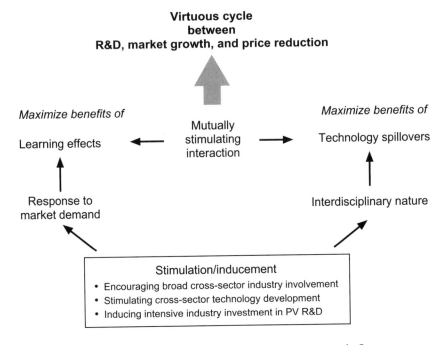

Figure 6.1. Conceptual Model of Furthering PV Development in Japan.

between R&D, market growth, and price reduction (Watanabe 1995b, 1997). The
achievement of a virtuous cycle (see Figure 6.1; see also Watanabe 1995a) has
significant techno-economic implications. However, to date only limited research
has been undertaken to elucidate Japan's policy framework in this area. Grübler
analyzed this virtuous cycle using the author's data and noticed the learning effect
triggered by MITI's inducement strategy (Grübler 1998). However, this analysis
was limited to the aggregate behavior of the entire PV industry and did not consider the impacts of inter-firm technology spillover effects. Although some pio-
neering studies have attempted to link learning and technology spillover effects
(e.g., Griliches 1957; Mansfield 1961; and Jovanovic and Lach 1989), further
research is necessary for analyzing the complicated mechanisms behind the creation of a "virtuous cycle" in the advancement of a new technology and the policy
framework conducive to such developments.

    This chapter presents an empirical analysis of Japan's firm-level PV technol-
ogy development over the past two decades. The analysis focuses on the impacts
of inter-firm technology spillovers to provide an empirical demonstration of the
industrial dynamism of this virtuous cycle created by a directed policy initiative. This chapter also illustrates the role of technology spillovers in stimulating these systemic positive feedbacks, resulting in an endogenous technological
innovation–development–diffusion process. A framework for relating technology
spillover effects and learning/experience curves in the PV innovation process is
developed in Section 6.2. The current state of cross-sector industry involvement in

PV technology development in Japan is reviewed in Section 6.3. In Section 6.4, the state of inter-firm technology spillovers in Japan's leading PV firms is analyzed. The virtuous cycle between R&D, market growth, and price reduction is assessed in Section 6.5. Finally, Section 6.6, discusses the implications of the present analysis for an induced technological change perspective and presents final observations and conclusions.

6.2   Analytical Framework

6.2.1   R&D, technology spillovers, and learning effects: Maximizing system efficiency on the market

R&D has become increasingly expensive. With such increased costs come opportunity costs, particularly in terms of choosing an optimal technology trajectory for sustainable development. Experience curves and learning effects often optimize a system that is linked to an external technological infrastructure that may not necessarily be the best choice from an environmental perspective. For these reasons, it is difficult for policy makers to choose which environmental technologies to support in using R&D incentives as an effective policy tool. Our work demonstrates the critical importance of "learning through the market," that is, making decisions based on market signals rather than trying to change the marketplace to achieve sustainable development.

In this context, technology spillovers not only alleviate some of the burden of huge R&D expenditures, but can also enhance the learning effects involved in assimilating environmentally friendly technologies and processes. The objective of the analytical framework presented here is to highlight the construction of a virtuous cycle in Japanese PV development. A learning/experience curve mechanism plays an important role in the policy-formation process. However, policy is only the tool that guides the technology trajectory by inducing a virtuous cycle. Its role is simply to stimulate learning and technology spillovers through—not against—the marketplace.

To illustrate these principles, an analytical framework is developed to elucidate empirically the development of PV technology in Japan. The framework incorporates not only technology spillovers within the PV technology sector, but also technology spillovers that have resulted in energy-efficiency improvements across industrial sectors, thereby further enhancing the positive effects of PV development. Figure 6.1 shows the links between these components. Japan's policy approach has created a virtuous development cycle in which energy-efficiency improvements from technology spillovers enhance learning effects across sectors, leading to effective innovation, particularly in PV technology. This effective innovation has contributed to economic growth, which in turn has stimulated technology spillovers. Thus, the positive feedback cycle continues. There is a loop not only between economic growth and technology spillovers, but also between technology spillovers and learning effects.

6.2.2   Correlation between technology improvement, technology spillovers, and learning effects

Technology Improvement and Learning Effects

In the following equation, a newly generated technology $[T]$ is a function $[\omega]$ of current technological level T and R&D investment R:

$$[T] = \omega(T, R) . \tag{6.1}$$

Equation (6.1) can be further developed into Equation (6.2) by using a learning function ϕ (Thomson 1993):

$$[T] = \phi(T) \times R . \tag{6.2}$$

The learning function $\phi(T)$ generally has the following characteristics (Arrow 1962):

$$\phi(T) > 0, \phi'(T) > 0, \phi''(T) < 0 . \tag{6.3}$$

With the generated $[T]$, the technology level will improve to T_2, as seen in Equation (6.4):

$$T_2 = T + [T] > T . \tag{6.4}$$

Since $\phi'(T) > 0$ and $\phi(T_2) > \phi(T)$, the learning effects increase as the technological level increases.

6.2.3   Technology spillover and learning effects

Based on Jaffe (1986), provided that technology Td spills over from the donor (D) to the host (H), assimilating technology in the host $[Td]h$ can be expressed using an assimilation capacity function θ as follows:

$$[Td]h = \theta(Th, \alpha) \times Td , \tag{6.5}$$

where $[Td]h$ is assimilating technology; $\theta(Th, \alpha)$ is assimilation capacity; Td is spillover technology flow; Th is technological level of host; and α is learning and assimilation capacity.

Equation (6.5) can be further developed as follows:

$$[Td]h = \phi(Th) \times f(\alpha) \times Td , \tag{6.6}$$

where Th is learning capacity, which is dependent on the host's technological level; and α is assimilation capacity. Equation (6.6) demonstrates that assimilating technology depends on learning capacity.

Technology newly generated by the host through assimilation of spillover technologies can be expressed as follows:

$$\begin{aligned}
[Th_2] &= \phi(Th) \times Rh + [Td]h = \phi(Th) \times Rh + \phi(Th) \times f(\alpha) \times Td \\
&= \phi(Th)[Rh + f(\alpha) \times Td] . \tag{6.7}
\end{aligned}$$

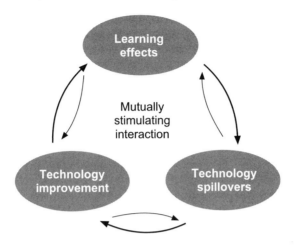

Figure 6.2. Correlation between Technology Improvement, Technology Spillovers, and Learning Effects.

Since the technology level of the host at this stage is $Th_2 = Th + [Th_2] > Th, \phi(Th_2) > \phi(Th)$, learning capacity increases as technology spillovers increase.

On this basis, we note a potential virtuous cycle between technology improvement, technology spillover, and learning effects, as illustrated in Figure 6.2. This concept is the basis of our analysis.

6.3 Cross-Sector Industry Involvement in PV Development

6.3.1 Intensive R&D by leading cross-sector firms

In line with its PV development strategy of encouraging the involvement of leading cross-sector firms, in 1990 MITI established an R&D consortium for PV development called the Photovoltaic Power Generation Technology Research Association (PVTEC). Table 6.1 lists the firms participating in PV development under the Sunshine Program and identifies the subset of companies that are also members of PVTEC. The table shows the broad spectrum of firms participating in PV development in Japan—from manufacturers of textiles, chemicals, petroleum and coal products, and ceramics to public institutes and the iron and steel, nonferrous metals, and electrical machinery industries. In addition to PVTEC, MITI has entrusted PV development to broad industrial sectors, including the electric power, housing, and construction industries. A total of 65 firms participated in MITI's PV development program in 1999.

Figure 6.3 illustrates the trends in Japan's R&D expenditures for PVs. Both public and private funding is shown; that is, the figure includes both expenditures

Table 6.1. Firms participating in the Sunshine Program and the Photovoltaic Power Generation Technology Research Association (PVTEC).

Textiles	Teijin Ltd	
Chemicals	**Kaneka Corp.**	Mitsubishi Chemical
	Mitsui Toatsu Chemicals	
	Shinetsu Chemical	
	Daido-hoxan Co.	
	Matsushita Battery Industrial Co., Ltd.	
Petroleum	Showa Shell Sekiyu K.K.	
and coal	Tonen Co.	
products	Japan Energy Co.	
Ceramics	**Kyocera Corp.**	
	Asahi Glass	
	Nippon Sheet Glass	
Iron and steel	Kawasaki Steel	Japan Steel Works
Nonferrous	Mitsubishi Materials	
metals and	Hitachi Cable	
products		
General		Kubota
machinery		
Electrical	**Sanyo Electric Co., Ltd.**	Sony
machinery	**Sharp Corp.**	Canon
	Fuji Electric Co., Ltd.	Anelva
	Hitachi, Ltd.	
	Mitsubishi Electric Corp.	
	Sumitomo Electric Industries, Ltd.	
	Matsushita Electric Industries Co., Ltd.	
	Oki Electric Industry	
Other		YKK
manufacturing		
Public	Japan Measurement and	Japan Quality Assurance Organisation
institutes	Inspection Institute	Shikoku Electric Power Research
	Central Research Institute	Institute
	of Electric Power Industry	Japan Electric Safety & Environ-
		ment Technology Laboratory
		Japan Weather Forecast Association
Electric		Okinawa Electric Power
power		Kashima North Joint Electric Power
Housing and		Misawa Homes
construction		National House Industry
		YKK Architectural Products
		Kajima

Note: Firms indicated in bold are discussed in this analysis; those listed in middle column are also members of PVTEC.

via MITI's Sunshine Program as well as those by Japan's PV industry over the 1974–1998 period. As the figure indicates, R&D expenditures increased dramatically after the early 1980s. This increase was in line with the recommendation by the Industrial Technology Council (one of MITI's advisory bodies) to accelerate PV R&D as a priority under the Sunshine Program (Industrial Technology Council 1979).[1] Although R&D expenditures began to decline in 1987—during

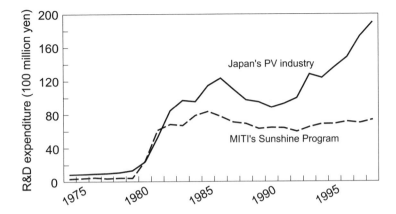

Figure 6.3. Trends in R&D Expenditures for PV in Japan, 1974–1998, in Constant 1985 Prices.

Japan's "bubble economy" period—they began to rise again as of 1993 and have since increased steadily, primarily from industry sources. This further increase can be attributed to MITI's new policy under the New Sunshine Program, an R&D program on energy and environmental technologies. The New Sunshine Program, which began in 1993, accelerated PV R&D activities further, particularly in light of the global environmental consequences of carbon dioxide emissions resulting from energy use (Watanabe 1995d, 1996).

6.3.2   The effect of MITI's R&D inducement strategy

Table 6.2 summarizes trends in PV R&D expenditures by eight leading Japanese PV firms and the government financial support provided under MITI's Sunshine Program. Approximately 40 percent of MITI's PV R&D budget was allocated to eight leading PV firms.[2] Table 6.3 summarizes the results of a correlation analysis of the PV R&D expenditures of these eight firms and MITI's financial support for PV R&D. The data indicate that, in all firms examined, MITI's financial support significantly induced private PV R&D expenditures with a time lag of one year. As demonstrated in Watanabe (1999) and (2000), MITI's energy R&D policy induced private R&D expenditures by firms in the following ways:

- It encouraged broad cross-sector industry involvement in national R&D programs such as the Sunshine Program (energy supply technologies) as well as the Moonlight Program (energy efficiency and conservation) (Watanabe and Honda 1992).

- It fostered cross-sector technology spillovers and inter-technology stimulation.

- It induced substantial industry activity in the broad area of energy R&D.

Table 6.2. Trends in Industry's PV R&D Expenditure and Government Support in Japan, 1981–1998, Billion Yen in Constant 1985 Prices.

	Firm								Other	
	A	B	C	D	E	F	G	H	firms	Total
1981	6.3	3.4	14.5	0.6	5.1	10.7	3.7	1.5	7.3	53.1
	(4.6)	(0.4)	(5.9)	(0.2)	(1.9)	(9.1)	(1.4)	(1.3)	[36.6]	[61.4]
1982	6.2	3.6	10.4	2.8	5.3	17.0	3.6	2.1	33.6	84.6
	(4.6)	(0.5)	(4.8)	(0.4)	(1.9)	(11.5)	(1.5)	(1.4)	[41.7]	[68.3]
1983	8.2	4.3	7.9	4.8	7.6	14.0	4.1	2.1	43.9	96.9
	(5.1)	(1.0)	(2.4)	(0.6)	(4.3)	(5.7)	(2.6)	(1.7)	[43.5]	[66.9]
1984	10.2	5.6	8.1	4.1	6.9	15.0	6.5	2.3	36.8	95.5
	(5.7)	(1.6)	(2.4)	(0.5)	(4.3)	(7.3)	(4.3)	(1.7)	[51.1]	[78.9]
1985	11.0	9.7	8.3	4.5	7.8	12.6	5.4	1.7	53.6	114.6
	(6.0)	(5.4)	(2.2)	(0.6)	(5.0)	(4.3)	(3.1)	(1.1)	[55.9]	[83.6]
1986	14.5	8.2	7.0	5.1	8.9	5.7	5.7	1.8	66.6	123.5
	(7.3)	(3.8)	(3.4)	(0.6)	(5.8)	(3.2)	(3.3)	(1.1)	[48.2]	[76.7]
1987	14.9	8.6	6.4	6.9	7.8	5.8	5.4	1.9	52.7	110.4
	(7.4)	(3.7)	(3.0)	(0.8)	(5.0)	(4.1)	(2.2)	(1.2)	[43.6]	[71.0]
1988	15.2	8.1	4.7	9.7	6.8	5.6	6.6	1.5	39.4	97.6
	(7.6)	(2.6)	(1.9)	(3.4)	(4.0)	(3.9)	(2.8)	(0.9)	[42.1]	[69.2]
1989	14.9	8.0	4.9	8.1	5.5	5.5	6.3	0.5	41.4	95.1
	(7.3)	(3.1)	(2.1)	(2.9)	(3.6)	(4.2)	(2.1)	(0.3)	[37.6]	[63.2]
1990	15.0	7.9	6.4	8.3	5.5	6.4	4.6	0.4	34.2	88.7
	(7.2)	(3.5)	(2.8)	(2.7)	(3.2)	(4.9)	(2.0)	(0.2)	[38.0]	[64.5]
1991	16.7	6.5	6.3	8.8	5.6	5.1	4.7	0.5	39.9	94.1
	(7.2)	(3.2)	(2.8)	(3.1)	(3.3)	(3.7)	(1.9)	(0.3)	[39.2]	[64.7]
1992	17.7	9.8	7.9	9.3	6.1	5.6	4.7	0.4	39.9	101.4
	(7.3)	(3.7)	(3.7)	(3.3)	(3.7)	(4.2)	(1.9)	(0.2)	[32.8]	[60.8]
1993	20.7	9.5	8.5	12.3	7.1	7.1	3.3	0.6	62.1	131.2
	(7.7)	(3.8)	(4.2)	(3.8)	(4.0)	(5.2)	(1.4)	(0.4)	[36.2]	[66.7]
1994	19.7	10.8	7.1	15.0	8.4	6.6	3.8	0.4	54.8	126.6
	(7.6)	(4.6)	(3.3)	(4.0)	(4.6)	(4.7)	(1.6)	(0.3)	[39.6]	[70.3]
1995	18.5	12.9	6.8	15.7	9.3	6.9	4.0	0.4	65.5	140.0
	(7.4)	(5.2)	(2.9)	(4.3)	(5.1)	(5.2)	(1.6)	(0.3)	[38.7]	[70.7]
1996*	20.0	13.9	7.5	16.5	9.5	7.5	4.5	0.5	68.6	148.5
										[71.7]
1997*	23.5	16.5	8.5	19.5	11.5	8.5	5.0	0.5	80.5	174.0
										[70.0]
1998*	25.5	18.0	9.5	21.5	12.5	9.5	5.5	0.6	87.4	190.0
										[73.6]

Note: Column headings are as follows: (A) Sanyo Electric Co., Ltd.; (B) Kyocera Corp.; (C) Sharp Corp.; (D) Kaneka Corp.; (E) Fuji Electric Co., Ltd.; (F) Hitachi, Ltd.; (G) Mitsubishi Electric Corp.; (H) Sumitomo Electric Industries, Ltd. Figures in parentheses indicate amount of financial support from government (MITI). Figures in square brackets include support to non-industry sectors (universities and national research institutes).
*Figures for 1996–1998 are estimated values based on interviews and statistics in the *Report on the Survey of Research and Development* (special issue on energy R&D), Agency of General Coordination and Management, Japanese Government, Tokyo, Japan.

- Enhanced R&D efforts led to an increase in industry's technology knowledge stock, which stimulated further research activities across technologies and sectors (Watanabe *et al.* 1991).

Table 6.3. Estimated Impact of Government PV R&D on Industry PV R&D Expenditure in Japan.

Regression model: $PVR = A \times SSPV_{t-1}^{\alpha}$.

	α (t-value)	adj.R^2	DW	Time period
Industry total	0.768 (29.35)	0.979	1.31	1976–1995
PV firm A	0.905 (8.60)	0.900	1.41	1980–1995
B	0.455 (12.50)	0.917	2.18	1981–1995
C	0.634 (13.89)	0.932	2.09	1976–1995
D	0.433 (12.57)	0.943	2.82	1982–1995
E	0.181 (4.51)	0.805	1.50	1981–1995
F	1.159 (33.14)	0.984	2.13	1976–1995
G	0.427 (6.23)	0.844	1.70	1981–1995
H	0.614 (5.98)	0.818	1.76	1981–1995

Note: $SSPV$ = PV R&D budget from Sunshine Program; PVR = industry's PV R&D expenditure; α = learning and assimilation capacity; DW = Durbin–Watson statistic. See Table 6.2 for firm names.

- This inducement mechanism played a catalytic role for industry's "technology substitution" for energy; that is, the substitution of efficiency improvements and conservation through the application of new technologies for energy as a factor input.

6.3.3 Creation of technology knowledge stock

The industry R&D expenditure induced by the Sunshine Program has furthered the PV-related technology knowledge stock. In line with previous research (Watanabe 1992), the technology knowledge stock of PVs is measured using the following equation:

$$T_t = R_{t-m} + (1 - \rho)T_{t-1} , \tag{6.8}$$

where T_t is technology knowledge stock of PV R&D at time t; R_{t-m} is PV R&D expenditure at time $t - m$; m is the lead time of (time lag between) PV R&D expenditures and PV commercialization; and ρ is the rate of obsolescence of PV technology.

To identify m and ρ, 19 leading PV firms were surveyed in 1993 with the support of MITI's Agency of Industrial Science Technology (AIST). Survey responses were received from 15 firms (79 percent of firms surveyed), including the 8 firms examined above. From these responses, 57 valid samples for estimating the time lag (m) and 28 for estimating the technology lifetime parameter (ρ) were obtained (see Appendix 6.A). As both samples are well balanced with respect to firms and technology life-cycle stages, the time lag and technology lifetime in leading Japanese PV firms over the past two decades were estimated by taking the average of the respective samples.

The average time lag between PV R&D and PV commercialization was estimated to be 2.8 years, and the average lifetime of PV technology was estimated to be 4.9 years. Assuming that technology depreciates and becomes obsolete over

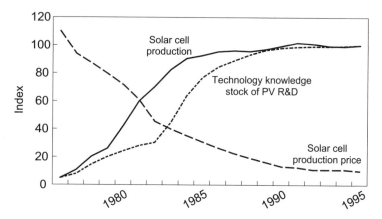

Figure 6.4. Trends in PV Development in Japan, 1976–1995: Index.

time, the annual rate of PV technology obsolescence was estimated to be 20.3 percent per annum (inverse of the lifetime of PV technology).[3] The estimated time lag and rate of technology obsolescence were evaluated by means of comparative statistical analysis using patent data; both proved to be statistically significant (see Appendix 6.A). Figure 6.4 illustrates the trends in the technology knowledge stock as measured by the lagged R&D expenditures as explained above and relates it to PV production volumes and declines in PV prices. Table 6.4 summarizes the same trends in eight leading PV firms. Figure 6.4 and Table 6.4 demonstrate that the technology knowledge stock as measured by PV R&D increased dramatically as of 1983/1984 in Japanese industry while PV R&D expenditures rose sharply from 1980/1981. This corresponds to a 2.8-year time lag between PV R&D and its translation into an increased technology knowledge stock enabling reduced costs and commercialization of PV technology in niche markets.

6.3.4   Solar cell production

Table 6.5 summarizes the development of solar cell production in Japan's PV industry. Japan's solar cell production in 1999 amounted to 80.0 MW, or 40 percent of a world production total of 200 MW (see Figures 6.5 and 6.6).[4] Figures 6.7 and 6.8 show the development of solar cell production by eight leading Japanese PV firms and their distribution. The shares of the top three firms accounted for nearly 90 percent of solar cell production in Japan in the 1995–1999 period.

A comparison of Tables 6.2 and 6.5 indicates that the cross-section of firms involved in PV R&D is broader than that of firms involved in solar cell production alone. Building on this observation, a comparison of Tables 6.4 and 6.5 indicates the following:

- Although the production levels of firms A and B are similar, the estimated technology knowledge stock of firm A is much greater than that of firm B.

Table 6.4. Trends in Japan's Industrial Technology Knowledge Stock of PV R&D, 1979–1998, Billion Yen at 1985 Fixed Prices.

| | Firm | | | | | | | | Other | |
	A	B	C	D	E	F	G	H	firms	Total
1979	4.8	10.5	4.2	0.4	7.8	1.8	2.8	0.3	1.0	133.6
1980	5.2	11.2	4.7	0.7	8.3	1.9	3.3	0.7	1.1	37.1
1981	5.9	11.5	5.2	1.1	9.2	2.0	4.1	1.0	1.2	41.2
1982	6.9	12.4	7.2	1.3	10.9	2.0	5.4	1.1	1.2	48.4
1983	9.8	13.1	10.5	1.4	13.1	4.5	7.7	1.4	7.0	68.5
1984	14.2	13.8	14.0	1.7	15.5	14.3	9.8	2.6	28.2	114.1
1985	17.5	14.8	18.6	4.1	17.7	30.9	11.5	4.1	59.2	178.4
1986	22.1	16.3	22.7	8.1	21.7	38.8	13.8	5.3	90.6	239.4
1987	27.9	19.5	26.3	10.6	24.2	47.0	17.3	6.6	111.4	290.8
1988	33.3	24.9	29.3	12.9	27.2	50.1	19.3	6.9	145.1	349.0
1989	41.2	28.2	30.4	15.4	30.6	45.7	21.0	7.3	180.2	400.0
1990	47.8	31.0	30.7	19.2	32.3	42.4	22.4	7.7	194.3	427.8
1991	53.3	32.8	29.3	24.9	32.6	39.5	24.3	7.5	195.1	439.3
1992	57.5	34.2	28.2	28.0	31.6	37.0	26.2	6.5	196.1	445.3
1993	61.0	34.9	28.9	30.7	30.7	36.0	26.0	5.6	192.0	445.8
1994	65.4	35.0	29.4	33.4	30.2	33.9	25.9	4.9	193.1	451.2
1995	70.0	37.5	31.3	36.0	30.2	32.7	25.7	4.3	198.8	466.5
1996	76.7	39.5	33.5	41.1	31.3	33.3	23.9	4.0	221.1	504.4
1997	81.1	42.4	33.9	47.9	33.4	33.2	22.9	3.6	231.7	530.1
1998	83.3	46.8	33.9	54.0	36.0	33.5	22.3	3.3	251.0	564.1

Note: See Table 6.2 for firm names.

- There are no substantial differences in the estimated technology knowledge stocks of firms B, C, D, E, F, G, and H; their production levels can be classified into three groups: firm B (production level much higher than those of the other firms considered); firms C and D (production levels similar to each other's, both firms among the top four); and firms E, F, G, and H (production levels much lower than those of other firms considered).

These observations suggest that broad inter-firm or cross-sector technology spillovers have prevailed in Japan's PV industry. The following subsection attempts to corroborate this hypothesis.

6.3.5   Correlation between technology potential and production

Figure 6.9 illustrates the relationship between PV production (SCP; average over the 1991–1995 period; vertical axis) and the estimated PV technology knowledge (TPV; horizontal axis). This map of the "techno-production structure" of leading Japanese PV firms can be divided into the following four clusters:

Table 6.5. Trends in Solar Cell Production in Japan's PV Industry, 1981–1999, in megawatts (MW).

	Firm								Other	
	A	B	C	D	E	F	G	H	firms	Total
1981	0.1	0.1	0.4	0.3	0.2	—	—	—	0.1	1.2
1982	0.9	0.2	0.5	0.3	0.7	—	—	—	0.1	2.7
1983	2.1	0.4	0.5	0.6	1.7	—	—	—	0.2	5.5
1984	3.0	1.0	0.6	0.9	2.1	0.1	—	—	1.2	8.9
1985	3.0	0.8	0.6	1.1	2.6	0.1	—	—	2.1	10.3
1986	3.4	0.9	0.6	1.2	2.5	0.1	—	—	3.9	12.6
1987	3.3	1.3	0.3	2.0	1.0	0.1	—	—	5.2	13.2
1988	3.1	1.7	0.3	2.4	0.5	0.1	—	—	4.7	12.8
1989	3.6	3.2	0.4	1.9	0.1	0.1	—	—	4.9	14.2
1990	4.4	4.6	0.6	1.8	0.1	0.1	—	—	5.2	16.8
1991	6.0	5.8	0.7	3.1	0.1	0.1	—	—	4.0	19.8
1992	6.5	5.1	1.0	3.0	—	—	—	—	3.2	18.8
1993	6.2	4.8	1.0	1.7	—	—	—	—	3.0	16.7
1994	5.5	5.3	2.0	1.8	—	—	—	—	1.9	16.5
1995	5.1	6.1	4.0	0.2	—	—	—	—	2.0	17.4
1996	4.6	9.1	5.0	0.0	—	—	—	—	2.5	21.2
1997	4.7	15.4	10.6	0.0	—	—	—	—	4.3	35.0
1998	6.3	24.5	14.0	0.0	—	—	—	—	4.2	49.0
1999	13.0	30.3	30.0	3.5	—	—	—	—	3.2	80.0

Note: See Table 6.2 for firm names.

- *Cluster 1*: Consists of firms A and B, which have the highest production levels in Japan. The technology knowledge stock level of firm A is much higher than that of firm B.

- *Cluster 2*: Includes the top four firms—firms A and B plus firms C and D—which together account for nearly 90 percent of solar cell production in Japan. Among these four firms, the technology knowledge stock levels of firms B, C, and D are almost the same, while that of firm A is much higher.

- *Cluster 3*: Consists of firms B, C, D, E, F, and G, which share similar technology knowledge stock levels. This cluster can be classified into two groups, firms B, C, and D, and firms E, F, and G. The former also belong to cluster 2 and rank higher in terms of production levels; the production levels of the latter are much lower.

- *Cluster 4*: Consists of firms E, F, G, and H. It includes firms of the lowest production levels. The technology knowledge stock levels of firms E, F, and G are reasonably high; that of firm H is extremely low.

An analysis of these clusters suggests the following possible technology spillovers among leading Japanese PV firms:

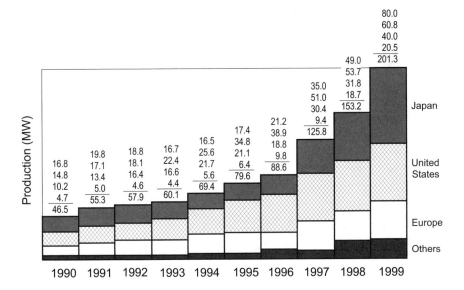

Figure 6.5. World Solar Cell Production, 1990–1999, Megawatts (MW).

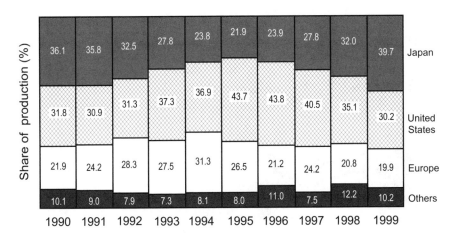

Figure 6.6. Regional Distribution of World Solar Cell Production, 1990–1999, Percent.

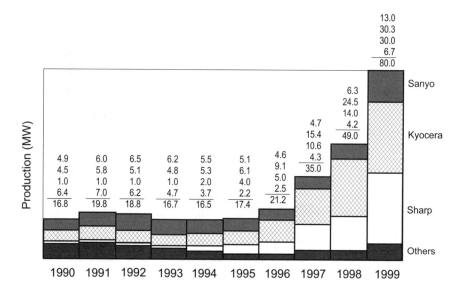

Figure 6.7. Solar Cell Production by Eight Leading Japanese PV Firms, 1990–1999, Megawatts (WM).

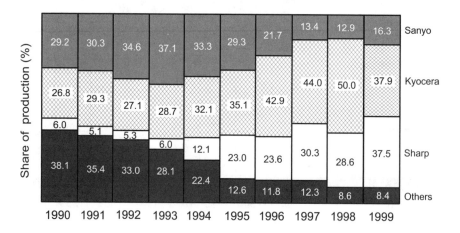

Figure 6.8. Solar Cell Production Shares for Eight Leading Japanese PV Firms, 1990–1999, Percent.

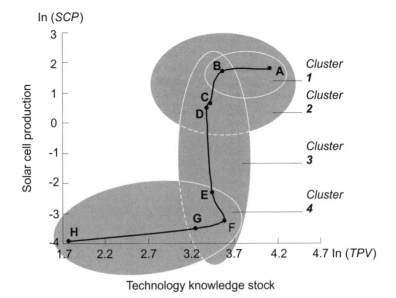

Figure 6.9. Correlation between Estimated Technology Knowledge Stock (via PV R&D Expenditures, log TPV) and Solar Cell Production (log SCP) in Leading PV Firms in Japan, Average, 1991–1995.

- Cluster 1 suggests that firm A plays the role of a technology donor while firm B enjoys the role of a technology receiver (host).

- Similarly, cluster 3 suggests that firms B, C, and D play the role of a technology donor, while firms E, F, and G enjoy the role of a technology receiver.

- Clusters 3 and 4 suggest that firm H acts as a technology receiver.

Even in the absence of detailed data on the mechanics and flows of PV technology knowledge between firms, the above mapping nonetheless suggests a high level of technology spillover between the firms analyzed.

6.4   Inter-Firm Technology Spillovers

6.4.1   Impact of inter-firm technology spillovers on PV innovation

Intensive PV R&D expenditures and the resulting improved technology knowledge stock are the sources of PV innovations. Consequently, R&D expenditures and the resulting increased technology stock will generate a number of patent

Table 6.6. Number of Patent Applications in Japan's PV Industry, 1981–1998.

	Firm								Other	
	A	B	C	D	E	F	G	H	firms	Total
1981	23	4	14	0	48	6	26	8	146	275
1982	28	11	16	7	14	20	35	15	214	360
1983	39	4	28	16	5	20	33	9	256	410
1984	61	14	45	9	19	51	40	14	249	502
1985	73	17	47	17	11	48	40	10	202	465
1986	116	21	78	23	52	29	54	18	97	488
1987	82	59	111	20	46	23	53	20	72	486
1988	134	47	63	26	27	29	74	15	82	497
1989	96	20	45	17	40	16	73	6	68	381
1990	73	15	41	18	26	11	62	3	88	337
1991	128	4	34	11	10	12	45	8	49	301
1992	108	12	42	5	13	15	35	1	171	402
1993	85	10	47	21	12	14	19	2	210	420
1994	71	11	34	25	20	7	12	1	328	509
1995	87	17	29	7	30	15	9	1	309	504
1996	49	13	34	12	27	5	8	1	326	475
1997	62	20	31	18	13	12	8	1	379	544
1998	36	9	39	4	19	6	11	0	359	483

Note: See Table 6.2 for firm names.

applications in the field of PVs. Table 6.6 and Figure 6.10 summarize trends in the number of patent applications in Japan's PV industry, including applications submitted by the eight leading PV firms. In line with previous research (e.g., Griliches 1984), the number of PV patent applications ($PVPA$) can be estimated by the following regression equation:

$$PVPA = F_1(PVRr, TPV),$$ (6.9)

where $PVRr$ is PV R&D expenditures at constant (1985) prices and TPV is the estimated PV technology knowledge stock.

Given that the essential requirement of a patent application is the novelty of an idea and that this novelty generally decreases as time passes (Freeman 1982), a third factor t representing a time trend should be incorporated into the regression equation for PV patent applications as follows:

$$PVPA = F_2(t, PVRr, TPV).$$ (6.10)

Furthermore, considering the interdisciplinary nature of PV R&D, MITI's PV R&D policy for stimulating cross-sector technology spillovers, and the resulting broad inter-firm technology spillovers [see Section 6.3.4 above and Watanabe (1999)], the PV technology knowledge stock TPV should be decomposed into proprietary knowledge (PV R&D performed by a given firm) and assimilated technology knowledge (spillovers of technology knowledge generated by the R&D of other firms). Provided that a firm makes every effort to maximize the

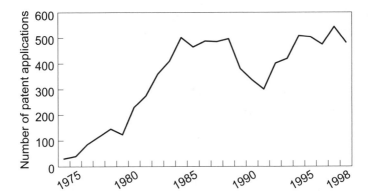

Figure 6.10. Number of Patent Applications in Japan's PV Industry, 1974–1998.

contribution of assimilated technology knowledge, TPV can be decomposed as follows:

$$TPV = TPV_i + Z(\sum_j TPV_j - TPV_i),$$ (6.11)

where TPV_i is the technology knowledge stock of PV R&D in firm i; $\sum_j TPV_j$ is the total industry technology knowledge stock of PV R&D; and $Z (0 < Z < 1)$ is a measure of the assimilation capacity (Cohen and Levinthal 1989; Watanabe and Griffy-Brown 1999; Watanabe *et al.* 2000).

On the basis of the above, Equation (6.9) can be estimated using the following simple Cobb-Douglas-type production function for Japan's PV industry patent applications over the 1976–1995 period:

$$PVPA = A \times e^{\lambda t} PVRr^{\alpha} \left[TPV_i + Z(\sum_j TPV_j - TPV_i) \right]^{\beta},$$ (6.12)

where A is a scale factor and λ, α, and β are elasticities.

In the case of leading PV firms that generate a reasonable portion of their own technology knowledge stock through proprietary R&D, and also considering that, generally, $Z \ll 1$, the ratio of assimilated spillover technology knowledge to proprietary technology knowledge stock is less than 1 $[Z(\sum_j TPV_j - TPV_i)/TPV_i - \ll 1]$, Equation (6.12) can be approximated by

$$
\begin{aligned}
\ln PVPA &= \ln A + \lambda t + \alpha \ln PVRr + \\
&\quad \beta \ln TPV_i (1 + Z \frac{\sum TPV_j - TPV_i}{TPV_i}) \\
&\approx \ln A + \lambda t + \alpha \ln PVRr + \\
&\quad \beta \ln TPV_i + \gamma (\frac{\sum TPV_j}{TPV_i} - 1),
\end{aligned}
$$ (6.13)

where $\gamma = Z\beta$.

Table 6.7. Factors Contributing to Changes in the Number of Patent Applications in Japan's PV Industry.

For model, see Equations (6.12) and (6.13) in text.

	λ	α	β	γ	adj.R^2	DW	Time period
Industry total	–0.02 (–2.09)	0.55 (16.19)	0.10 (1.89)		0.982	2.31	1976–1995
PV firm A	–0.16 (–1.97)	1.38 (10.52)	0.75 (1.64)	0.03 (0.57)	0.982	2.50	1976–1995
B	–0.22 (–2.48)	2.67 (4.00)	1.37 (2.28)	0.05 (1.11)	0.927	1.73	1976–1995
C	–0.14 (–3.03)	0.11 (0.45)	0.87 (4.86)	0.10 (2.28)	0.812	1.56	1976–1995
D	–0.46 (–6.89)	1.52 (7.73)	0.98 (3.73)	0.01 (0.87)	0.908	1.77	1981–1995
E	–0.32 (–3.42)	1.73 (3.11)	1.82 (2.64)	0.33 (2.49)	0.875	2.56	1976–1995
F	–0.14 (–3.05)	1.41 (8.24)	1.10 (3.26)	0.04 (0.83)	0.937	2.65	1980–1995
G	–0.18 (–1.39)	1.49 (2.09)	1.40 (1.44)	0.03 (1.03)	0.942	2.00	1976–1995
H	–0.18 (–1.37)	1.13 (4.64)	1.06 (3.12)	0.01 (0.76)	0.941	2.56	1980–1995

Note: See Table 6.2 for firm names; λ, α, and β are elasticities; $\gamma = Z\beta$, with Z being the assimilation capacity [see Equation (6.13)]; DW = Durbin–Watson statistic; t-values are given in parentheses.

Table 6.8. Factors Contributing to Changes in Solar Cell Production in Japanese Industry.

For model, see Equations (6.14) and (6.15) in text.

	α	β	γ	adj.R^2	DW	Time period
Industry total	5.94 (13.60)	2.19 (35.48)		0.985	1.75	1976–1995
PV firm A	12.40 (7.45)	3.53 (10.35)	0.32 (3.80)	0.935	1.92	1980–1995
B	6.82 (13.41)	5.05 (18.06)	0.14 (4.93)	0.985	1.98	1978–1995
C	2.94 (2.92)	1.01 (9.07)	0.11 (1.91)	0.893	1.19	1976–1995
D	0.10 (0.08)	0.77 (3.76)	0.02 (2.77)	0.902	1.64	1981–1995
E	14.52 (7.85)	2.19 (1.97)	0.25 (1.49)	0.877	1.77	1980–1990
F	6.20 (13.05)	0.88 (6.10)	0.04 (1.10)	0.956	2.57	1976–1990

Note: For variables and firm names, see Tables 6.2 and 6.7.

In Table 6.7, Equations (6.12) and (6.13) are used to identify factors governing patent applications by leading PV firms over the 1976–1995 period. The table suggests that PV R&D expenditures make the most statistically significant contribution to PV patent applications (except for firm C, where the correlation is insignificant), followed by our measure of the technology knowledge stock. In many firms, the technology knowledge stock of proprietary R&D and assimilated spillover technology knowledge are statistically significant contributors to PV patent applications.[5] The elasticity of the time trend (λ) is negative and statistically significant for all firms examined. These findings suggest the following interpretation of the factors governing patent applications in Japan's PV industry over the past two decades:

- R&D expenditures—representing R&D activities at the forefront—make the most significant contribution.[6]

- R&D expenditure flows and the estimated technology knowledge stock of proprietary R&D make an additional contribution. In many firms, the assimilated technology knowledge stock also makes a significant contribution toward explaining PV patent applications.

- Considering the general downward trend of novel ideas worthy of patent applications, a general decrease in the number of patents over time was observed in all firms examined.

6.4.2 Impacts of inter-firm technology spillovers on solar cell production

Trends in PV production depend *inter alia* on the improved technology knowledge stock (arising from R&D). As suggested in Section 6.3.4, this knowledge stock consists of the technology knowledge stock from proprietary PV R&D plus the assimilated technology stock acquired via spillovers from the technology knowledge generated by other firms. Meyer-Krahmer (1992) suggests that the extent of internal (R&D) and external knowledge acquisition (assimilative capacity) also depends on price signals.

On the basis of these observations, an equation describing the governing factors of solar cell production in Japanese industry over the past two decades is estimated using the following simple Cobb-Douglas-type production function:

$$SCP = A \times Pey^{\alpha} \left[TPV_i + Z(\sum_j TPV_j - TPV_i) \right]^{\beta} , \qquad (6.14)$$

where SCP is solar cell production and Pey represents relative energy prices.

Similar to Equation (6.13), Equation (6.14) can be approximated by

$$
\begin{aligned}
\ln SCP &= \ln A + \alpha \ln Pey + \\
&\quad \beta \ln TPV_i (1 + Z \frac{\sum TPV_j - TPV_i}{TPV_i}) \\
&\approx \ln A + \alpha \ln Pey + \\
&\quad \beta \ln TPV_i + \gamma(\frac{\sum TPV_j}{TPV_i} - 1) , \qquad (6.15)
\end{aligned}
$$

where $\gamma \equiv Z\beta$.

In Table 6.8, Equations (6.14) and (6.15) are used to identify factors governing solar cell production of leading PV firms over the past two decades. The table indicates that the technology knowledge stock—both from proprietary PV R&D as well as from assimilated technology knowledge spillovers—contributes significantly to increases in solar cell production. In addition, an increase in relative energy prices contributes significantly to a production increase (except for firm D, where the influence of this variable is statistically insignificant). A comparison of Tables 6.7 and 6.8 reveals that inter-firm technology spillovers have a more

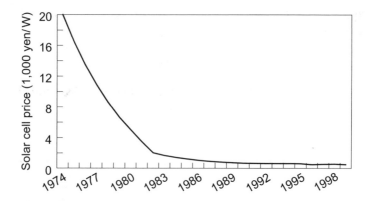

Figure 6.11. Trends in Solar Cell Prices in Japan, 1974–1999: Current Prices, 1,000 Yen/Watt.
Source: Table 6.9.

significant impact on solar cell production than on patent applications. This suggests that technology spillovers contribute to solar cell production directly rather than by stimulating PV innovation. Therefore, technology spillovers play a critical role in determining the trajectory of technological change of PVs, linking it to experience curves and the existing web of technological infrastructures.

As is commonly pointed out, any system of equations should be carefully estimated by multiple regression analysis using ordinary least squares, because this sometimes demonstrates statistical coincidental correlations. Although the interpretation presented here is based on a cross-evaluation of empirical observations, further statistical tests are necessary before drawing definitive conclusions.

6.4.3   Impact of inter-firm technology spillovers on change in solar cell prices

Increased solar cell production resulting from an increase in the PV technology knowledge stock and induced by high energy prices can be expected to lead to falling costs; that is, learning curve effects for both solar cell producers and customers. In addition, economies of scale effects can be expected in solar cell production operations. Higher production levels have led to a decline in solar cell prices, as shown in Table 6.9 and Figure 6.11. The solar cell production price in 1974, the year the Sunshine Program was started, was 20,000 yen per watt (W); by 1999 it had decreased by a factor of 40 to 490 yen/W (in current prices). In constant 1985 yen, prices decreased from 26,120 yen/W to 590 yen/W between 1974 and 1999; that is, by a factor of more than 44. This process can be explained by two factors: (1) an improved technology knowledge stock coupled with inducement mechanisms through changing relative energy prices, and (2) effects due to learning and economies of scale.

Table 6.9. Trends in Solar Cell Prices in Japan, 1974–1999: Government Purchasing Prices, Yen/Watt.

	Current prices	1985 constant prices
1974	20,000	26,120
1975	16,500	20,960
1976	13,500	16,230
1977	10,900	12,790
1978	8,600	10,100
1979	6,600	7,390
1980	5,000	4,980
1981	3,450	3,400
1982	2,000	1,960
1983	1,650	1,630
1984	1,400	1,390
1985	1,200	1,200
1986	1,030	1,090
1987	900	980
1988	800	880
1989	720	780
1990	650	690
1991	635	675
1992	620	670
1993	615	680
1994	610	695
1995	595	690
1996	475	550
1997	520	610
1998	540	640
1999	490	590

Source: MITI/NEDO.

Solar cell production prices in Japanese industry over the past two decades are therefore estimated using the following Cobb-Douglas-type production function:

(1) Inducement by Technology Knowledge Stock and Energy Prices

$$PSC = A \times Pey^{\alpha} \times [TPV_i + Z\left(\sum_j TPV_j - TPV_i\right)]^{\beta}, \qquad (6.16)$$

where PSC is solar cell production price (in constant prices).

Similarly, Equation (6.16) can be approximated by

$$\ln PSC \quad = \quad \ln A + \alpha \ln Pey +$$
$$\beta \ln TPV_i \left(1 + Z(\frac{\sum TPV_j - TPV_i}{TPV_i}\right)$$

Table 6.10. Factors Contributing to Changes in Solar Cell Production Prices in Japanese Industry.

A. Inducement by technology knowledge stock and energy prices
For model, see Equations (6.16) and (6.17) in text.

	α	β	γ	adj.R^2	DW	Time period
Industry total	−1.04 (−5.59)	−0.97 (−39.51)		0.988	1.53	1976–1995
PV firm A	−0.30 (−2.56)	−0.29 (−7.35)	−0.01 (−0.34)	0.992	2.79	1980–1990
B	−1.83 (−4.29)	−1.25 (−5.29)	−0.06 (−5.04)	0.961	2.91	1979–1990
C	−2.77 (−7.42)	−0.23 (−2.87)	−0.06 (−1.70)	0.873	1.80	1976–1990
D	−0.80 (−1.29)	−0.68 (−6.73)	−0.02 (−4.80)	0.963	2.40	1981–1990
E	−0.87 (−2.74)	−0.62 (−3.42)	−0.14 (−4.89)	0.988	1.62	1980–1990
F	−0.22 (−0.75)	−0.60 (−6.64)	−0.07 (−4.54)	0.966	1.98	1980–1990

B. Effects due to learning and economies of scale
For model, see Equation (6.18) in text.

	λ	α	adj.R^2	DW	Time period
Industry total	−0.06 (−9.32)	−0.31 (−18.87)	0.995	1.38	1976–1995
PV firm A	−0.03 (−5.75)	−0.09 (−9.00)	0.979	2.30	1980–1990
B		−0.29 (−22.61)	0.983	2.38	1979–1990
C		−0.52 (−25.34)	0.979	1.57	1976–1990
D		−0.84 (−34.20)	0.993	1.74	1981–1990
E	−0.19 (−19.25)	−0.06 (−2.29)	0.977	2.73	1980–1990
F	−0.12 (−18.61)	−0.06 (−2.40)	0.973	2.42	1980–1990

Note: See Table 6.2 for firm names; λ, α, and β are elasticities; $\gamma = Z\beta$, with Z being the assimilation capacity [see Equation (6.13)]; DW = Durbin–Watson statistic; t-values are given in parentheses.

$$\approx \ln A + \alpha \ln Pey +$$

$$\beta \ln TPV_i + \gamma \left(\frac{\sum TPV_j}{TPV_i} - 1 \right) , \qquad (6.17)$$

where $\gamma = Z\beta$.

(2) Effects due to Learning and Economies of Scale

$$PSC = A \times e^{\lambda t} SCP^{\alpha} . \qquad (6.18)$$

In Table 6.10, Equations (6.16), (6.17), and (6.18) are used to identify factors governing the decline in solar cell prices by leading firms. Table 6.10a shows that technology knowledge stock—represented both by proprietary PV R&D and assimilated technology knowledge through spillovers—contributed significantly to a dramatic decrease in solar cell production prices (except for firm A, where the statistical influence is insignificant). In addition, increases in relative energy prices contributed significantly to solar cell price decreases (except for firm F). Table 6.10b suggests that over the past two decades the effects of economies of scale have contributed significantly to a decline in solar cell prices in all the

Table 6.11. Learning Coefficients of Solar Cell Development in Japan.
Model: $PSC = A \times CMSCP^{\eta}$.

	η	t-value	adj.R^2	DW	Time period
Industry	–0.35	–22.80	0.981	1.42	1979–1999
total	–0.37	–73.88	0.997	1.60	1976–1995
PV firm A	–0.12	–28.31	0.988	2.09	1979–1999
B	–0.25	–25.92	0.988	2.63	1979–1999
C	–0.29	–7.47	0.871	1.49	1976–1990
D	–0.69	–18.99	0.978	2.53	1981–1990
E	–0.41	–9.34	0.911	1.77	1980–1990
F	–0.31	–14.46	0.962	1.52	1979–1999

Note: PSC = solar cell production price (fixed price); A = constant; $CMSCP$ = cumulative solar cell production; DW = Durbin–Watson statistic. See Table 6.2 for firm names.

leading Japanese PV firms examined, whereas the impacts of learning effects seem to have been rather limited.[7]

Because the lack of data precluded an estimate of all variables of the production function for all firms, and because the estimated influence of learning and economies of scale effects is extremely variable among the limited sample of firms analyzed (see discussion below), definitive conclusions await further empirical and statistical corroboration.

6.4.4 Impact of inter-firm technology spillovers on learning effects

Learning effects can be clearly observed at the aggregate industry level in line with the increase in PV R&D technology knowledge stock and its embodiment in production facilities, as illustrated in the learning curve shown in Figure 6.12. In Table 6.11, the learning coefficient among leading PV firms in Japan over the 1979–1999 period is compared with that over the 1980–1990 period. A comparison of the interdependency of technology spillover and solar cell price decreases (Table 6.10a) and the difference in learning coefficients (rates) in Table 6.11 indicates a clear correlation between the two. Figure 6.13 illustrates this correlation in leading PV firms. The figure suggests that firms with a higher dependency on technology spillovers, as indicated by the higher "potentiality of technology spillover assimilation" indicator in Figure 6.13, also seem to demonstrate better performance with respect to learning curve effects. This seems to corroborate the theoretical proposition set forth in Section 6.2 that there is a mutually stimulating interaction between technology spillovers and enhancement of learning curve effects.

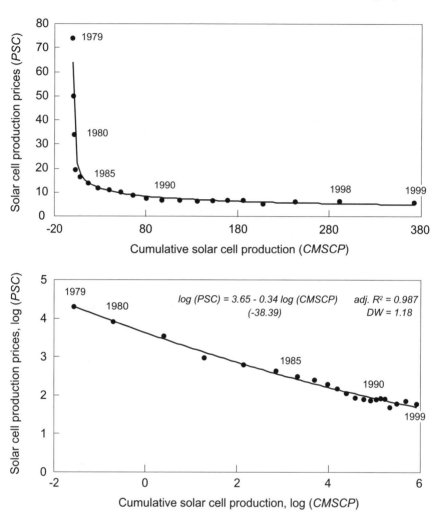

Figure 6.12. PV Learning Curve in Japanese Industry, 1979–1999, Linear (Top) and Double-Logarithmic (Bottom) Representations.

6.5 Virtuous Cycle between R&D, Market Growth, and Price Reduction

6.5.1 Feedback loop to a further production increase

As demonstrated in Table 6.9 and Figure 6.11, a dramatic decrease in solar cell prices induces further production increases. This can be attributed to both a demand- and a supply-side response. Demand for PV cells increases with falling prices. Suppliers aim to maintain sales volumes; that is, to compensate for price

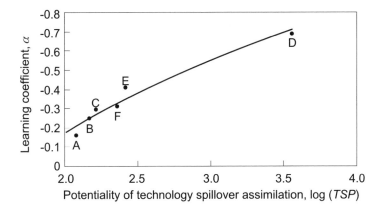

Figure 6.13. Correlation between Potentiality of Technology Spillover Assimilation and Learning Coefficient in Solar Cell Production in Japan, 1980–1990.[8]

Table 6.12. Factors Contributing to Changes in the Solar Cell Production Feedback Loop in Japanese Industry.
Model: $SCP = A \times PSC_{t-1}^{\alpha} Pey_{t-1}^{\beta}$.

	α	β	adj.R^2	DW	Time period
Industry total	−2.02 (−58.73)	2.98 (13.61)	0.995	1.25	1976–1995
PV firm A	−4.07 (−14.75)	2.03 (4.97)	0.974	2.82	1981–1990
B	−2.02 (−7.14)	0.80 (0.79)	0.902	2.01	1980–1990
C	−1.12 (−5.17)	2.10 (2.31)	0.967	2.48	1976–1990
D	−1.63 (−32.62)	2.54 (5.69)	0.990	1.79	1978–1990
E	−1.48 (−6.88)	11.10 (13.36)	0.963	1.62	1981–1990
F	−1.31 (−6.56)	6.65 (15.35)	0.975	1.94	1981–1991

Note: SCP = solar cell production; A = constant; PSC = solar cell production price (fixed price); Pey = relative energy prices; α and β are elasticities; DW = Durbin–Watson statistic; t-values are given in parentheses. See Table 6.2 for firm names.

declines with increased production volumes. In addition, changing relative prices (in the case of increasing energy prices) also induces a production increase. Equation (6.19) depicts this behavior. In this simple Cobb-Douglas-type production function, both solar cells and aggregate energy prices in the previous year are used as explanatory variables:

$$SCP = A \times PSC_{t-1}^{\alpha} Pey_{t-1}^{\beta} . \qquad (6.19)$$

In Table 6.12, Equation (6.19) is used to demonstrate this feedback loop and the factors that have induced it in leading PV firms over the past two decades. As the table indicates, for any given year both a decrease in solar cell prices and an increase in energy prices in the previous year provide significant inducement for a production increase in all firms examined (except for the energy price inducement effect for firm B, which turns out to be statistically insignificant).

Table 6.13. Impact of Inducement of Solar Cell Production Increase on PV R&D in Japanese Industry.

Model: $PVR = A \times SCP^{\alpha}$.

	α (*t*-value)	adj.R^2	DW	Time period
Industry total	0.43 (25.96)	0.973	1.03	1976–1995
PV firm A	0.24 (13.61)	0.952	1.46	1980–1990
B	0.24 (10.12)	0.932	1.57	1979–1990
C	0.62 (15.62)	0.946	1.17	1976–1990
D	1.19 (10.04)	0.908	2.21	1980–1990
E	0.10 (6.10)	0.933	1.53	1980–1990
F	0.66 (6.32)	0.829	2.04	1979–1991

Note: *PVR* = industry PV R&D expenditure; *A* = constant; *SCP* = solar cell production; α = learning and assimilation capacity. See Table 6.2 for firm names.

6.5.2   Feedback loop to further R&D expenditures

Similar to the feedback loop between falling PV prices and rising PV production, stepped-up PV production induces further R&D. Equation (6.20) illustrates this inducement mechanism:

$$PVR = A \times SCP^{\alpha} .$$
(6.20)

In Table 6.13, Equation (6.20) is used to demonstrate this feedback loop in leading PV firms over the past two decades. The table illustrates that solar cell production simultaneously induces PV R&D in all firms examined.[9]

6.5.3   Creation of a virtuous cycle

The analysis presented above demonstrates the creation of a virtuous cycle in PV development in Japan. The simultaneous involvement of cross-sector industry collaboration, MITI's inducement policies for R&D (and niche market incentives), the creation of a continuously rising PV technology knowledge stock, technology spillovers, and the interaction between these factors leading to the formation of a PV technological trajectory characterized by dramatically falling costs were analyzed. Figure 6.14 illustrates the virtuous cycle between R&D triggered by the Sunshine Program, steady market growth, and the resulting dramatic price reduction in Japan's PV industry over the 1976–1995 period. A noteworthy element of this cycle is the "double boost" to solar cell production from the increased technology knowledge stock resulting from PV R&D and from falling solar cell prices. A similar "double boost" effect can be observed in PV R&D—the source of the increasing technology knowledge stock—arising from both increased solar cell production volumes and MITI's PV R&D budget's stimulation of further private PV R&D. This virtuous cycle of PV development in Japan suggests that a variety of policy mechanisms exist for inducing endogenous technological change.

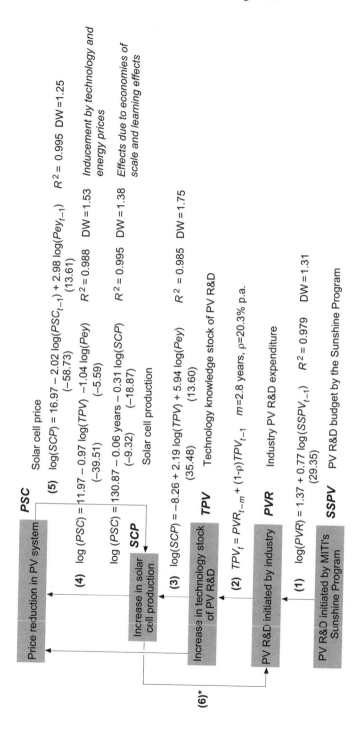

PSC Solar cell price

(5) $\log(SCP) = 16.97 - 2.02 \log(PSC_{t-1}) + 2.98 \log(Pey_{t-1})$ $R^2 = 0.995$ DW $= 1.25$
          (-58.73) (13.61)

Inducement by technology and energy prices

(4) $\log(PSC) = 11.97 - 0.97 \log(TPV) - 1.04 \log(Pey)$ $R^2 = 0.988$ DW $= 1.53$
          (-39.51) (-5.59)

$\log(PSC) = 130.87 - 0.06 \text{ years} - 0.31 \log(SCP)$ $R^2 = 0.995$ DW $= 1.38$
          (-9.32) (-18.87)

Effects due to economies of scale and learning effects

SCP Solar cell production

(3) $\log(SCP) = -8.26 + 2.19 \log(TPV) + 5.94 \log(Pey)$ $R^2 = 0.985$ DW $= 1.75$
          (35.48) (13.60)

TPV Technology knowledge stock of PV R&D

(2) $TPV_t = PVR_{t-m} + (1-\rho)TPV_{t-1}$ $m=2.8$ years, $\rho=20.3\%$ p.a.

PVR Industry PV R&D expenditure

(1) $\log(PVR) = 1.37 + 0.77 \log(SSPV_{t-1})$ $R^2 = 0.979$ DW $= 1.31$
          (29.35)

SSPV PV R&D budget by the Sunshine Program

Price reduction in PV system

Increase in solar cell production

Increase in technology stock of PV R&D

PV R&D initiated by industry

PV R&D initiated by MITI's Sunshine Program

(6)*

*$\log(PVR) = 3.59 + 0.43 \log(SCP)$ $R^2 = 0.973$ DW $= 1.03$
          (25.96)

Figure 6.14. Steps for Creating a Virtuous Cycle for PV Development in Japan, 1976–1995; t-Values Given in Parentheses.[10]

6.6   Implications for Techno-Economics

On the basis of the empirical analysis presented in this chapter, the following policy implications emerge:

- Institutional and technological demonstrations of PV development and utilization should be carried out.

- Technology improvements and organizational learning should be accelerated.

- R&D on renewable energy technologies (RETs) and the development of market incentive structures that promote the industrial dynamic mechanism of a virtuous cycle involving market growth and price reductions for technically proven advanced RETs should be intensified (Williams *et al.* 1996).

- Coherent systemic policies aimed at unleashing the industrial dynamic mechanism of a virtuous cycle involving market growth and price decreases for RETs should be formulated.

- Market opportunities for RETs developed with private-sector resources should be pursued and viable RETs industries should be established.

As these suggestions have been extracted from an empirical analysis of the relationships between PV R&D and production in Japan, they may not be directly applicable to other technologies or countries. However, the following general observations could be helpful in considering the relationships between inter-firm technology spillovers, R&D, market growth, and price reductions for the development and diffusion of innovative environmental technologies.

Creating a virtuous cycle in the development of a new technology trajectory depends on a number of factors within the endogenous technological innovation process. The analysis presented here has demonstrated the complex and important role policy can play in inducing technological change if critical success factors are acting in concert. In the case of PV development in Japan, it was critical that the targeted technology was embedded in a complex web of related technologies. Like semiconductors, PVs can maximize the benefits of an improved technology knowledge stock resulting from both public and private R&D, as well as from proprietary knowledge generation combined with knowledge assimilation (spillover effects), learning effects (via experience curves), and economies of scale. This induced technological change appears particularly successful because of the interdisciplinary nature of its development, which maximized the benefits of technology spillovers.

The creation of a virtuous cycle has promising policy applications in terms of induced technological change and the environment. The case of a virtuous cycle in PV technology development demonstrates how critical factors can alter technology trajectories in ways that benefit the environment and how policy, market forces, and R&D can work together to this end. Interestingly, our analysis also indicates the existence and importance of network externalities arising from technological interrelatedness. This highlights the critical importance of considering and coordinating entire "technology networks" as targets for technology policy. Of particular importance in this context is interpreting and responding to market signals rather than creating "false" signals from a policy perspective. In view of a multitude of positive effects on the economy and the environment, further examination of these issues is certainly worth continued study and policy consideration.

Appendix 6.A
Measurement of Time Lag between PV R&D and Commercialization and Rate of Obsolescence of PV Technology

Measuring the technology knowledge stock via PV R&D expenditures requires reliable up-to-date estimates of the time lag between R&D expenditures and increases in the technology knowledge stock, and of the rate of obsolescence of that knowledge. However, no reliable survey exists estimating these factors for PV R&D. Therefore, in 1993, with the support of AIST of MITI, the authors prepared a questionnaire for 19 leading PV firms from the 26 member firms of the Photovoltaic Power Generation Technology Research Association (PVTEC). The survey included questions related to the time lag between R&D and commercialization and the lifetime of PV technology. Responses were received from the following 15 firms (including the 8 firms examined here): Sanyo Electric Co., Ltd.; Kyocera Corp.; Sharp Corp.; Kaneka Corp.; Fuji Electric Co., Ltd.; Hitachi, Ltd.; Mitsubishi Electric Corp.; Sumitomo Electric Industries, Ltd.; Daido-hoxan Co.; Matsushita Battery Industrial Co., Ltd.; Showa Shell Sekiyu K.K.; Tonen Co.; Japan Energy Co.; Osaka Titanium Co., Ltd.; and Matsushita Electric Industries Co., Ltd.

The distribution of replies to the survey from the 15 firms is given in Tables 6.A1 and 6.A2.

Table 6.A1. Number of Observations Concerning Time Lag between PV R&D and Commercialization.

Time lag (years)	Number of Observations	
1	13	
2	18	
3	11	
4	5	
5	4	
6	3	
7	2	
8	1	
Total	57	Average: 2.8 years

Table 6.A2. Number of Observations Concerning Lifetime of PV Technology.

Lifetime (years)	Number of Samples	
2	3	
3	4	
4	5	
5	6	
6	4	
7	3	
8	2	
9	1	
Total	28	Average: 4.9 years (20.3 percent p.a.)

To evaluate the estimated time lag ($m = 2.8$ years) and the rate of obsolescence of technology ($\rho = 20.3$ percent), a comparative evaluation was made using the following equation:

$$PVPA = A \times e^{\lambda t} PVRr^{\alpha} \times TPV^{\beta} ,$$
$$TRV = TPV(PVRr, m, \rho) , \hspace{2cm} (6.A1)$$

where $PVPA$ is the number of PV patent applications, $PVRr$ is PV R&D expenditure (in constant prices), and TPV is technology stock of PV R&D.

TPV is a function of $PVRr, m$, and ρ, with $m(2.8 \pm \varepsilon)$ and $\rho(20.3 \pm \varepsilon)$. Results of a sensitivity analysis varying m and ρ are reported in Table 6.A3. The case with $m = 2.8$ and $\rho = 20.3$ is the most statistically significant.

Table 6.A3. Comparative Evaluation of m and ρ.

m (years)	ρ (%)	λ	α	β	adj.R^2	DW	AIC
2.8	20	−0.02	0.54	0.11	0.983	2.06	−77.97
		(−2.55)	(17.72)	(2.28)			
	15	−0.02	0.54	0.12	0.982	2.05	−77.66
		(−2.47)	(17.66)	(2.20)			
	25	−0.02	0.54	0.11	0.956	1.98	−70.27
		(1.91)	(11.30)	(1.57)			
2.5	20	−0.03	0.54	0.13	0.958	11.98	−70.57
		(2.00)	(10.79)	(1.67)			
3.0	20	−0.02	0.55	0.10	0.982	2.04	−77.35
		(2.93)	(18.03)	(2.12)			

Note: m = lead time of PV R&D and its commercialization; ρ = rate of obsolescence of PV technology; λ, α, and β = elasticities; DW = Durbin–Watson statistic; AIC = Akaike Information Criterion; t-values are given in parentheses.

Acknowledgments

This chapter is based on work originally presented at the International Workshop on Induced Technological Change and the Environment (IIASA, Laxenburg, Austria, 1999).

Notes

1.   Facing a second energy crisis in 1979, the minister of MITI consulted with the Industrial Technology Council about a priority policy menu. In response, the Council prepared a recommendation entitled "Strategy for Acceleration of the Sunshine Program," which identified certain R&D priority areas, including PV R&D. In response to this recommendation, MITI introduced new policies in 1980, including a Law for the Promotion of Development and Introduction of Oil Alternative Energy; created a new funding system by means of special accounts for energy security; and established NEDO (the New Energy Development Organization, MITI's affiliate responsible for energy R&D). Consequently, R&D activities, particularly in priority areas such as PVs, were accelerated.

2.   The remaining 60 percent was appropriated to national research institutes, to universities for basic research, and to other industries such as the electric power, housing, and construction industries for application-oriented research.

3.   The rate of technology obsolescence for energy R&D in Japan's manufacturing industries was estimated to vary between 14.5 percent (1970) and 22.2 percent (1994), while total R&D intensity was estimated to be between 8.2 percent (1970) and 12.1 percent (1994) (Watanabe 1999). Table 6.A3 in Appendix 6.A gives more details on these estimates.

4.   The world's total solar cell production in 1999 was 201.3 MW, including Japan with 80.0 MW (39.7 percent); United States, 60.8 MW (30.2 percent); Europe, 40.0 MW (19.9 percent); and other countries, 20.5 MW (10.2 percent).

5.   A significance of 2.5 percent for two firms, 15 percent for two firms, and 20–25 percent for three firms. The contribution is insignificant for firm A's technology stock of proprietary R&D, which is much greater than that of the other firms.

6.   An exception is the case of firm C, where the influence of factors on PV patent applications is statistically insignificant. Looking at Tables 6.3 and 6.6, we note that, contrary to its recent increase in solar cell production, firm C's PV R&D expenditure share has decreased. This suggests that firm C depends on its proprietary technology knowledge stock based on its previous R&D and the assimilation of technology knowledge via spillovers rather than by proprietary R&D for PV patent applications. Table 6.8 supports this view.

7.   Contrary to the significant contribution of learning effects to a decrease in solar cell production prices in firms A, E, and F, contributions in firms B, C, and D are not statistically significant. This is due to a rapid production increase in later years in firms B, C, and D that provided significant opportunities for benefiting from economies of scale with limited opportunities for learning effects.

8.   The learning coefficient α can be defined by the following equation: $PSC = A \times CMSCP^\alpha$, where PSC is the solar cell price (in constant money), and $CMSCP$ is cumulative solar cell production. The "potentiality of technology spillover assimilation" indicator in firm i, TSP_i, can be defined as follows: $TSP_i = \frac{\sum TPV_j}{TPV_i} - 1$, where TPV_i is the technology stock of PV R&D in firm i. Firms shown are as follows: (A) Sanyo Electric Co., Ltd.; (B) Kyocera Corp.; (C) Sharp Corp.; (D) Kaneka Corp.; (E) Fuji Electric Co.,

Ltd.; (F) Hitachi, Ltd.; (G) Mitsubishi Electric Corp.; (H) Sumitomo Electric Industries, Ltd.

9. This can be imputed by the following simple identity: $\Delta R = \Delta R/Y + \Delta Y$, where R is R&D investment, R/Y is R&D intensity, and Y is production.

10. PVR is industry PV R&D expenditure; $SSPV$ is PV R&D budget from the Sunshine Program; TPV is technology knowledge stock of PV R&D; m is time lag of PV R&D to commercialization; ρ is the rate of obsolescence of PV technology; SCP is solar cell production; Pey is relative energy prices; and PSC is the solar cell price (in constant 1985 prices).

References

Arrow, K.J., 1962, The economic implication of learning by doing, *Review of Economic Studies*, **29**:155–173.

Cohen, W.M., and Levinthal, D.A., 1989, Innovation and learning: Two faces of R&D, *Economic Journal*, **99**:569–596.

Freeman, C., 1982, *The Economics of Industrial Innovation*, The MIT Press, Cambridge, MA, USA.

Griliches, Z., 1957, Hybrid corn: An exploration in the economics of technical change, *Econometrica*, **25**:501–522.

Griliches, Z., ed., 1984, *R&D, Patents, and Productivity*, The University of Chicago Press, Chicago, IL, USA.

Grübler, A., 1998, *Technology and Global Change*, Cambridge University Press, Cambridge, UK.

Industrial Technology Council of MITI, 1979, Strategy to Accelerate the Sunshine Program, Tokyo, Japan.

Jaffe, A.B., 1986, Technological opportunity and spillovers of R&D: Evidence from firm's patents, profits, and market value, *The American Economic Review*, **76**(5):984–1001.

Jovanovic, B., and Lach, S., 1989, Entry, exit, and diffusion with learning by doing, *American Economic Review*, **79**:690–699.

Mansfield, E., 1961, Technical change and the rate of imitation, *Econometrica*, **29**:741–766.

Meyer-Krahmer, F., 1992, The German R&D system in transition, *Research Policy*, **21**(5):423–436.

MITI, 1970–1990, Annual Report on MITI's Policy, Ministry of International Trade and Industry, Tokyo, Japan.

Thomson, R., 1993, *Learning and Technological Change*, St. Martin's Press, New York, NY, USA.

Watanabe, C., 1992, Trends in the substitution of production factors to technology, *Research Policy*, **21**(6):481–505.

Watanabe, C., 1995a, The feedback loop between technology and economic development: An examination of Japanese industry, *Technological Forecasting and Social Change*, **49**(2):127–145.

Watanabe, C., 1995b, Identification of the role of renewable energy: A view from Japan's challenge, *Renewable Energy*, **6**(3):237–274.

Watanabe, C., 1995c, The Interaction between Technology and Economy: National Strategies for Constrained Economic Environments: The Case of Japan 1955-1992,

WP-95-16, International Institute for Applied Systems Analysis, Laxenburg, Austria.

Watanabe, C. 1995d, Mitigating global warming by substituting technology for energy: MITI's efforts and new approach, *Energy Policy*, **23**(4/5):447–461.

Watanabe, C., 1996, Choosing energy technologies: The Japanese approach, in *Comparing Energy Technologies*, OECD/IES, Paris, France.

Watanabe, C., 1997, A Technometric Approach to the Dynamic Mechanism of Technological Innovation, Abstract of the Annual Conference of the Japan Society for Science Policy and Research Management, Tsukuba, Japan.

Watanabe, C., 1999, Systems option for sustainable development: Effect and limit of MITI's efforts to substitute technology for energy, *Research Policy*, **28**(7):719–749.

Watanabe, C., 2000, MITI's Policy as a System to Substitute Technology for Energy: Lessons, Limits and Perspective, Paper presented to a Joint Meeting of the Energy Modeling Forum, International Energy Agency and International Energy Workshop, Stanford, CA, USA.

Watanabe, C., and Clark, T., 1991, Inducing technological innovation in Japan, *Journal of Scientific & Industrial Research*, **50**(10):771–785.

Watanabe, C., and Griffy-Brown, C., 1999, Inter-firm Technology Spillover and the Creation of a "Virtuous Cycle" between R&D, Market Growth and Price Reduction: The Case of Photovoltaic Power Generation (PV) Development in Japan, Paper presented at the International Workshop on Induced Technological Change and the Environment, International Institute for Applied Systems Analysis, Laxenburg, Austria.

Watanabe, C., and Honda. Y., 1992, Inducing power of Japanese technological innovation, mechanism of Japan's industrial science and technology policy, *Japan and the World Economy*, **3**(4):357–390.

Watanabe, C., Santoso, I., and Widayanti, T., 1991, *The Inducing Power of Japanese Technological Innovation*, Pinter Publishers, London, UK.

Watanabe, C., Wakabayashi, K., and Miyazawa, T., 2000, Industrial dynamism and the creation of a "virtuous cycle" between R&D, market growth and price reduction, *Technovation*, **20**(6):299–312.

Williams, R., Karekezi, S., Parikh, J., and Watanabe, C., 1996, The Outlook for Renewable Energy Technologies, Public Policy Issues, and Roles for Global Environment Facility, STAP/GEF, Nairobi, Kenya.

Chapter 7

Technological Change and Diffusion as a Learning Process

Nebojsa Nakicenovic

7.1 Introduction

Energy and carbon intensities of aggregate economic activities, as measured by the gross domestic product, have generally been declining since the onset of industrialization two centuries ago (Grübler and Nakicenovic 1996). This historical tendency can be observed for most countries, and for some throughout the industrialization process during the past two centuries, as will be shown for the United States (Nakicenovic 1996). This contrasts significantly with the perspective provided by disaggregated energy and carbon intensities of individual economic sectors and activities, and even with short-term intensity increases in some countries (Schmalensee *et al.* 1998). An important part of the secular decline of energy and carbon intensities is the result of technological change. Technologies that are more energy efficient have replaced less efficient ones, and technologies that are less carbon intensive have replaced those that are more carbon intensive. In this way, technological change has made a major contribution to these long-term improvements in the productivity of energy. In particular, the decarbonization of energy—namely, the reduction of the specific carbon content of energy—can be represented by a learning curve and thus interpreted as a long-term learning process. In this chapter it is argued that an important component of the dynamics of technological change and diffusion is a cumulative process of learning by doing. Surely technology diffusion also takes place as a result of changes (decreases) in the price of a technology or changes (increases) in the price of a saved input (energy), neither of which need be directly driven by a learning-by-doing process. To the extent that it is a result of cumulative learning processes, technological change is not an "autonomous" process, although it is often represented as such in energy and economic models.

A number of implications will be considered with reference to the mitigation of carbon dioxide (CO_2) emissions. Various mitigation strategies for countering the possibility of climate change have been proposed. Recently, research has

This chapter was originally published in *Perspectives in Energy*, Volume 4, Number 2:173–189, 1997. ©Moscow International Energy Club and International Academy of Energy.

begun to focus on the formulation of global CO_2 emissions profiles that would lead to the stabilization of atmospheric concentrations at some negotiated level in accordance with Article 2 of the Framework Convention on Climate Change (UN/FCCC 1992). For example, all of the CO_2 emissions profiles that lead to stabilization of concentrations that were analyzed by the Intergovernmental Panel on Climate Change (IPCC 1996, 2001) require the eventual elimination of global carbon emissions sometime during the next two centuries. In view of the increasing need for energy services in the world, especially in developing countries, such emissions reductions will require a substantial increase in the decarbonization rate. This, in turn, implies a larger future role for new technologies with lower CO_2 emissions. Thus, there is an increasing recognition in the literature that abatement of CO_2 emissions requires a sustained commitment to research, development, and demonstration (RD&D) today that could lead to diffusion of new, less carbon-intensive technologies in the future (see, e.g., Wigley *et al.* 1996).

It will be shown that, in conjunction with RD&D, timely investment in new technologies with lower CO_2 emissions might be a more cost-effective strategy for reducing global emissions than postponing investment decisions in the hope that mitigation technologies might somehow become more attractive through "autonomous" RD&D improvements and cost reductions in step with the natural turnover of capital. It has been argued that the latter strategy is superior to a timelier introduction of lower-emission technologies, because at present these technologies are generally costlier than the alternatives (see, e.g., Wigley *et al.* 1996). In some cases, there is a trade-off between the cost savings that may be brought about by rapid technological change and the cost increases that may thereby be brought about by prematurely rendering parts of the capital stock obsolete. Although this is true, postponement in itself will bring few additional benefits.

While the costs and performance of technologies are generally modeled as if they were exogenous, they are not. Costs of new technologies have been shown to decline and performance to increase with accumulated experience and improvements. Unless there is dedicated, timely, and pronounced investment in these technologies, they are unlikely to be developed and thus become commercially viable and competitive in the marketplace. Learning by doing is a prerequisite for performance improvements, cost reductions, and eventual technology diffusion. Postponing investment decisions will not by itself bring about the technological change required to reduce CO_2 emissions in a cost-effective way. Even worse, under unfavorable conditions it might bring about further "lock-in" of energy systems and economic activities along fossil-intensive development paths.

The implication is that there may be great leverage in policies and measures that accelerate the accumulation of experience in new technologies with lower environmental impacts, for example, through early adoption and development of special niche markets. This leverage can be important, particularly if these policies can minimize the "deadweight" loss to society associated with the foregone exploitation of cheaper fossil fuels and possible reductions of RD&D in other parts of the economy. It is important to note that the approach taken here does not consider potential welfare losses associated with moving resources away from

RD&D efforts, for example, in other sectors. That is, an acceleration of energy-related technical progress may be accompanied by reduced levels of RD&D activities in other sectors, leading to a slowdown in labor and capital productivity. These are some of the problems and issues that must be resolved before technological change can become a truly endogenous component in standard modeling approaches. In the meantime, an increasing number of models are being adapted to explore alternative ways of incorporating endogenous technological change. In this chapter, we will explore the nature of the relationship between technological change, costs and performance of new technologies, and resulting emissions profiles from the global electricity generation system with the MESSAGE model.

7.2   Decarbonization

Through decarbonization, energy services can be provided with lower carbon emissions. The process can be expressed as a product of two factors: decarbonization of energy and reduction of the energy intensity of economic activities, for example, as measured by gross domestic product (GDP). Figure 7.1 shows the decarbonization of GDP; Figures 7.2 and 7.3 show the decarbonization of energy and the reduction of energy intensity of GDP, respectively. The example for the United States is shown in the three figures primarily because the data are of relatively good quality; however, available data allow the assessment of decarbonization trends with reasonable confidence for other major energy-consuming regions and countries, such as France and the United Kingdom, and for the world as a whole (see, e.g., Nakicenovic 1996; Grübler and Nakicenovic 1996). Over shorter time periods similar decarbonization trends can be obtained for many developed and industrializing countries, such as India and China. In Figure 7.1, the decarbonization rate is expressed in kilograms of carbon (kgC) per unit of GDP in US dollars measured at 1990 prices. The average annual rate of decline is about 1.3 percent, meaning that every year about 1.3 percent less carbon is emitted to generate one dollar of value added.

Today, about a quarter of a kilogram of carbon is emitted per dollar value added in the United States, and about half that amount is emitted per dollar value added in Europe and Japan. However, the amount of carbon emitted per dollar value added is significantly greater in most developing and many re-forming countries. Thus, it is evident there are different paths of economic development that lead to similar levels of affluence at quite different levels of CO_2 emissions. The prime objective of possible mitigation strategies is to reduce these emission levels by increasing the rate of decarbonization throughout the world. At an average decarbonization rate of 1.3 percent per year, global CO_2 emissions will increase about 1.7 percent annually, assuming the economic growth rate remains at about 3 percent per year. This increase will lead to a doubling of emission levels in about 40 years. Thus, to stabilize global emissions at some (higher) level in the future, the decarbonization rate would have to at least double to offset the current

Figure 7.1. Decarbonization of Economic Activities in the United States.
Expressed in kilograms of carbon per unit of GDP at constant 1990 prices [kgC/US(1990)$].

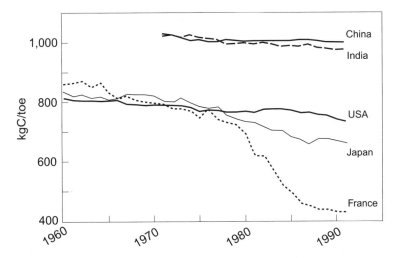

Figure 7.2. Decarbonization of Primary Energy in the United States and Selected Countries.
Expressed in kilograms of carbon per ton oil equivalent (kgC/toe).

rate of economic growth. The second alternative, maintaining lower rates of economic growth, is clearly undesirable in light of the existing widespread poverty and deprivation throughout the world.

Figure 7.4 portrays another image of the dynamics of decarbonization. The data from Figure 7.1 are now shown as a learning or experience curve. The ratio of carbon emissions to GDP is shown versus the cumulative emissions in a double

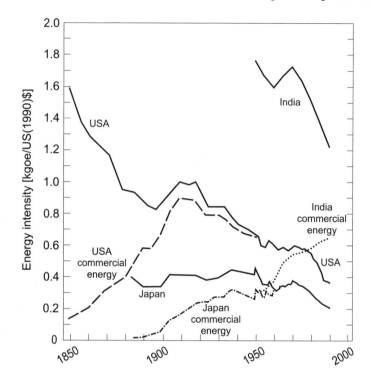

Figure 7.3. Primary Energy Intensity of Economic Activities in the United States and Selected Countries.

Expressed in kilograms of oil equivalent per unit GDP at constant 1990 prices [kgoe/US(1990)$].

logarithmic diagram. There is an exponential decline (linear on double logarithmic scales) in specific carbon emissions per doubling of cumulative emissions. Apparently, the more we emit, the more we learn about how to emit less per unit value. The progress ratio is actually quite high at about 76 percent (representing a 24 percent cost reduction in specific emissions) per doubling of cumulative emissions. This figure compares with progress ratios in the range of 70–90 percent across a number of energy technology learning curves reported in the literature (see, e.g., Christiansson 1995).

The fact that decarbonization of the US economy can be represented as a learning curve suggests that at least a part of the carbon reductions could be due to a process of technological learning resulting from cumulative experience. At the highly aggregated level of the relationship between cumulative emissions and decreased emissions per unit of value for a whole country, it is difficult to identify the component of decarbonization that is due to learning by doing, as opposed to other mechanisms. The process of cumulative learning may be no more than a small part of the explanation, but it may also be the dominant part. However,

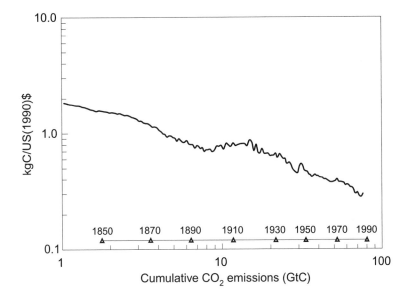

Figure 7.4. Decarbonization of Economic Activities in the United States as a Learning Curve.

Accumulated experience is represented by cumulative CO_2 emissions, expressed in kilograms of carbon per unit of GDP at constant 1990 prices [kgC/US(1990)$] versus cumulative CO_2 emissions in gigatons of carbon (GtC) on double logarithmic axes.

determining its contribution would require an in-depth analysis of the underlying processes that is not possible at this time, especially because of the lack of detailed engineering and microeconomic data for such long periods even for a fairly well-documented country such as the United States.

As a kind of thought experiment, assume a hypothetical case where this rate of decarbonization continues for another century. In this case, one could expect the specific carbon emissions to continue to decline. To date, the United States has emitted about 100 gigatons of carbon (GtC, or billion tons of carbon), slightly less than half the cumulative global emissions, estimated at about 250 GtC. If the rate of decarbonization were to remain the same, another 100 GtC would be emitted before the specific emissions could be reduced by another 24 percent. This rate is clearly too slow for a transition to the post-fossil era within a century or two. Thus, for a more drastic increase of decarbonization, substantially higher progress ratios would be required.

Before discussing the process of endogenizing technological learning, let us first consider the technology dynamics behind the historical rates of decarbonization and the implications decarbonization carries for the possible diffusion of less carbon-intensive energy technologies in the future. Figure 7.5 shows the hierarchy of replacements of old energy sources with new ones in the United States.

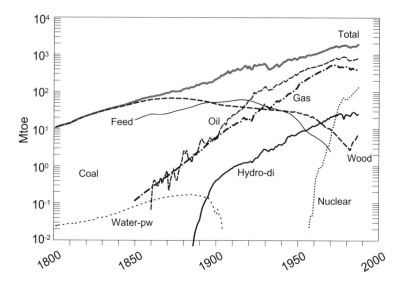

Figure 7.5. Primary Energy Consumption by Major Energy Sources in the United States.

Expressed in million tons of oil equivalent (Mtoe).

This dynamic process of technological substitution is the driving force behind the historical rates of decarbonization.

Traditional energy forms such as animal feed and wood have a high carbon content, both per unit of energy and per unit of economic activity, because of the relatively low efficiency with which they deliver demanded energy services. Draft animals and open fire have very low energy-conversion efficiencies compared with contemporary prime movers and furnaces. It is true that some of the released carbon can be reabsorbed by new plant growth and new trees, and by the replanting of animal feed, but quite often the land is not used in a sustainable fashion. For example, because many of these activities are associated with deforestation and land degradation, they often lead to net carbon flux to the atmosphere. The carbon intensity of fuelwood and animal feed is substantially higher than that of coal. Moreover, coal can be used with generally higher efficiencies and often much greater convenience for the consumer. For these reasons, coal eventually supplanted traditional energy forms. This progress toward energy sources with lower carbon contents and higher conversion efficiencies has continued, with shifts from coal to oil to natural gas, and more recently to nuclear energy and new renewable sources of energy, both of which have minimal carbon emissions. Natural gas in itself brings enormous reductions in carbon emissions (with half the carbon emissions of coal) as well as higher efficiencies.

Using the available data, the historical replacement of coal with oil and later with natural gas can be illustrated for most countries and major energy-consuming regions, as well as for the world as a whole (Marchetti and Nakicenovic 1979;

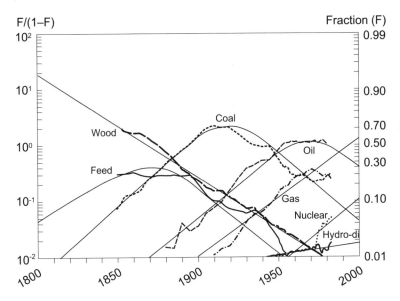

Figure 7.6. Primary Energy Substitution in the United States.

Historical data and model projections for the future, expressed in fractional market shares (F) and transformed as F/(1–F) on logarithmic axes.

Nakicenovic 1979). If all energy sources are considered, the replacement process is very intricate and complex, as can be seen from Figure 7.5. Similar dynamics of technological substitution have been studied for other systems, such as transport and steel making (Grübler and Nakicenovic 1988; Nakicenovic 1990). It is a process with long transition periods from older to newer technologies, especially in the areas of energy systems and infrastructure. The competitive struggle between the six main sources of primary energy—wood, animal feed, coal, oil, gas, and nuclear materials—has proved to be a process with regular dynamics that can be described by relatively simple rules. This process is shown in Figure 7.6 for the United States, based on the data from Figure 7.5.

A glance reveals the dominance of coal as the principal energy source between the 1880s and the 1950s, after a long period during which fuelwood, animal feed, and other traditional energy sources were predominant. The mature coal economy meshed with the massive expansion of railroads and steamship lines, the growth of steel making, and the electrification of factories. During the 1960s, oil assumed a dominant role in conjunction with the development of automotive transport, the petrochemical industry, and markets for home heating oil. If this substitution continues to progress at similar rates in the future, natural gas will be the dominant source of energy during the first decades of the twenty-first century, although oil is likely to maintain the second-largest share until the 2020s. Such an exploratory look into the future requires additional assumptions to describe the subsequent competition of potential new energy sources such as nuclear, solar,

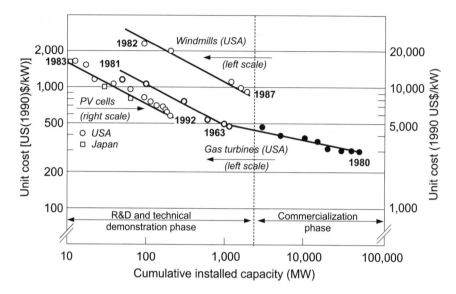

Figure 7.7. Reductions of Investment Costs for Three Representative New and Advanced Technologies as a Learning Process.

Expressed in US dollars at constant 1990 prices per unit installed capacity [US(1990)$/kW] versus cumulative installed capacity (MW) on double logarithmic axes.

and other renewables, which have not yet captured sufficient market shares to allow an estimation of their penetration rates and market potentials. Because all of these alternative energy sources have only minimal CO_2 emissions and natural gas has the lowest emissions of all fossil fuels, the unfolding of primary energy substitution implies a continuation of gradual energy decarbonization throughout the world.

7.3   Technological Learning

The replacement of old technologies with new ones occurs gradually. The performance of new technologies improves and their costs decrease with increases in production and use. Accumulated experience and learning can be assumed to increase with increases in the market shares of a new technology. As technologies mature, their improvement potentials decrease. A somewhat stylized difference between new and old technologies is that the former are costlier at the time of their introduction, but their costs can be assumed to decrease with increases in their market share so that at some point the cost curves might cross, making them a more attractive choice than the old technology. Learning curves capture this process. Figure 7.7 presents a number of illustrative examples (Grübler *et al.* 1996; Nakicenovic and Rogner 1996; Nakicenovic *et al.* 1998). It shows rapid declines in investment costs with every doubling of cumulative installed capacity of gas combustion turbines and wind and photovoltaic (PV) systems. This pattern of performance improvement and cost reductions with accumulated experience and learning is common to most technologies, although its specific shape depends on the technology. Typical progress ratios listed in the literature range between 65 percent and 95 percent for all technologies and between 70 percent and 90 percent for energy technologies (Christiansson 1995). There are significant cost improvements during the RD&D phases of technological development. For example, in Figure 7.7 an 18 percent reduction in investment costs per doubling of cumulative production (a progress ratio of 88 percent) is shown for the case of gas combustion turbines. These improvements during the RD&D phase are followed by more modest improvements after commercialization, 7 percent per production doubling for combustion turbines, for example. If such cost reductions were to continue in the future for the PV systems, these systems could become commercially viable in a few decades, with cost reductions of about a factor of five to one order of magnitude compared with today's costs [from between US$10,000/kW and US$5,000/kW to as little as US$1,000/kW; see Ishitani *et al.* (1996); Nakicenovic *et al.* (1996); Nakicenovic *et al.* (1998)].

Technological learning is reflected in most energy and emission scenarios and their underlying assumptions. New and emerging technologies are assumed to have better performance and lower costs in the future compared with current levels. Figure 7.8 reflects a range of such assumptions for some new and emerging energy-conversion technologies. It is based on the International Institute for Applied Systems Analysis (IIASA) inventory of mitigation technologies, CO2DB (Messner and Strubegger 1991; Messner and Nakicenovic 1992; Schäfer *et al.* 1992). This database currently includes characterization of about 1,600 energy technologies, from energy extraction and conversion to energy end use. The database includes current and future technologies based on information from the literature for a number of countries and representative world regions. A large share of technology descriptions come from various energy modeling efforts. Most of the information is available for energy-conversion technologies. In many cases, there are a sufficient number of data points for a given type of technology, such as for gas combustion turbines or PV systems, so that sample mean and standard deviation can be meaningfully derived. Figure 7.8 shows such statistics for 10 representative conversion technologies and gives the mean and standard deviation for current and future (about 2020) investment costs (Strubegger and Reitgruber 1995). A glance reveals a clear pattern: current costs are higher than the assumed future costs. The less mature a technology is today (such as the PV systems), the higher the future cost reductions and the higher the uncertainty, as evidenced by the wider distribution of cost estimates. This is indeed consistent with the phenomenon of cost reductions associated with learning, assuming that the installed capacities of these technologies will increase in the future, making them more competitive compared with current alternatives.

Equivalent assumptions are made in most modeling efforts and scenarios about future energy and emissions. Over time, new technologies become more

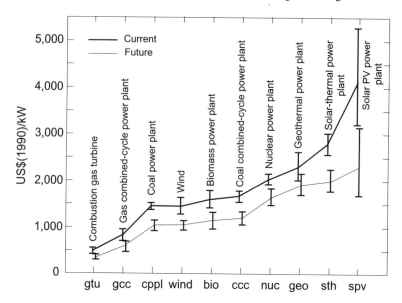

Figure 7.8. Mean and Standard Deviation of Investment Costs for 10 Representative Conversion Technologies, Current and Future (about 2020).

Based on data in the IIASA technology inventory CO2DB. *Source*: Strubegger and Reitgruber (1995).

attractive as their costs decrease and their performance improves. Sometimes such new technologies are called "backstops." Originally, Nordhaus (1973) formulated the concept of a backstop to mean a technology that has a virtually infinite resource base (e.g., PV systems). Generally it is assumed that backstop technologies require RD&D and that they are too costly to be competitive at the present time. Alternatively, if the costs of other technologies increase, the backstops may become competitive at some point in the future. There is, of course, a fundamental difference between the two approaches. In the first approach, it is assumed that new technologies will become cheaper and have better performance through RD&D and "autonomous" technological change, without, however, explicitly accounting for RD&D and appropriability issues. In the second approach, backstops become more attractive as supply limitations of currently competitive technologies lead to increases in their costs compared with those of the alternatives. In either case, technological change either is assumed to occur implicitly through specified market increases or takes the form of an exogenous parameter. This is a standard view of technological change in most economic modeling approaches. In some manner technologies are "ready" before entering the economic world and the entrepreneurs can choose among them according to their costs and relative performance so that they do have incentives to postpone investment in new technologies. In general, the problem is that new technologies appear as "manna from heaven" in the standard approaches to modeling technological change: as time passes, new technologies become the best choices without any explicit RD&D effort or investment and without any of the risks that entrepreneurs usually face. This is why these models are said to have an "autonomous" rate of technological change.

Models that employ autonomous technological change portray exogenous improvement of technologies over time. Because these models employ market allocation algorithms, the technologies gradually penetrate the market. This kind of simulation can emulate the introduction of new technologies and their subsequent diffusion. The employment of autonomous technological change assumptions can lead to either too much or too little technological change relative to an endogenous model, unless the nature of the autonomous path of technological change is known *a priori* as a scenario assumption.

The exogenous specification of costs of new technologies and their decrease over time implies that later adoption would be cheaper than early adoption. Thus, it is evident that in a model where a given autonomous rate of technological change is assumed, it is a cost-effective strategy to postpone investment in low-carbon technologies until they become cheaper and until the current vintages become obsolete. In reality, such results are misleading. If such mitigation strategies were to be adopted, there would be no investment in new technologies: all agents would wait for them to become more attractive, and no one would risk an early investment. Consequently, the technologies would not enter the marketplace and there would be no backstops in the future to reduce emissions. Instead, an emissions-intensive development path would be adopted that might prove difficult if not impossible to change midcourse. Even worse, there is some evidence that technological "forgetting by not doing" can occur (Rosegger 1991). Figure 7.6 illustrates how important inertia is in the energy systems: it takes decades to achieve a transition from old to new technologies through active innovation and diffusion of new technologies, and for each success there are many failures. It is in this light that the policy-relevant assessments of cost-optimal time paths of emission reductions should be considered.

7.4   Endogenizing Technological Change

The lack of technological realism and dynamics in most energy modeling work obviously must be rectified. This has been recognized for a long time. For example, Nordhaus and van der Heyden (1977) attempted to endogenize technological change in an energy model of the United States two decades ago. They included RD&D and learning by doing in the form of cost reductions as a function of the cumulative output of a technology. In the meantime, mathematical programming and computing techniques have improved so that it is now possible to capture RD&D and learning processes in greater detail, although computation requirements are still quite challenging.

A new research effort currently under way at IIASA aims at endogenizing technological change into the energy systems mathematical programming model MESSAGE (Messner 1995, 1997; Grübler and Messner 1998; and Chapter 11

Table 7.1. Reductions of Investment Costs as a Learning Process for Electricity Generation by Six New and Advanced Technologies.
Expressed in US dollars at constant 1990 prices per unit installed capacity [US(1990)$/kW].

Technology	1990	2050	Progress ratio
Advanced coal	1,650	1,350	0.93
Gas combined cycle	730	400	0.85
New nuclear	2,600	1,800	0.93
Wind	1,400	600	0.85
Solar thermal	2,900	1,200	0.85
Solar PV	5,100	1,000	0.72

in this volume) and introducing uncertainty into the characteristics of new and emerging technologies (Messner *et al.* 1996; Gritsevskyi and Nakicenovic 2000; and Chapter 10 in this volume).

Messner (1997) introduced technological learning into MESSAGE in terms of investment-cost reductions as a function of cumulative installations for six new and emerging electricity-generating technologies: advanced coal, natural gas combined cycle, new nuclear, wind, solar thermal, and PV systems. The learning process starts at present costs and can reach much lower and more competitive costs by accumulating experience. For example, for solar PV systems the assumed learning curve can lead to cost reductions of a factor of five between the base year (1990) and 2050 (from US$5,100 to US$1,000 per kW installed); the reduction potential for gas combined-cycle systems is approximately 45 percent (from US$730 to US$400 per kW installed). The technological learning assumptions for all six conversion technologies are shown in Table 7.1, reproduced from Messner (1997). In the model, RD&D activities and investments must be made in expensive new technologies if the technologies are to become cheaper through accumulated experience, represented by cumulative increase in installed capacity.

The representation of endogenous RD&D and technological learning in the energy systems model MESSAGE requires so-called mixed integer programming techniques, because the constraint set is nonconvex. Computationally, this approach is very demanding, so that only six new technologies are explicitly modeled as a single-region world model of the electricity sector. The next research tasks will include the extension of the approach to the whole energy system and inclusion of other downstream technologies in addition to electricity generation [see Gritsevskyi and Nakicenovic (2000) and Chapter 10 in this volume]. Among the shortcomings of the approach are that the shape of the learning curves is specified exogenously (including the RD&D phase) and that the uncertainty of technological change is not yet captured in this particular model.

To compare the technological learning case with alternative ways of modeling technological change, Messner (1997) developed two additional cases. The first variant, the "static" case, is the least realistic of the three cases. In this variant, it is assumed that the investment costs of the new technologies remain at their 1990 levels over the entire time horizon. The "dynamic" variant assumes the same degree of cost reductions given in Table 7.1, but the reductions are exogenous

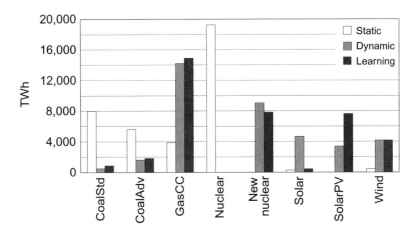

Figure 7.9. World Electricity Generation (TWh) in 2050.

Figure shows eight generation technologies for three alternative cases: the "static" case, with constant investment costs; the "dynamic" case, with exogenously declining costs; and the "technological learning" case, with endogenously declining costs.

("autonomous"), occurring at continuous rates between the base year (1990) and 2050. The dynamic case emulates the most common approach to modeling technological change in energy systems. In fact, it corresponds to Case A presented in the joint IIASA and World Energy Conference (WEC) study *Global Energy Perspectives to 2050 and Beyond* (Nakicenovic *et al.* 1998).

Figure 7.9 shows the mix of global electricity generation in 2050 from eight different conversion technologies, including the six selected new and emerging technologies. The static variant relies primarily on established technologies such as standard coal and nuclear power plants, and to a more limited degree on less costly advanced coal and natural gas combined-cycle technologies. With the exception of some coal, the new and advanced technologies are hardly used, because of the relatively high investment costs. In comparison, the dynamic cost profile does indeed lead to greater investment in new and advanced technologies. The roles of coal and standard nuclear technologies diminish compared with the static case; they are replaced by natural gas combined-cycle, new nuclear, solar, and wind technologies. Because in the dynamic case these technology improvements are exogenous, the shift in investments from traditional to new and advanced technologies changes in step with the cost reductions. In contrast to the dynamic case, with technological learning investments in new technologies must be made up front, when these technologies are much costlier than the conventional alternatives, if they are to become cheaper with cumulative experience as installed capacity increases. With technological learning, the structure of electricity production in 2050 is not all that different from the dynamic alternative, with the exception of a slight shift from new nuclear to solar PV systems.

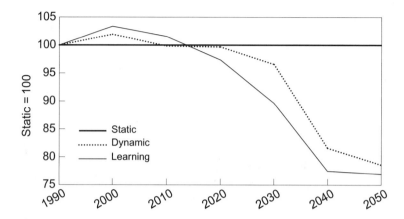

Figure 7.10. Annual Investment Requirements for Global Electricity Generation. Investment is shown for three alternative cases: the "dynamic" case, with exogenously declining costs, and the "technological learning" case, with endogenously declining costs, compared with the "static" case, with constant costs (index = 100), expressed as an index.

Messner (1997) has analyzed the different dynamics of investment paths in new and advanced technologies in the two alternative cases—the dynamic case with exogenous cost reductions and the technological learning case with endogenous cost reductions. Figure 7.10 presents her findings for global annual investments in electricity generation in the technological learning and dynamic cases compared with the static case. The most striking difference is that the case with endogenous learning shows higher up-front investment costs but has lower discounted systems costs than the dynamic case with exogenous cost reductions. Both cases lead to roughly the same investment costs in 2050, because there is sufficient cumulative investment in new and advanced technologies to reduce the costs along the learning curve to the level of exogenous reductions in the dynamic case. Over the entire time period (1990–2050), cumulative discounted investments are 6.6 percent lower in the dynamic case with exogenous learning and 9.7 percent lower in the case with endogenous learning than in the static case (Messner 1997). The difference in the investments is particularly large between 2020 and 2050. The discounted investment costs in the case with technological learning are 50 percent below the discounted investment costs of the dynamic case.

This single example illustrates some of the generic differences between the two approaches to modeling future technology costs and performance. In the dynamic case it pays to postpone some investment in new technologies until the costs are reduced (exogenously). In the case of technological learning there is no time to waste. Higher levels of costly investments are made immediately to accrue sufficient experience to be able to reap the benefits of cost reductions at some point further along the learning curve. If these costly investments are not made, the technology stays expensive. Nonetheless, despite high initial investments, the overall discounted costs are lower in this example than in the other cases. This result means that early RD&D expenditures and development of niche markets for new technologies may be able to reduce the overall discounted costs of long-term mitigation strategies, even if similar rates of "autonomous" technology improvement are assumed in the case without learning. In reality, however, the exogenous cost reductions are unlikely to occur unless someone else invests instead. At the global level this is of course a contradiction, because even in the dynamic case such investments must be included in the calculations if cost reductions are to occur.

7.5 Conclusion

Incorporating the concept of technological learning into the energy model MESSAGE led to lower CO_2 mitigation costs compared with an alternative model employing a fixed rate of autonomous technological change, as is usually done in studies of future energy and emissions perspectives. The costs were also lower although exactly the same rates of performance improvements and cost reductions were assumed to occur over the study time horizon in both approaches. Compared with the case of endogenized learning, the "autonomous" case leads to the postponement of investment decisions until lower-emission technologies "become" cheaper. This means that initially the investments are somewhat lower. In the case with endogenous technological learning, initial investments are higher. However, this higher investment is offset later through the possibilities of reducing emissions at substantially lower costs when installed capacities and emission levels are higher. Even with discounting at 5 percent per year, the endogenous learning case leads to lower total costs in the global electricity sector. Of course, these results are sector specific and do not reflect any of the deadweight loss or intersectoral trade-offs stipulated by Goulder (1996). In other words, the analysis does not consider the potential loss of welfare associated with the costly initial market penetration of the new technologies or the transfer of resources away from other technology development toward the development of new technologies. The results, however, do shed light on the process by which new technologies enter and penetrate the market, which has important implications for both the cost and timing of policy interventions designed to achieve emission mitigation.

Endogenization of technological change through technological learning captures some of the positive externalities generated by RD&D and early investment in new technologies. This means that not only will a given technology be improved through RD&D and learning, but other technologies of the same "family" will improve, as well. Knowledge spillover is often assumed to be determined by the combination of processes by which knowledge diffuses and by which it becomes obsolete. It has a positive impact on the social return of the technological learning development strategies.

The introduction of technological learning into the model does not solve all the problems associated with understanding technological change or the future costs of alternative energy technology strategies. Some basic problems also encountered in the autonomous technological change approach are still unsolved.

Technical performance and cost profiles of learning curves must be specified *a priori*. In the real world the performance improvement rates of new technologies are not known *a priori*, which is reflected in the risks that entrepreneurs usually face when they make new technology adoption decisions. It should be acknowledged that technical change is only one of several factors that determine technology costs and performance and thus ultimately also emissions paths.

Including this "stylized" treatment of technological change in the model captures some of the dynamic patterns common to the cost reductions and improvement in performance of almost all technologies that are successful in the marketplace. Initially, costs are high owing to batch-production methods that require highly skilled labor. Performance optimization and cost minimization are rarely important; the overriding objective is the demonstration of technical feasibility. When the technology seeks entry into a market niche, costs begin to matter, although usually what is of central importance is the technology's ability to perform a task that cannot be accomplished by any other technology. Examples are fuel cells in space applications, PV systems for remote and unattended electricity generation, gas turbines for military aircraft propulsion, and drill-bit steering technology in oil and gas exploration. Including in the model the more costly new and advanced technologies with the promise of lower costs and better performance through accumulated learning captures these effects of early and pre-commercial technology development and entry into specialized niche markets.

A technology's success in a niche market, however, does not ensure its successful commercialization. Improvements must be made in reliability, durability, and efficiency, and, even more important, costs must be reduced. Any RD&D devoted to these objectives creates a supply push. This supply push must be complemented by a demand pull, by which initial markets are expanded sufficiently to further reduce costs through economies of scale. The demand pull may be policy driven. Technically feasible technologies that are not yet economically competitive might benefit from environmental or energy security policies that increase their competitors' costs. For example, other electricity generation options benefit from requirements for flue gas desulfurization in coal-fired plants, or from bans on electricity generation from natural gas that restrict combined-cycle gas technology. New technologies may also benefit from economies of scale and market dominance already achieved by older technologies. The existing transmission infrastructure, for example, can be readily used by new electricity-generating technologies (Nakicenovic *et al.* 1998). Including such effects in the model by initially introducing new and advanced technologies only in some niche markets and later in more widespread applications as their costs decrease captures some of these complex phenomena associated with innovation diffusion and technological change.

Thus, the rate of technological change depends on the diffusion of innovations and the dynamics of their adoption. The replacement of carbon-intensive technologies with zero- or low-carbon alternatives can be expressed as the process of energy decarbonization. Scenarios with high shares of coal actually lead to a reversal of the historical trends toward decarbonization. Other scenarios that

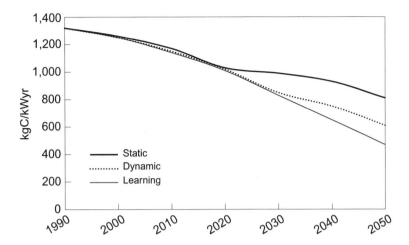

Figure 7.11. Decarbonization of Global Electricity.
Rates are shown for three alternative cases: the "static" case, with constant investment costs; the "dynamic" case, with exogenously declining costs; and the "technological learning" case, with endogenously declining costs. Rates are expressed in kilograms of carbon per kilowatt-year of electricity (kgC/kWyr).

envisage that the transition to the post-fossil era will occur during the twenty-first century portray decarbonization rates similar to, or sometimes even higher than, historical rates. Decarbonization must continue if CO_2 emissions and eventually also concentrations are to stabilize in the future. Quite high rates would be required to actually reduce global CO_2 emissions, as would be required to achieve stabilization of atmospheric concentrations at some negotiated level in accordance with Article 2 of the Framework Convention on Climate Change (UN/FCCC 1992). Figure 7.11 captures the differences in the decarbonization of global electricity generation with and without technological learning presented in this chapter.

Without improvements in technological performance or cost reductions compared with the present situation, the static case actually leads to a reversal of historical trends toward decarbonization after the 2020s as the global electricity generation is "locked-in" the carbon-intensive generation technologies. Decarbonization occurs in the dynamic case, indicating a high degree of structural change in electricity-generating capacity. However, the rate slows down after the 2030s compared with the technological learning case. The more dynamic interplay in the learning case among different electricity-generating technologies leads to the highest degree of decarbonization, and yet here the total discounted costs are the lowest of all three alternatives. That the costs are lower than in the static case is not at all surprising, as the static case does not include any reduction in costs, and thus older and cheaper technologies are generally chosen, leading to relatively high emissions and high costs.

An interesting result of this analysis is that technological learning leads to lower emissions and costs compared with the dynamic case, even though costs and emission-reduction potentials are the same as the exogenously assumed improvement rates in the dynamic case by the end of the time horizon. The additional degree of freedom of initially introducing promising technologies in the niche markets although they are still too costly leads to overall cost reductions, because cumulative learning allows for significant cost reductions later on, when installed capacities and emission levels are high. In contrast, the dynamic case does not lead to early market entry of new and advanced technologies. These technologies diffuse as they become more attractive, but by that time the system's inertia and the still-high shares of older technologies in the vintage structure do not allow a more dynamic transition toward lower emissions.

The "stylized" treatment of RD&D and technological learning in the model requires further improvement. Endogenous technological change is captured only for six new technologies in the presented example. This is seriously deficient and clearly needs to be extended to other technologies in the energy system and other sectors of the economy [see also Gritsevskyi and Nakicenovic (2000) and Chapter 10 in this volume]. High computational requirements are a serious barrier to such extensions, so that new research is required. There are serious methodological shortcomings to the approach, as it captures RD&D and learning only for low-carbon-emitting technologies. According to Goulder (1996), knowledge-generating resources are generally scarce, so that expansion of technological progress in one industry often implies a reduction in the rate of technological progress in others, even if the policy in question does not intend to discourage any industry's rate of technological progress. Another critical issue is that endogenization of technological change through learning by doing means that the energy system will be "locked-in" a few technologies that have high progress ratios. But variety has a value in itself. This means that a number of speculative projects should be funded in any case, with the idea that this will enlarge the stock of future possibilities.

This first result of endogenizing technological change indicates that the postponement of investments in new and advanced technologies in itself will bring few additional benefits to future CO_2 mitigation strategies. In other cases there might be benefits from delay. The costs of some technologies might decrease as a result of "exogenous" improvement of other technologies. For example, improvements in information technologies might benefit energy technologies so that postponement might be attractive. The main result of the analysis, however, is robust: unless there is dedicated, timely, and pronounced investment in CO_2 mitigation technologies, they are less likely to be developed and thus become commercially viable and competitive in the marketplace. Learning by doing is a prerequisite for performance improvements, cost reductions, and eventual diffusion. Postponement of investment decisions will not bring about the technological change required to reduce CO_2 emissions in a cost-effective way. Even worse, it might bring about further "lock-in" of energy systems and economic activities along fossil-intensive development paths.

Acknowledgments

This chapter reports on some results of introducing increasing returns through technological learning into the model MESSAGE based on the analysis presented in greater detail by Golodnikov et al. (1995), Messner (1997), Messner et al. (1996), Grübler and Messner (1998), Grübler and Gritsevskyi (1998 and Chapter 11 in this volume), and Gritsevskyi and Nakicenovic (2000 and Chapter 10 in this volume). It is a substantively revised version of an earlier paper by the author (Nakicenovic 1997). Helpful comments and suggestions were received from Jae Edmonds, Dominique Foray, Bill Nordhaus, and Rich Richels. The views presented are solely those of the author.

References

Christiansson, L., 1995, Diffusion and Learning Curves of Renewable Energy Technologies, WP-95-126, International Institute for Applied Systems Analysis, Laxenburg, Austria.

Golodnikov, A., Gritsevskyi, A., and Messner, S., 1995, A Stochastic Version of the Dynamic Linear Programming Model MESSAGE III, WP-95-94, International Institute for Applied Systems Analysis, Laxenburg, Austria.

Goulder, L.H., 1996, Notes on the Implications of Induced Technological Change for the Attractiveness and Timing of Carbon Abatement, Paper presented at the Workshop on Timing and Abatement of Greenhouse-Gas Emissions, 17–18 June, Paris, France.

Gritsevskyi, A., and Nakicenovic, N., 2000, Modeling uncertainty of induced technological change, *Energy Policy*, **28**:907–921.

Grübler, A., and Messner, S., 1998, Technological change and the timing of mitigation measures, *Energy Economics*, **20**:495–512.

Grübler, A., and Nakicenovic, N., 1988, The dynamic evolution of methane technologies, in T.H. Lee, H.R. Lindin, D.A. Dreyfus, and T. Vasko, eds, *The Methane Age*, Kluwer Academic Publishers, Dordrecht, Netherlands.

Grübler, A., and Nakicenovic, N., 1996, Decarbonizing the global energy system, *Technological Forecasting and Social Change*, **53**(1):97–110.

Grübler, A., Jefferson, M., and Nakicenovic, N., 1996, Global energy perspectives: A summary of the joint study by the International Institute for Applied Systems Analysis and World Energy Council, *Technological Forecasting and Social Change*, **51**(3):237–264.

IPCC (Intergovernmental Panel on Climate Change), 1996, Second Assessment Synthesis of Scientific-Technical Information Relevant to Interpreting Article 2 of the UN Framework Convention on Climate Change, Climate Change 1995, IPCC, Geneva, Switzerland.

IPCC (Intergovernmental Panel on Climate Change), 2001, *Climate Change 2001: Mitigation. Contributions of the Working Group III to the Third Assessment Report of the Intergovernmental Panel on Climate Change*, Cambridge University Press, Cambridge, UK.

Ishitani, H., Johansson, T., convening lead authors; Al-Khouli, S., Audus, H., Bertel, E., Bravo, E., Edmonds, J.A., Frandsen, S., Hall, D., Heinloth, K., Jefferson, M., de Laquil III, P., Moreira, J.R., Nakicenovic, N., Ogawa, Y., Pachauri, R., Riedacker, A., Rogner, H.-H., Saviharju, K., S. B., Stevens, G., Turkenburg, W.C., Williams,

R.H., Fengqi, Z., principal lead authors, 1996, Energy supply mitigation options, in R.T. Watson, M.C. Zinyowera, and R.H. Moss, eds, *Climate Change 1995: Impacts, Adaptations and Mitigation of Climate Change: Scientific-Technical Analyses. Contribution of Working Group II to the Second Assessment Report of the Intergovernmental Panel on Climate Change*, Cambridge University Press, Cambridge, UK.

Marchetti, C., and Nakicenovic, N., 1979, *The Dynamics of Energy Systems and the Logistic Substitution Model*, RR-79-13, International Institute for Applied Systems Analysis, Laxenburg, Austria.

Messner, S., 1995, Endogenized Technological Learning in an Energy Systems Model, WP-95-114, International Institute for Applied Systems Analysis, Laxenburg, Austria.

Messner, S., 1997, Endogenized technological learning in an energy systems model, *Journal of Evolutionary Economics*, **7**(3):291–313.

Messner, S., and Nakicenovic, N., 1992, A comparative assessment of different options to reduce CO_2 emissions, *Energy Conversion and Management*, **33**(5–8):763–771.

Messner, S., and Strubegger, M., 1991, User's Guide to CO2DB: The IIASA CO_2 Technology Data Bank, Version 1.0, WP-91-31a, International Institute for Applied Systems Analysis, Laxenburg, Austria.

Messner, S., Golodnikov, A., and Gritsevskyi, A., 1996, A stochastic version of the dynamic linear programming model MESSAGE III, *Energy*, **21**(9):775–784.

Nakicenovic, N., 1979, *Software Package for the Logistic Substitution Model*, RR-79-12, International Institute for Applied Systems Analysis, Laxenburg, Austria.

Nakicenovic, N., 1990, Dynamics of change and long waves, in T. Vasko, R. Ayres, and L. Fontvielle, eds, *Life Cycles and Long Waves*, Springer-Verlag, Berlin, Germany.

Nakicenovic, N., 1996, Freeing energy from carbon, *Daedalus*, **125**(3):95–112.

Nakicenovic, N., 1997, Technological change and learning, *Perspectives in Energy*, **4**:173–189.

Nakicenovic, N., and Rogner, H.-H., 1996, Financing global energy perspectives to 2050, *OPEC Review*, **XX**(1):1–24.

Nakicenovic, N., Grübler, A., Ishitani, H., Johansson, T., Marland, G., *et al.*, 1996, Energy primer, in R.T. Watson, M.C. Zinyowera, and R.H. Moss, eds, *Climate Change 1995: Impacts, Adaptations and Mitigation of Climate Change: Scientific-Technical Analyses. Contribution of Working Group II to the Second Assessment Report of the Intergovernmental Panel on Climate Change*, Cambridge University Press, Cambridge, UK.

Nakicenovic, N., Grübler, A., and McDonald, A., eds, 1998, *Global Energy Perspectives*, Cambridge University Press, Cambridge, UK.

Nordhaus, W.D., 1973, The allocation of energy resources, Brookings Institution, Washington, DC, USA, *Brookings Papers on Economics Activity*, **3**:529–576.

Nordhaus, W.D., and van der Heyden, L., 1997, Modeling Technological Change: Use of Mathematical Programming Models in the Energy Sector, Cowles Foundation Discussion Paper No. 457, New Haven, CT, USA.

Rosegger, G., 1991, Diffusion through interfirm cooperation: A case study, in N. Nakicenovic and A. Grübler, eds, *Diffusion of Technologies and Social Behavior*, Springer-Verlag, Berlin, Germany.

Schäfer, A., Schrattenholzer, L., and Messner, S., 1992, Inventory of Greenhouse-Gas Mitigation Measures: Examples from IIASA Technology Data Bank, WP-92-85, International Institute for Applied Systems Analysis, Laxenburg, Austria.

Schmalensee, R., Stoker, T.M., and Judson, R.A., 1998, World carbon dioxide emissions: 1950–2050, *The Review of Economics and Statistics*, **80**:15–27.

Strubegger, M., and Reitgruber, I., 1995, Statistical Analysis of Investment Costs for Power Generation Technologies, WP-95-109, International Institute for Applied Systems Analysis, Laxenburg, Austria.

UN/FCCC (United Nations Framework Convention on Climate Change), 1992, Convention Text, IUCC, Geneva, Switzerland.

Wigley, T.M.L., Richels, R., and Edmonds, J.A., 1996, Economic and environmental choices in the stabilization of atmospheric CO_2 concentrations, *Nature*, **379**:240–243.

Chapter 8

Modeling Induced Innovation in Climate-Change Policy

William D. Nordhaus

8.1 Introduction

Studies of environmental and climate-change policy—indeed of virtually all aspects of economic policy—have generally sidestepped the thorny issue of induced innovation, which refers to the impact of economic activity and policy on research, development, and the diffusion of new technologies. This omission arises both because of the lack of a firm empirical understanding of the determinants of technological change and because of the inherent difficulties of economic modeling processes with externalities and increasing returns to scale. While we suspect that we know the direction of this effect—toward overestimates of the cost of emissions reductions and the trend increase in climate change—we have little sense of the magnitude of the effect or the importance of this omission. Would including induced innovation have a large or small impact on climate change and on climate-change policies? This is a major open question.

One way of remedying this omission is to draw on research in the area of induced innovation. The formal theory of induced innovation arose in the 1960s in an attempt to understand why technological change appears to have been largely labor saving (see Nelson 1959; Arrow 1962; Kennedy 1962; and von Weizsäcker 1966). More recently, theories of induced technological change have been resuscitated as the "new growth theory," pioneered by Lucas (1986) and Romer (1990).[1] The thrust of this research is to allow for investment in knowledge and human capital. Such investments improve society's technologies, and a higher level of investment in knowledge will change society's production possibilities and may improve the long-run growth rate of the economy.

Virtually all studies of induced innovation have been theoretical.[2] With few exceptions, they do not lay out a set of hypotheses that can be tested or used to model the innovation process at an industrial level.[3] The present study sets out an approach that draws on the theory of endogenous technological change as well as empirical results in this area and applies the theory to the issue of induced innovation in climate-change policy.

8.2 Theory of Induced Innovation as Applied to the Environment

It will be useful to describe the forest terrain before getting immersed in tree mathematics. This chapter presents a model of induced innovation that can describe the impact of changes in prices or regulations on the innovations in different sectors. At any given time, society has a stock of existing general and sector-specific basic knowledge and applied and engineering knowledge. By investing in improvements in the stock of knowledge, society can improve the productivity of its resources.

Inventive activity in market economies is fundamentally a private sector activity, so the decisions about the allocation of inventive activity depend on private sector incentives. The allocation to particular sectors will depend on the relative sizes of different sectors, the degree of appropriability, and the underlying "innovation-possibility frontier" (the production function for producing new knowledge). These functions are derived from or calibrated to empirical studies that analyze the inventive process. The calibrated innovation production function is then embedded into a model of the economics of global warming to determine the impact of innovation on the important variables, such as the time path of greenhouse-gas emissions and concentrations, and climate change, along with the policy variables.

The analytical background for the model is developed in detail below. The discussion in this section lays out the model of technological change to be used in the simulations that follow. In describing the analytical and modeling framework, five issues are discussed: (1) the underlying model of technology, (2) the firm's decision framework, (3) the divergence between social and private return, (4) the functional specification of the induced innovation function, and (5) the welfare implications of induced innovation.

8.2.1 The underlying model of technology

We begin with a discussion of the underlying view of technology and innovation. At any time, there is a stock of basic, applied, and engineering knowledge—both general and sector specific. The state of fundamental knowledge at each time is represented by H_t. The state of fundamental knowledge is assumed to proceed exogenously from the point of view of individual sectors.

Within an individual sector, there is a level of sector-specific technological knowledge. In each sector, new knowledge is generated by combining sector-specific research with general and sector-specific knowledge. Resources can be applied to improve the state of knowledge, and improvements in knowledge are called "innovation." These resources comprise a wide variety of activities, including basic research, applied research and development (R&D), development, and commercialization. In the discussion that follows, the inputs into the process of technological change are labeled as "research," or R_i. The outputs of research—the innovations—should be thought of in the broad sense of "new combinations" or new products, processes, or ways of doing business. $A_{i,t}$ denotes the level of technological knowledge in sector i in year t. Hence, technological change (i.e., changes in $A_{i,t}$) is denoted by $\partial A_{i,t}/\partial t = \dot{A}_{i,t}$, where a dot over a variable represents the time derivative of that variable, or as $\Delta A_{i,t}/A_{i,t-1}$ in a discrete, period model.

Production in sector i is given by $X_{i,t} = A_{i,t} f_i(\cdot)$, where $f_i(\cdot)$ is a function of inputs of capital, labor, etc. Useful new innovations increase technology, as represented by the following *innovation-possibility frontier*:

$$\dot{A}_{i,t}/A_{i,t} = \phi_i(R_{i,t}, H_t),\tag{8.1}$$

where $\dot{A}_{i,t}/A_{i,t}$ is the rate of technological change (either discrete or continuous depending on the context).

8.2.2   The firm's decision framework

Although research takes place in a wide variety of institutions, we examine research that takes place in profit-oriented enterprises. For this purpose, we assume that research is an investment activity that is pursued for commercial purposes and has well-determined private and social returns. This assumption allows us to model induced innovation in a conventional economic framework once we take into account the externalities of research.

This basic framework contains two major assumptions. The first is that inventive activity is undertaken with an eye to increasing profits rather than for prestige, Nobel prizes, not-for-profit altruism, or curiosity. While much academic and government research does not fit this pattern, the focus here is on the profit-oriented activity in firms that is directed to improving energy and environmental production technology. The second major assumption is that the innovation-possibility schedule is deterministic. This could easily be modified to include a risk component to inventive activity as long as the risks were independent and reasonably nicely distributed. This second assumption is worrisome because of the evidence that the distribution of returns to inventive activity is highly skewed.[4]

To understand the allocation of research, we need to examine the way innovation affects profitability of *private* individual agents. This can be treated in a number of ways, but the following is consistent with current findings in the industrial organization literature.[5]

We view the production process as consisting of a large number of elemental goods and production processes (e.g., sending bits of information, producing Barbie dolls, lighting rooms, or producing a kilowatt-hour of electricity). For simplicity, we assume that these elemental processes have constant returns to scale. At any time, there is a dominant technology for each process. The dominant technology in one period becomes widely available in the next period and sets the upper limit on prices. In the model, the period is a decade and represents the effective life of a new invention. In the preceding period, this dominant technology costs $C_{i,t-1}$ and has a market price of $P_{i,t-1} = C_{i,t-1}$. A single new innovation in period t then lowers the cost to $C_{i,t} < C_{i,t-1}$. We assume that the inventor

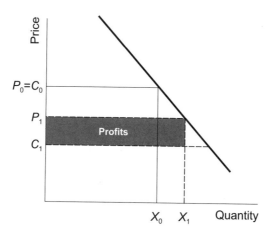

Figure 8.1. Impact of Innovation on Price, Cost, and Profits.

can appropriate the fraction α of the cost savings from the innovation. Then (for "run-of-the-mill" innovations), the inventor maximizes profit by setting the price at $C_{i,t} + \alpha[C_{i,t-1} - C_{i,t}]$.[6]

Inventive activity in a sector leads to an improvement of the basic technology, increasing the level of productivity, $A_{i,t}$. The improvement reduces the cost of production from $C_{i,t-1}$ in the preceding period to $C_{i,t}$ in the current period. The inventor captures the fraction α of the innovation (this being the appropriability ratio). Figure 8.1 shows the initial competitive price, new cost, and new price under these assumptions. The shaded profit region is "Schumpeterian" or innovational profits. As is shown in the figure, the current period price (P_1) therefore lies between the competitive cost of the old technology (C_0) and the new, lower cost of the innovation (C_1). The extent to which the current price is above the cost depends on the appropriability ratio.

The inventor's profits are equal to $(P_1 - C_1)X_1$, which can be approximated by $\alpha(C_0 - C_1)X_0 = \alpha[(C_0 - C_1)/C_0](P_0X_0) = \alpha[\Delta A_1/A_0](P_0X_0)$. In other words, the value of the innovation is approximately equal to the appropriability ratio times the improvement in technology times the value of output.

The firm's decision is then to determine the profit-maximizing amount of research given all the parameters and constraints of the market. The major parameters are the degree of appropriability, the form of the innovation-possibility function, the size of the market, and the cost of research.

8.2.3   The divergence between social and private return

A major issue raised by the presence of technological change is the likelihood that invention involves external economies or, more precisely, that R&D has a social rate of return that greatly exceeds the private rate of return on those investments. This is of course the primary economic justification for government intervention in the allocation of research funds. The divergence between social and private return also poses thorny problems for economic modeling. If research had equal private and social returns, it could be treated as simply another investment.

While estimates differ from sector to sector, there is a substantial body of evidence that, for the United States, research has a social return of 30 to 70 percent per annum, versus private returns on capital in the 6 to 15 percent range.[7] A government summary of the literature stated that the "benefits to industries which purchase new or improved products from innovating firms [have] rates of return ... estimated to range from 20 to 80 percent" (National Science Board 1977:126). A summary by the leading practitioner in this area, Edwin Mansfield, stated that "practically all of the studies carried out to date indicate that the average social rate of return from industrial R&D is very high. Moreover, the marginal social rate of return also seems high, generally in the neighborhood of 30 to 50 percent" (Mansfield 1996:191).

An alternative approach would be to assume that all residual economic growth is determined by "human ingenuity." This approach would imply a social rate of return of around 150 percent per annum rather than 50 percent per annum. This calculation assumes that about 1 percent of output is devoted to the "human ingenuity sector" and that the rate of total factor productivity after removing all other inputs is 1.5 percent per annum.

8.2.4   Functional specification of the induced innovation function

Given these background observations, we now discuss the specification of the induced innovation function employed in the present study. We assume that production takes place with fixed proportions. Under this view, substitution itself is a costly process, and research and engineering costs must be devoted to employing knowledge to develop and deploy new techniques.

In the specific application, we investigate the role of "carbon-energy-saving technological change," or CESTC. The model used here simulates the energy system by defining a new input into production called "carbon energy." Carbon energy is the carbon equivalent of energy consumption and is measured in carbon units. Carbon energy is one of the inputs (along with capital and labor) used to produce output. Carbon dioxide (CO_2) emissions and output are therefore a joint product of using inputs of carbon energy. This can be interpreted as a world in which energy is produced either by pure carbon or by substitutes that are some blend of pure carbon and capital and labor.

Technological change takes two forms: economy-wide technological change and CESTC. Economy-wide technological change is Hicks neutral, while CESTC reduces the carbon energy coefficient of the fixed-proportions production function. In other words, CESTC reduces the amount of carbon emissions per unit output at given input prices.

The next step is to find an appropriate functional form for the innovation-possibility function in Equation (8.1). For this purpose, we concentrate only on the CESTC and assume that the Hicks-neutral component of technological change is exogenous. We further assume that there are diminishing returns to research, so that there is an interior solution to the representative firm's profit-maximizing decision. Moreover, we allow for depreciation of knowledge, either because of economic obsolescence (as new products replace old ones) or because of natural factors (such as adaptive pests) that reduce the value of the innovation over time.

In this spirit, it will be convenient to assume that the innovation-possibility frontier is generated as a constant-elasticity function of the level of research in the sector:

$$\dot{A}_{i,t}/A_{i,t} = \phi_i(R_{i,t}, H_t) = \kappa_i R_{i,t}^{\beta_i} - \delta_i \, , \qquad \cdot \qquad (8.2)$$

where the roman-letter variables are as defined in Equation (8.1) and the Greek letters are parameters. Equation (8.2) is a standard approach to modeling the economic impact of inventive activity (see, e.g., Griliches 1998).

There is one assumption that is hidden in the specification in Equation (8.2), concerning the "recharge" of the innovational potential. This question involves how rapidly the pool of potential innovations is recharged or replenished. There are two interesting polar cases. The first, which we follow, is the "building-on-shoulders" model of innovation, in which current innovations build on past innovations. As modeled in Equation (8.2), there are diminishing returns to inventive activity in any particular period, but in the next period the innovation possibilities are replenished. This implies that a high rate of innovation in one period leads to a larger stock of knowledge in the next period, and innovation in the next period builds on the "shoulders" of the larger stock of knowledge. Hence, in the building-on-shoulders approach, high innovation in one period leads to automatic recharge of the pool of potential innovations in the next period. This approach leads to technological drift in which the long-run level of technology depends on the past intensity of research.

An alternative approach, called the "depletable-pool" approach, is one in which the stock of potential innovations is an exogenous pool that is determined by activities outside the sector. In this approach, potential innovations are like a pool of fish that breed at a given rate so that heavy research fishing today reduces the number of invention-fish available in the next period. In other words, the stock of innovations is exogenously given, so more rapid innovation today leads to depleted opportunities tomorrow. The maximum level of technology in this approach is essentially exogenous, although a sector can be closer to or farther from the frontier; research can move the industry closer to the exogenous frontier but it does not advance the frontier.

It would seem appropriate to use the building-on-shoulders model when the relevant technology is largely self-contained and the research process is intrinsically cumulative. The depletable-pool model would be appropriate for processes like diffusion, or when the sector is small and much of the relevant technology is determined outside the sector.

8.2.5   The welfare implications of induced innovation

The present study does not consider the general-equilibrium impacts of induced innovation through its impact on other sectors. In the modeling of induced innovation presented below, aggregate output, labor, capital, and factor prices are taken as given. However, a few observations about the welfare impacts of induced innovation are in order.

In considering the impact of induced innovation in the carbon-energy sector, we must consider the possibility that higher inventive activity in one sector will reduce inventive activity in other sectors. That is, as research is reallocated from other areas to the carbon-energy sector, there may be more rapid technological change in the climate-change sector and less rapid technological change in other sectors.

In analyzing the welfare impact of changing conventional inputs, we generally put a social price tag of US$1 on inputs that cost US$1. For research, this would be appropriate if the higher research spending simply required more bricks and scribes without diminishing research in other sectors. For inventive activity, however, the higher inventive activity in the carbon-energy sector may reduce inventive activity in other sectors. This would occur under either of two conditions. First, under the commercial view described in the last section, increasing inventive incentives in one sector reduces incentives in others. The decline in other sectors occurs because the increased market value of output in the carbon-energy sector through carbon taxes or emissions constraints raises the relative cost of production in the carbon-energy sector and lowers the relative cost of production in other sectors, both of which tend to raise returns and reallocate inventive activity to the now-larger sector. Hence, if total output is unchanged, and if the technological parameters of the innovation-possibility frontiers in different industries are similar, a unit increase in research in the carbon-energy sector would accompany a unit decrease of research in the rest of the economy.

The second condition under which the social opportunity cost of reduced research in other sectors would be more than US$1 would be where the supply of research is less than perfectly elastic. For example, consider the case where there is a fixed stock of researchers (e.g., a fixed stock of people who are inclined to tinker, invent, and try new combinations). As some of the tinkerers are attracted to the carbon-energy sector to try to solve the world's environmental problems, they will pay less attention to unsolved problems in other sectors. If John von Neumann had been attracted to solving problems of the environment rather than those of computers, the cost to society would have been much more than a fractional lifetime of labor inputs into computation.

From a modeling perspective, we handle this issue by assuming that the opportunity cost of higher research in the carbon-energy sector is a multiple of its dollar costs; for the numerical calculations we take this multiple to be four. This assumption implies that the social return to research is four times the private return and is a convenient way of handling the externality of research discussed in Section 8.2.3.[8] In other words, this assumption implies that the present-value opportunity cost of research is US$4 of output per US$1 of research, so that when we calculate the social costs of increasing the research in the environmental sectors, we must reckon the loss in research outputs in the non-environmental sectors. It seems likely that the first of these assumptions is well supported by existing evidence. The second is more speculative, but it is important only for the welfare implications and not for the findings about the dynamics of technological change or the impact of induced innovation on the energy sector.

8.3 Application in the R&DICE Model

Here, we estimate the impact of induced innovation by putting the model of technological change described in the previous section in a new version of the globally aggregated DICE model, referred to here as the R&DICE model.[9] This section provides a brief description of the DICE-99 model and describes the modifications that are incorporated for the R&DICE model.

8.3.1 The DICE-99 model

The version of the model used for this study is the DICE-99 model, a globally aggregated model of the economics of global warming that includes the economy, carbon cycle, climate science, and impacts. This highly aggregated model allows a weighing of the costs and benefits of taking steps to slow greenhouse warming.

The latest versions are the RICE-99 and DICE-99 models. The RICE-99 model is an eight-region model of the world economy with a climate module. The DICE-99 model is a globally aggregated version of the RICE-99 model calibrated to match the major features of the larger model. While losing the regional detail of the RICE-99 model, the DICE-99 model has several advantages. It is more useful for understanding the basic structure of economic policy issues posed by greenhouse warming because it is small enough that researchers can understand the individual linkages in an intuitive way. It is more easily modified because the number of parameters is much smaller. It is much faster, so that alternative experiments can be tested more easily. In addition, it is available in a spreadsheet format, which can be more easily understood and manipulated by researchers.

The basic approach in the DICE model is to consider the trade-off between consumption today and consumption in the future. By taking steps to slow emissions of greenhouse gases (GHGs) today, the economy reduces the amount of output that can be devoted to consumption and productive investment. The return for this "climate investment" is reduced damages and therefore higher consumption in the future.

The DICE-99 model has an objective function which is the discounted utility of consumption. More precisely, the objective function to be maximized is the sum across periods of the discounted utilities, where the utility function is the population times the logarithm of per capita consumption. This objective function is maximized subject to a number of economic and geophysical constraints. The decision variables that are available to the economy are consumption, the rate of investment in tangible capital, and the climate investments, primarily reductions of GHG emissions.

The model contains both a traditional economic sector found in many economic models and a climate sector designed for climate-change modeling. In the economic sector there is a single commodity, which can be used for either consumption or investment. The world is endowed with an initial stock of capital and labor and an initial level of technology.

The environmental part of the model contains a number of geophysical relationships that link together the different forces affecting climate change. This part contains a carbon cycle, a radiative forcing equation, climate-change equations, and a climate-damage relationship. In DICE-99, endogenous emissions are limited to industrial CO_2, which is a joint product of carbon energy. Other contributions to global warming are taken as exogenous. DICE-99 contains a new structural approach to carbon-cycle modeling that uses a three-reservoir model calibrated to existing carbon-cycle models. Climate change is represented by global mean surface temperature, and the relationship uses the consensus of climate modelers and a lag derived from coupled ocean–atmospheric models. The DICE-99 model also contains new estimates of the damage function from climate change.

The major equations of the model along with modifications (discussed in the next section) are provided in Appendices 8.A and 8.B.

8.3.2   Modifications for the R&DICE-99 model

In the basic DICE model, carbon intensity is affected by *substitution* of capital and labor for carbon energy. The economic mechanism at work is substitution in that increases in the price of carbon energy relative to other inputs induce users to purchase more fuel-efficient equipment or employ less energy-intensive products and services. In the Cobb-Douglas production function, for example, a 10 percent increase in the price of carbon energy relative to other prices, when markups and other factors are taken into account, will lower the carbon/output ratio by approximately 4 percent.

In the R&DICE model, carbon intensity is affected by *technological change*. Here, the mechanism at work is induced innovation, which works in a manner quite different from substitution. A rise in the price of carbon energy will give an inducement to firms to develop new processes and products that are less carbon intensive than existing products. They therefore invest in new knowledge—through research, development, and informal investments. The new knowledge produces new processes and products, which lower the carbon intensity of output. To simplify the analysis, we have removed the processes of substitution from the R&DICE model to focus the analysis on the induced-innovation mechanism.

The difference between the standard substitution approach and the induced-innovation approach is illustrated in Figures 8.2 and 8.3. Figure 8.2 illustrates how an increase in the relative price of carbon energy would induce substitution from point A to point B. By contrast, in the R&DICE model, an investment in research

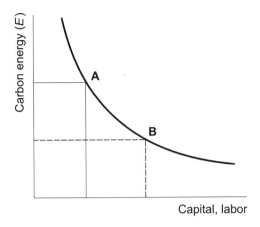

Figure 8.2. Substitution Approach.

In the standard substitution approach, an increase in the price of carbon energy leads to a substitution of capital and labor for carbon energy, moving along the production isoquant from A to B. In the usual approach, substitution is costless and reversible.

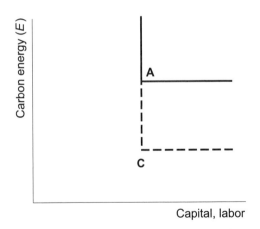

Figure 8.3. Induced-Innovation Approach.

In the induced-innovation approach, we assume that there is no costless substitution. Initially, the economy is at point A. After an investment in research, a new technique is developed with the same output as at A but with inputs shown by point C. We assume that technological change takes place in the baseline solution, and the R&DICE model investigates the role of carbon-saving technological change. The primary difference between this and the substitution approach is that the movement from A to C is both costly and irreversible.

leads to purely carbon-saving technological change, moving the technique from point A to point C (see Figure 8.3).

Analytical Changes

More precisely, the implementation of the R&DICE model takes the following steps:

- We begin with the DICE-99 model as described above and shown in Appendices 8.A and 8.B.

- We then take the capital, labor, and interest rate variables from the DICE model and set these as exogenous. This gives the exogenously determined level of output, $Q^*(t)$ in Equation (1-R&DICE) in Appendix 8.B.

- We then adjust the output balance equation in Equation (2-R&DICE) by making two further corrections. First, we subtract the cost of carbon energy from output to reflect the fixed-proportions assumption. Second, because research is now an endogenous variable, we subtract the cost of research. Note that the cost of research is multiplied by the factor of 4 to reflect the divergence of social and private costs discussed in Section 8.2.

- We next adjust the emissions equation to reflect the assumption that all changes in carbon intensity occur through induced innovation rather than substitution. In the standard DICE model, shown in Equation (4-DICE), substitution is reflected in the parameter μ. This is removed in Equation (4-R&DICE), and industrial carbon emissions are a function only of output and the endogenous carbon intensity, $\sigma(t)$.

- The final change comes in determining the carbon intensity. In the DICE model, the no-controls carbon intensity (i.e., the carbon intensity with no climate-change policy in the form of regulations or carbon taxes) is determined by estimates of technological trends. In the R&DICE model, we assume that the carbon intensity is endogenous and determined by the equation for induced innovation described in Section 8.2 and shown as Equation (5-R&DICE). In this specification, the change in the carbon intensity is a function of research spending on carbon-saving technological change.

Equations (6) through (11) in Appendix 8.B are identical in the DICE and R&DICE models.

Calibration

The final issue is the calibration of the parameters of the R&DICE model. More precisely, we need to specify the parameters of the induced-innovation function in Equation (5-R&DICE). Given the functional form, there are three parameters to estimate: Ψ_1, the productivity of research; Ψ_2, the elasticity of technology with respect to research; and Ψ_3, the depreciation rate of technology.

Table 8.1. R&D and Output for Fossil Fuels, United States, Billions of Dollars per Year.

	1987	1988	1989	1990	1991
Gross output, fossil fuels	77.0	86.9	83.3	99.6	87.8
R&D, fossil fuels	1.5	1.6	1.6	1.8	2.0
R&D/output (percent)	2.0	1.8	1.9	1.8	2.3

Sources: Bureau of Economic Analysis and National Science Foundation.

There is a great deal of research in the existing literature on both the elasticity and the depreciation rate. The elasticity has been estimated to lie in the range of 0.05 to 0.20 depending on the specification, industry, and time period. The depreciation rate is variously estimated at between 1 and 10 percent per year depending on the specific technology and the time period. We use these broad bounds to ensure that the actual calibration produces sensible estimates.

Within these broad ranges, the model produces three variables that can be used for calibration:

- *Criterion 1*. The first calibrating variable is the rate of return on R&D. From earlier discussion, we determined that a major feature of induced innovation is the divergence between social and private return. On the basis of existing studies, we impose a ratio of social to private return of 4.

- *Criterion 2*. The second calibrating variable is the trend in the carbon intensity of production. We assume that at the level of privately optimized R&D that occurs with *no climate damages*, the trend of carbon intensity would be as close as possible to the baseline (zero-carbon-tax) emissions path of the DICE-99 model. The numerical criterion to be minimized is the sum through 2150 of the squared deviations between the baseline projections of carbon intensity of the DICE-99 model and the baseline carbon intensity projected by the R&DICE model.

- *Criterion 3*. The final calibration is optimized R&D. Table 8.1 shows recent data for the United States on energy R&D for fossil fuels along with total gross output originating in the oil, gas, and coal industries. The data indicate that this industry has had a ratio of R&D to output of about 2 percent of sales in recent years. We therefore assume that R&D for carbon energy is 2 percent of output at the optimized level. For the calibration, we assume that the R&D/output ratio is constant for the entire period.

We then select the three parameters of the model so that they meet the calibration targets. The actual procedure is as follows: First, the level of R&D is optimized for an initial set of parameters $[\Psi_1^{(i)}, \Psi_2^{(i)}, \Psi_3^{(i)}]$. The calibration criterion 1 is automatically met through this optimization. We then adjust the parameter values so as to meet calibration criterion 3 exactly. We then set the parameters so that criterion 2 is best achieved (in the sense that the sum of the squared deviations of the carbon/output ratio from its target is minimized). In summary, we set the parameters so that the depreciation rate and the R&D/output ratio are exactly equal to their estimated values, while the remaining parameters are set to fit the path of the estimated CO_2 output trajectory. In these estimates, calibration criterion 2 has a standard deviation of 15 percent.

The actual form of the calibrated equation is

$$[\sigma(t) - \sigma(t-1)]/\sigma(t-1) = .415R(t)^{0.139} - 0.20 , \tag{8.3}$$

where $\sigma(t)$ is global industrial CO_2 emissions per unit of world output and where the time period t is 10 years.

The DICE and R&DICE model runs are calculated for 40 periods (400 years). The optimization in the R&DICE model estimates the optimal research inputs for the first 10 periods, but for the subsequent periods the ratio of research spending to spending on carbon energy is assumed to be constant for computational simplicity.

Model Runs

We will analyze three simulations:

- *Case A: No controls.* In the market run, there is no climate-change policy. Carbon-energy research is determined endogenously to maximize global income according to the procedure described above.

- *Case B: Climate policy with substitution but exogenous technology.* In the second experiment, we determine the optimal policy with an exogenous technology but with price-induced substitution. The mechanism is a carbon tax (determined by the discounted value of the marginal damages from emissions) that induces substitution away from carbon energy. This is estimated using the DICE model with adjustments for the baseline technology so that the base conforms to case A.

- *Case C: Climate policy with endogenous technology but no substitution.* In the third experiment, there is induced innovation but no substitution. Here, carbon taxes are again determined by the marginal damages from climate change, and the carbon tax induces research and technological change. Research is determined according to the criterion that the social rate of return is four times the private rate of return.

8.4   Results

8.4.1   Overall trends

One of the major issues is the impact of induced innovation on carbon intensity. We would expect that induced innovation would lead to a slower initial reduction in carbon intensity than would substitution, but the cumulative reduction in carbon intensity might well be larger in the long run. Figure 8.4 shows the carbon intensities of the baseline, substitution, and optimal-research (or induced-innovation)

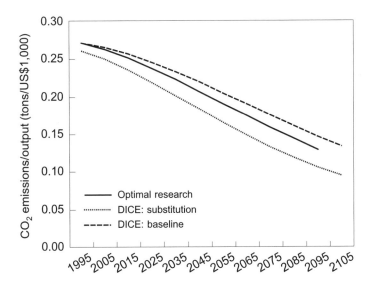

Figure 8.4. Carbon Intensity of Output under Different Approaches.

scenarios. Figure 8.5 shows the percentage reduction in emissions for the two policy scenarios. The first important conclusion is that the reduction in carbon intensity in the induced-innovation strategy is quite modest in the early decades. The reduction in emissions from induced innovation is about 6 percent over the first half century and about 12 percent after a century.

Second, it is interesting to note that the reduction in emissions from substitution is substantially larger than the reduction from induced innovation at the beginning (see Figure 8.5). The reduction in emissions from substitution is 12 percent by 2050 and 22 percent by 2100. The reductions from substitution are larger than those from induced innovation through the twenty-second century. Indeed the "crossover point" between induced innovation and substitution does not come until about 2230—although the exact timing is sensitive to the specification. This is perhaps the greatest surprise of the present study. We might have thought that allowing induced innovation with the supernormal rates of return would have led to quite rapid reductions in carbon intensities, but this proved not to be the case. We discuss the reasons for this result below.

Figure 8.6 shows the impact of optimizing emissions of industrial CO_2 in both the baseline and substitution cases of the standard DICE model and in the case of induced innovation in the R&DICE model. Figure 8.7 shows the impact on projected global mean temperature. Induced innovation produces more emissions and less temperature reduction than does substitution. After two centuries, induced innovation reduces global mean temperature by $0.25°C$ while substitution reduces global mean temperature by $0.54°C$. The crossover point where induced innovation beats out substitution is not reached at any point in the entire

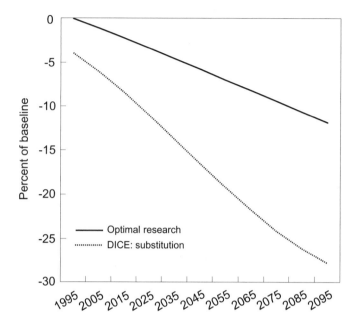

Figure 8.5. Emissions Relative to Baseline under Different Approaches.

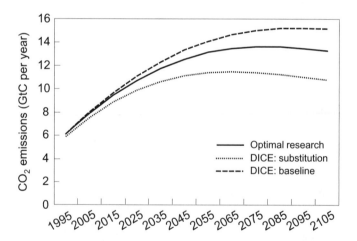

Figure 8.6. Industrial CO_2 Emissions under Different Approaches.

simulation period, although the crossover point for CO_2 concentrations is reached after about four centuries.

We next turn to the impact on research. Global R&D on carbon energy in 1995 in the baseline R&DICE run is US$11 billion in 1990 prices. This compares with estimates of US$2 billion in 1991 for the United States alone. There are no comparable data for most other countries, but we can extrapolate to determine whether

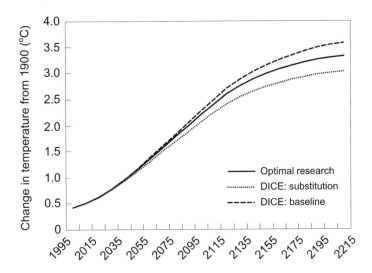

Figure 8.7. Global Temperature Change under Different Approaches.

the numbers in the R&DICE model are reasonable. The United States accounted for 23 percent of carbon use; with growth in the economy and the same ratio of R&D to output for the world, this would yield a global R&D total of US$10 billion in 1995. This probably overstates the actual world total given the higher research intensity in high-income countries, but the numbers are close enough to allow us to conclude that the model captures the basic numbers accurately.

The impact of induced innovation on research is shown in Figure 8.8. Total research spending increases by 40 percent when the climate externality is internalized through appropriate pricing. For this graph, we also show the impact on R&D spending of a doubling of carbon prices: a doubling of carbon-energy prices leads to an increase in R&D spending of 80 percent in the model. The elasticity of research with respect to the price of carbon energy is therefore around 80 percent, which is consistent with existing studies (see Popp 1997).

As Figure 8.9 shows, the optimal carbon taxes for both the induced-innovation (R&DICE) and the substitution (DICE) cases are virtually identical. The reason they are so similar is that the optimal carbon tax equals the present discounted value of the damage from emissions. The optimal carbon tax is basically the same in the two approaches because there is so little impact on the path of climate change.

The welfare impacts of the two policies are shown in Table 8.2. As noted above, the welfare interpretation in the R&DICE model is somewhat tricky because of the externality. In calculating the welfare gains, we have assumed that the research reallocated to the carbon-energy sector has an offset in other sectors that just balances the externality. That is, a US$1 increase in research in the energy sector leads to a US$4 increase in the value of production in the energy sector (in present value terms); but the increase in value in the energy sector is just offset

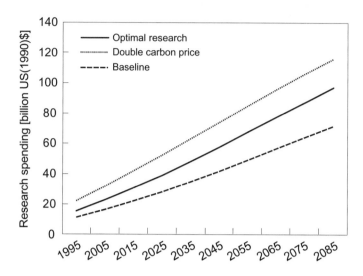

Figure 8.8. Research and Development on Carbon Energy under Different Scenarios.

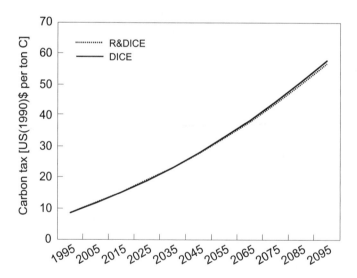

Figure 8.9. Optimal Carbon Taxes in DICE and R&DICE Models.

by a US$4 decrease in the value of research in the non-energy sector. This would be the case if either the total stock of research is fixed or the gap between social and private returns is the same in different sectors and the induced effect in other sectors is equal and opposite to the induced effect in the carbon-energy sector.

Table 8.2. Welfare Implications of Alternative Models.

	Difference from no controls [billions of US(1990)$]
DICE	
No controls	0
Optimal controls	585
R&DICE	
No induced innovation	0
Induced innovation	238

Under these assumptions, the introduction of induced innovation increases the discounted value of world consumption by US$238 billion. This is about 40 percent of the welfare gain from substitution policies, which is $585 billion (1990 US dollars). Basically, induced innovation has less welfare potential because the quantitative impact on research is smaller than that on substitution.

8.4.2 Discussion

A major surprise of the present study is how little impact induced innovation has in the near term. A simple example illustrates why this outcome is, on reflection, quite sensible. For this calculation, we take 10 periods (100 years) starting in the year 2000. We calculate that the optimized level of R&D on carbon energy with no climate-change policies would average about 2 percent of total expenditures on carbon energy. This would average about US$30 billion per year.

If we internalize climate damages into carbon-energy prices, the wholesale carbon price (including the carbon tax) would increase about US$22 per ton on average, an increase of 24 percent. Firms operating in the energy sector would see the cost of their energy fuels rise under a carbon tax and would increase their carbon-energy R&D. Suppose that the R&D/output ratio remains constant. Assuming for simplicity that output is unchanged, a 24 percent increase in the price would therefore induce a 24 percent increase in R&D. (The actual result over the first 10 periods is 37 percent, which is above the static calculation because the carbon tax is rising.) This would lead to an increase in R&D of US$7.7 billion per year on average.

Next consider the returns on this investment. For this calculation, we ignore depreciation. The return on this investment is 14.2 percent per year in carbon emissions savings (four times the rate of return on capital). The annual costs of carbon energy during the first 10 periods are US$1,490 billion per year. With these costs and an average increase of US$7.7 billion in R&D, a 14.2 percent annual rate of return would lead to a decline in the carbon/output ratio of $0.142 \times$ US$7.7/$1,490 = 0.073$ percent per year.

The decline in the carbon/output ratio over these 10 decades induced by the higher carbon-energy prices in the R&DICE model is 0.126 percent per year.

Table 8.3. Sensitivity Analysis for Major Parameters: Impact upon Emissions Decline Rate, Percent per Year, First 10 Periods.

	R&D/sales ratio (percent)			
	1	2	3	
Externality factor				
2	–0.03	–0.06	–0.09	
4	–0.06	–0.13	–0.19	
8	–0.13	–0.25	–0.38	
Decline rate:				
In baseline				–0.62
Necessary to stabilize				
emissions from 2000 to 2100				–0.91

The divergence between the simple calculation (0.073) and the R&DICE model (0.126) arises because the estimated R&DICE elasticity of research with respect to carbon prices is 1.5 rather than 1.0. With this correction, the simple example in the previous paragraph yields an estimated decline of 0.109 percent per year. The small remaining difference between the simple example and the R&DICE model results from depreciation of technology and timing factors.

8.4.3 Sensitivity analysis

One of the difficulties with using complex numerical methods like the R&DICE model is that it is always difficult to know how sensitive the conclusions are to alternative specifications or alternative values of the key parameters. We have undertaken extensive sensitivity analyses in previous studies and limit the analysis here to asking how sensitive the key results are to alternative parameter values. Some preliminary tests indicated that the key important parameters are the baseline level of R&D and the externality factor (e.g., the ratio of social to private returns to research).

   Table 8.3 shows the impact of alternative specifications of these two key parameters on the rate of decline in the emissions/output ratio. For these calculations, we have used a reduced-form model rather than the full model because of the complexity of re-estimating the parameters and re-running the model. The central case is shown in the middle of the three-by-three matrix of numbers. It states that under the central assumptions, the decline in the emissions/output ratio from induced innovation is 0.13 percent per year over the first 10 periods. Looking at the columns to the left and right, we see what the impact of a higher or lower R&D/sales ratio is. If we have underestimated the current level of R&D on energy by 50 percent, then the impact of induced innovation will be to reduce the emissions rate by 0.19 rather than 0.13 percent per year. Reading vertically, we see that if the externality factor is 8 rather than 4, then the induced decline rate is 0.25 percent rather than 0.13 percent. Neither of these seems particularly likely. There is no evidence that the externality ratio is a factor of 8, and we believe that the R&D/output ratio of 2 percent is likely to be high when developing countries are included.

The last two lines of Table 8.3 put these sensitivity analyses into perspective. The first shows the projected decline in the baseline emissions/output ratio. The last line shows our estimate of the required *further* reduction in the emissions/output ratio necessary to stabilize emissions over the twenty-first century. According to our estimates, in the central case, induced innovation is likely to reduce emissions about one-eighth of the amount necessary to stabilize emissions. Even in the most optimistic case at the lower right, induced innovation will produce less than half the amount required to stabilize emissions.

8.5 The US Program to Promote Climate-Friendly Technologies

8.5.1 The US program for induced innovation

In response to the Kyoto Protocol, the United States, under the Clinton administration, proposed both an undefined future program for emissions limitations and a current technological initiative to improve the carbon efficiency of energy use. The rationale of the program was described in the 1999 *Economic Report of the President*, which stated that

> [A]chieving climate change policy goals will require improving the energy efficiency of the economy. In addition to policies affecting energy prices directly, the Administration believes that a strong argument can be made for policies to stimulate innovation and diffusion through R&D and appropriate fiscal incentives. (Council of Economic Advisers 1999:206)

The administration's budget for fiscal year 2000 included US$4 billion in a Climate Change Technology Initiative (CCTI). This program was designed to provide fiscal incentives "offsetting in part the appropriability problems associated with [climate change] R&D" (Council of Economic Advisers 1999:207).

The CCTI for 2000 contained two kinds of programs:

- *Tax credits for low-carbon energy and energy efficiency* amounting to US$3.6 billion (although only US$0.4 billion is actually in the budget proposal). These credits are for fuel-efficient vehicles, solar equipment, energy-efficient building equipment and homes, biomass, and other similar investments.

- *R&D investments* for the major sectors (buildings, industry, transportation, and electric power) totaling US$1.4 billion.

These programs are not currently under consideration by the George W. Bush administration, but given the propensity of governments to propose these kinds of measures, it will be useful to analyze their impacts.

8.5.2   Analysis

It is difficult to analyze the CCTI because of the lack of programmatic detail and because we know relatively little about the rate of return on federal investments in this area. We will estimate the impact based on the following assumptions:

- The tax credits are generally ones to promote energy saving rather than carbon saving. We therefore assume that they reduce the energy/output ratio rather than the carbon/output ratio. For this purpose, we assume that one-half of the expenditures go to reducing carbon and one-half go to reducing capital and labor.

- The tax credits are assumed to have an impact of US$1 of increased investment per US$1 of federal outlay. This is a common finding in studies of the impact of investment tax credits. For modeling purposes, these are assumed to have no impact on induced innovation.

- Expenditures on federal R&D are assumed to have a social rate of return that is one-half that of the return on private R&D. While there is no firm evidence on this issue, the rationale for this assumption is that there is no "bottom line" check on federal research and that there are many examples of persistent waste in federal programs. Even with this assumption, the overall social rate of return is well above that of conventional investment.

- We assume that the investments continue at the same real level indefinitely.

8.5.3   Results

We have analyzed the CCTI by augmenting the baseline investment and R&D expenditures in the R&DICE model according to the assumptions made above. The investments are assumed to be carbon reducing with a normal rate of return, while the R&D component is assumed to be carbon saving with a rate of return one-half that of private R&D. We then solve the model with the augmented R&D expenditures.

The first result is that the US technology program is unlikely to have a major impact on emissions. According to our estimates, the program will reduce emissions by about 30 million tons per year over the next half century. The emissions reduction is far less than would be induced by optimal substitution or induced innovation (see Figure 8.10).

Figure 8.11 compares the impact on emissions from the US technology program with the emissions reductions called for under the Kyoto Protocol.[10] The upper line in Figure 8.11 gives the total emissions reductions (which would be required were there no emissions trading), while the lower two trading lines show the estimated US emissions reductions with Annex I trading and with global trading. The US technology program would lead to only a small fraction of the total US emissions reductions called for under the Kyoto Protocol—around 10 percent

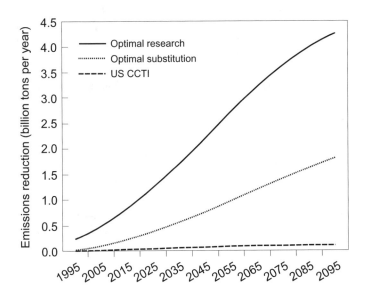

Figure 8.10. Comparison of Impact of US Technology Program with That of Other Policies on Global Emissions.

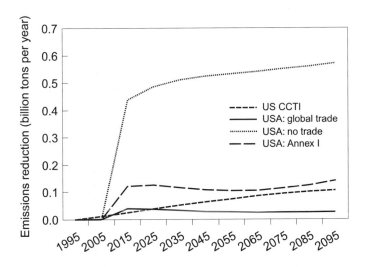

Figure 8.11. Comparison of Impact of US Technology Program with Emissions Reductions for the United States under the Kyoto Protocol.

of scheduled total emissions reductions over the next century. However, the technology program is approximately the same order of magnitude as the emissions reductions under the trading programs. The reduction in global temperature from the CCTI after a century is estimated to be 0.006°C.

We should emphasize that these estimates are highly speculative and are likely to be quite optimistic. They rely on the assumption that the technology program would be effective in promoting efficient investments in low-carbon energy and technologies and in energy efficiency. Should the program turn out to be a wasteful pork-barrel program or become hung up on exotic pet projects, the emissions reductions would be much smaller or even negligible.

One final point is that the technology initiative is a relatively high-cost program. The discounted cost is US$150 per discounted ton of carbon (in 1990 prices), whereas we estimate that the same carbon benefits can be attained with an average cost of less than US$10 per ton carbon. This figure overestimates the costs, however, because approximately one-half the benefits would come in reducing non-carbon inputs.

8.6   Conclusions

The primary conclusion of the present study is that induced innovation is likely to be a less powerful factor in implementing climate-change policies than substitution. A rough calculation is that the reductions in CO_2 concentrations and in global mean temperature resulting from induced innovation are approximately one-half those resulting from substitution.

All of this is a major surprise, at least to this author. But in hindsight, the results are intuitive. The primary reason for the small impact of induced innovation on the overall path of climate change is that the investments in inventive activity are too small to make a major difference unless the social returns to R&D are much larger than the already-supernormal returns. R&D is about 2 percent of output in the energy sector, while conventional investment is close to 30 percent of output. Even with supernormal returns, the small fraction devoted to research is unlikely to outweigh other investments.

What are the likely shortcomings in the model? One issue comes from the assumption of no substitution. If in fact substitution occurs, this would tend to reduce the value of induced innovation because the value of induced innovation would decline as non-carbon energy is substituted for carbon energy. A second issue lies in the oversimplified specification of the underlying technology. The R&DICE model assumes that carbon emissions and carbon energy are joint products, with no alternative fuels available. Hence, the possibility of using low-carbon fuels instead of high-carbon fuels is omitted. A richer specification of the technology would allow induced innovation in technologies that lower the carbon intensity of carbon-energy fuels. The difficulties of modeling this richer technology set are formidable, however, because of the lack of any data on which to base the innovation-possibility function. It is difficult to guess at the potential bias here, as the impact of induced innovation in this broader set of fuels could go either way.

If these results are confirmed for other models, they suggest our climate-change problem is unlikely to be solved by induced innovation. The interpretation is tricky, however. It definitely does not say that technology is unimportant. A technological miracle of a low-cost and environmentally benign backstop technology may indeed by just around the corner and put the climate-change worry-warts out of business. Rather, the proper interpretation is that we should not look to regulatory stringency or high emissions taxes as a way of forcing inventors to solve our global environmental problems. Necessity may indeed by the mother of invention, but there is limited payoff in inducing the delivery through regulations or high taxes.

Appendix 8.A
Definition of Variables for R&DICE Model

The variables are defined as follows. We have omitted certain obvious equations such as the capital balance equation and the definition of per capita consumption. In the listing, t always refers to time ($t = 1990, 2000, \dots$).

Exogenous variables

$A(t)$ = level of technology
$E^L(t)$ = land-use CO_2 emissions
$L(t)$ = population at time t, also equal to labor inputs
$O(t)$ = radiative forcings of exogenous greenhouse gases
$Q^*(t)$ = gross world output
t = time

Parameters

β = marginal atmospheric retention ratio of CO_2 emissions
γ = elasticity of output with respect to capital
δ_K = rate of depreciation of the capital stock
λ = feedback parameter in climate model (inversely related to temperature-sensitivity coefficient)
$\rho(t)$ = pure rate of social time preference
θ_i = parameters of climate damage function
Ψ_i = parameters of innovation-possibility function
η_i = parameters of climate system
ϕ_i = parameters of carbon cycle

Endogenous variables

$C(t)$ = total consumption
$c(t)$ = per capita consumption
$D(t)$ = damage from greenhouse warming

$E(t)$ = carbon-energy inputs, also equal to industrial CO_2 emissions
$F(t)$ = radiative forcing from all greenhouse gas concentrations
$I(t)$ = investment
$K(t)$ = capital stock
$M_{AT}(t)$ = mass of CO_2 in atmosphere (deviation from pre-industrial level)
$M_{UP}(t)$ = mass of CO_2 in upper reservoir (deviation from pre-industrial level)
$M_{LO}(t)$ = mass of CO_2 in lower ocean (deviation from pre-industrial level)
$Q(t)$ = gross world product
$T_{UP}(t)$ = atmospheric temperature relative to base period
$T_{LO}(t)$ = deep ocean temperature relative to base period
$U(t) = U[c(t), L(t)]$ = utility of consumption
$\sigma(t)$ = industrial carbon/output ratio
$\Omega(t)$ = output scaling factor due to damages from climate change

Policy variables

$R(t)$ = R&D inputs into the carbon-energy sector (in induced-innovation approach)
$\mu(t)$ = control rate (in substitution approach)

Appendix 8.B
Equations of DICE-99 and R&DICE model

Objective function
$$\max_{\{c(t)\}} = \sum_t U[c(t), L(t)]\rho(t) = \sum_t L(t)\{\log[c(t)]\}\rho(t) \qquad (0)$$
Production function
$$Q(t) = \Omega(t)A(t)K(t)^\gamma L(t)^{1-\gamma} \qquad \text{(1-DICE)}$$
$$Q(t) = \Omega(t)Q^*(t) \qquad \text{(1-R\&DICE)}$$

Consumption equals output less investment and production costs
$$C(t) = Q(t) - I(t) \qquad \text{(2-DICE)}$$
$$C(t) = Q(t) - I(t) - 4R(t) - p(t)E(t) \qquad \text{(2-R\&DICE)}$$

Capital accumulation
$$K(t) = (1 - \delta_K)K(t-1) + 10I(t-1) \qquad \text{(3-DICE)}$$
$$K(t) \text{ exogenous} \qquad \text{(3-DICE)}$$

Emissions
$$E(t) = [1 - \mu(t)]\sigma(t)A(t)K(t)^\gamma L(t)^{1-\gamma} + E^L(t) \qquad \text{(4-DICE)}$$
$$E(t) = \sigma(t)Q^*(t) + E^L(t) \qquad \text{(4-R\&DICE)}$$

Carbon intensity
$$\sigma(t) = \sigma(0)\exp(-\upsilon t) \qquad \text{(5-DICE)}$$
$$[d\sigma(t)/dt]/\sigma(t) = \Psi_1 R(t)^{\Psi_2} - \Psi_3 \qquad \text{(5-R\&DICE)}$$
Per capita consumption
$$c(t) = C(t)/L(t) \qquad (6)$$

Carbon cycle

$$M_{AT}(t) = 10 \times E(t) + \phi_{11}M_{AT}(t-1) - \phi_{12}M_{AT}(t-1)+ \qquad (7a)$$
$$\phi_{21}M_{UP}(t-1)$$

$$M_{UP}(t) = \phi_{22}M_{UP}(t-1) + \phi_{12}M_{AT}(t-1) - \phi_{21}M_{UP}(t-1)+ \qquad (7b)$$
$$\phi_{32}M_{LO}(t-1) - \phi_{23}M_{UP}(t-1)$$

$$M_{LO}(t) = \phi_{33}M_{LO}(t-1) - \phi_{32}M_{LO}(t-1) + \phi_{23}M_{UP}(t-1) \qquad (7c)$$

Radiative forcings

$$F(t) = \eta\{\log[M_{AT}(t)/M_{AT}^*]/\log(2)\} + O(t) \qquad (8)$$

Climate equations

$$T_{UP}(t) = T_{UP}(t-1) + \eta_1\{F(t) - \lambda T_{UP}(t-1)- \qquad (9a)$$
$$\eta_2[T_{UP}(t-1) - T_{LO}(t-1)]\}$$

$$T_{LO}(t) = T_{LO}(t-1) + \eta_3[T_{UP}(t-1) - T_{LO}(t-1)] \qquad (9b)$$

Damage equation

$$D(t) = \theta_1 T(t) + \theta_2 T(t)^2 \qquad (10)$$

Damage parameter

$$\Omega(t) = [1 - b_1\mu(t)^2]/[1 + D(t)] \qquad (11)$$

For those familiar with the DICE model, the major changes are in Equations (B.2), (B.5), (B.11), and (B.12) of the original model (see Nordhaus 1994).

Acknowledgments

This research was supported by the National Science Foundation and the US Department of Energy. The author would like to thank Joseph Boyer, Larry Goulder, Arnulf Grübler, Adam Jaffe, Dale Jorgenson, Charles Kolstad, Alan Manne, Nebojsa Nakicenovic, Leo Schrattenholzer, Robert Stavins, Paul Waggoner, and John Weyant for their thoughtful comments.

Notes

1. See also the extensive survey in Jorgenson (1996).
2. For a recent overview, see Ruttan (2001) as well as the related chapters in this volume.
3. One example of incorporating technological change in policy analysis is the work of Jorgenson and his colleagues; see, for example, Jorgenson and Wilcoxen (1991). See also Goulder and Schneider (1996) and Goulder and Mathai (1997). This literature is surveyed by Weyant and Olavson (1999) and by Clarke and Weyant (Chapter 12 in this volume). A major problem with current approaches is that in these models there is no explicit linkage between innovation or technological change and inventive inputs.
4. For a discussion of this point and its implications, see Nordhaus (1984).
5. This approach is inspired by Arrow (1962).
6. Run-of-the-mill innovations are those that are sufficiently small that the price of the innovation is determined by the limit price rather than the monopoly price.
7. There is an extensive literature on this subject. In particular, Nelson (1959); Griliches (1973); Mansfield *et al.* (1975, 1977); National Science Board (1977); Nathan

Associates (1978); Mansfield (1980, 1985); Pakes (1985); Jaffe (1986); Levin *et al.* (1987); Jaffe *et al.* (1993); and Hall (1995).

8.   Strictly speaking, the two assumptions are equivalent if the rate of return is constant and if there is no depreciation of knowledge. Given the actual data and assumptions in the DICE model, using the precisely correct shadow price would lead to a time-varying factor between three and four. However, it does not seem fruitful to calculate the exact time-varying factor for the different depreciation rates.

9.   Full documentation of the RICE-99 and DICE-99 models is provided in Nordhaus and Boyer (2000). The spreadsheet version of the model is available on the Internet at http: //www.econ.yale.edu/~nordhaus/home page/dicemodels.htm.

10.   The estimates for the impact from the Kyoto Protocol are from Nordhaus and Boyer (1999).

References

Arrow, K.J., 1962, Economic welfare and the allocation of resources for invention, in R. Nelson, ed., *The Rate and Direction of Inventive Activity*, Princeton University Press for National Bureau of Economic Research, Princeton, NJ, USA.

Council of Economic Advisers, 1999, *Economic Report of the President*, US Government Printing Office, Washington, DC, USA.

Goulder, L.H., and Mathai, K., 1997, Optimal CO_2 Abatement in the Presence of Induced Technological Change, working paper, June, Stanford University, Stanford, CA, USA.

Goulder, L.H., and Schneider, S.L., 1996, Induced Technological Change, Crowding Out, and the Attractiveness of CO_2 Emissions Abatement, working paper, October, Stanford University, Stanford, CA, USA.

Griliches, Z., 1973, Research expenditures and growth accounting, in M. Brown, ed., *Science and Technology in Economic Growth*, Wiley, New York, NY, USA.

Griliches, Z., 1986, Productivity, R&D, and basic research at the firm level in the 1970s, *American Economic Review*, **76**:141–154.

Griliches, Z., 1998, *Patents and Technology*, Harvard University Press, Cambridge, MA, USA.

Hall, B.H., 1995, The private and social returns to research and development, in B. Smith and C. Barfield, eds, *Technology, R&D, and the Economy*, Brookings Institution, Washington, DC, USA.

Jaffe, A., 1986, Technological opportunity and spillover of R&D: Evidence from firms' patents, profits, and market value, *American Economic Review*, **76**:984–1001.

Jaffe, A., Trajenberg, M., and Henderson, R., 1993, Geographical localization of knowledge spillovers as evidence by patent citations, *Quarterly Journal of Economics*, **108**(3):577–598.

Jorgenson, D., 1996, Technology in growth theory, in J.C. Fuhrer and J. Sneddon Little, eds, *Technology and Growth, Conference Proceedings*, Federal Reserve Bank of Boston, Boston, MA, USA, pp. 45–77.

Jorgenson, D., and Wilcoxen, P.J., 1991, Reducing US carbon dioxide emissions: The cost of different goals, in J.R. Moroney, ed., *Energy, Growth, and the Environment*, JAI Press, Greenwich, CT, USA.

Kennedy, C., 1962, The character of improvements and technical progress, *Economic Journal*, **72**(288):899–911.

Levin, R., Klevorick, A., Nelson, R., and Winter, S.G., 1987, Appropriating the returns from industrial research and development, *Brookings Papers on Economic Activity*, 3:783–820.

Lucas, R.E., Jr., 1986, On the mechanics of economic development, *Journal of Monetary Economics*, **22**(January):3–42.

Mansfield, E., 1980, Basic research and productivity increase in manufacturing, *American Economic Review*, **70**:863–873.

Mansfield, E., 1985, How fast does new industrial technology leak out?, *Journal of Industrial Economics*, **34**:217–223.

Mansfield, E., 1996, Macroeconomic policy and technological change, in J.C. Fuhrer and J. Sneddon Little, eds, *Technology and Growth, Conference Proceedings*, Federal Reserve Bank of Boston, Boston, MA, USA.

Mansfield, E., *et al.*, 1975, *Social and Private Rates of Return from Industrial Innovations*, National Technical Information Services, Washington, DC, USA.

Mansfield, E., *et al.*, 1977, Social and private rates of return from industrial innovations, *Quarterly Journal of Economics*, **91**:221–240.

National Science Board, 1977, *Science Indicators, 1977*, National Science Foundation, Washington, DC, USA.

Nathan Associates, 1978, Net Rates of Return on Innovation, Report to the National Science Foundation, NSF, Washington, DC, USA.

Nelson, R.R., 1959, The simple economics of basic scientific research, *Journal of Political Economy*, **67**:297–306.

Nordhaus, W.D., 1984, Comments on Griliches, *Brookings Papers on Economic Activity*, 1984:1.

Nordhaus, W.D., 1994, *Managing the Global Commons: The Economics of the Greenhouse Effect*, MIT Press, Cambridge, MA, USA.

Nordhaus, W.D., and Boyer, J., 1999, Requiem for Kyoto: An economic analysis of the Kyoto Protocol, *Energy Journal* (special issue: Costs of the Kyoto Protocol): pp. 93–103.

Nordhaus, W.D., and Boyer, J., 2000, *Warming the World: Economic Models of Global Warming*, The MIT Press, Cambridge, MA, USA.

Pakes, A., 1985, On patents, R&D, and the stock market rate of return, *Journal of Political Economy*, **93**:390–409.

Popp, D., 1997, Induced Innovation in the Energy Industries, Ph.D. dissertation, Yale University, New Haven, CT, USA.

Ramsey, F.P., 1928, A mathematical theory of saving, *The Economic Journal*, (December):543–559.

Romer, P., 1990, Endogenous technological change, *Journal of Political Economy*, **98**(October, Part 2):S71–S102.

Ruttan, V.W., 2001, *Technology, Growth, and Development*, Oxford University Press, New York, NY, USA.

von Weizsäcker, C.C., 1966, Tentative notes on a two-sector model with induced technical progress, *Review of Economic Studies*, **33**(95):245–251.

Weyant, J., and Olavson, T., 1999, Issues in modeling induced technological change in energy, environment, and climate policy, *Environmental Modeling and Assessment*, **4**:67–85.

Chapter 9

Optimal CO_2 Abatement in the Presence of Induced Technological Change

Lawrence H. Goulder and Koshy Mathai

9.1 Introduction

Over the past decade considerable efforts have been directed toward evaluating alternative policies to reduce the atmospheric accumulation of greenhouse gases, particularly carbon dioxide (CO_2). Initial assessments tended to disregard interconnections between technological change and CO_2-abatement policies, treating the rate of technological progress as autonomous—that is, unrelated to policy changes or associated changes in relative prices. Recently, however, several researchers have emphasized that CO_2 policies and the rate of technological change are connected: to the extent that public policies affect the prices of carbon-based fuels, they affect incentives to invest in research and development (R&D) aimed at bringing alternative fuels on-line earlier or at lower cost. Such policies may also prompt R&D oriented toward the discovery of new production methods that require less of *any* kind of fuel. Moreover, climate policies can affect the growth of knowledge through impacts on learning by doing: to the extent that these policies affect producers' experience with alternative energy fuels or energy-conserving processes, they can influence the rate of advancement of knowledge.

Thus, through impacts on patterns of both R&D spending and learning by doing, climate policy can alter the path of knowledge acquisition. What does this connection imply for the design of CO_2-abatement policy? In particular, how do the optimal timing and extent of carbon emissions abatement, as well as the optimal time path of carbon taxes, change when we recognize the possibility of induced technological change (ITC)?

Policy makers and researchers are divided on these questions. Wigley *et al.* (1996) have argued that the prospect of technological change justifies relatively

This chapter was originally published in the *Journal of Environmental Economics and Management*, Volume 39:1–38, 2000. ©Academic Press.

little current abatement of CO_2 emissions: better to wait until scientific advances make such abatement less costly. In contrast, Ha-Duong *et al.* (1996) have maintained that the potential for ITC justifies relatively more abatement in the near term, in light of the ability of current abatement activities to contribute to learning by doing. Yet another possibility is that ITC makes it optimal to increase abatement in all periods and thus achieve more ambitious overall targets for atmospheric CO_2 concentrations.

In addition to these disagreements on the optimal profile for abatement, there are differing viewpoints concerning the optimal carbon tax profile. One frequently heard claim is that ITC justifies a higher carbon tax trajectory than would be optimal in its absence. The argument is that in the presence of ITC, carbon taxes not only confer the usual environmental benefit by forcing agents to internalize the previously external costs from CO_2 emissions, but also yield the benefit of faster innovation, particularly in the supply of alternative energy technologies.[1] Another possibility, however, is that with technological progress, a lower carbon tax profile is all that is needed to achieve desired levels of abatement.

This chapter aims to clarify the issues underlying these controversies. We derive analytical expressions characterizing the optimal paths of emission abatement and carbon taxes under different specifications for the channels through which knowledge is accumulated, considering both *R&D-based* and *learning-by-doing-based* knowledge accumulation. We examine each of these specifications under two different optimization criteria: the *cost-effectiveness* criterion of obtaining by a specified date and thereafter maintaining, at minimum cost, a given target for the atmospheric CO_2 concentration; and the *benefit–cost* criterion under which we also choose the optimal concentration target, thus obtaining the path of carbon abatement that maximizes the benefits from avoided climate damages net of abatement costs.[2] To gain a sense of plausible magnitudes, we also perform illustrative numerical simulations.

Our analysis is in the spirit of two studies by Nordhaus (1980a, 1980b)—the first to obtain analytical expressions for the optimal carbon tax trajectory—as well as more recent work by Farzin and Tahvonnen (1996), Farzin (1996), Peck and Wan (1996), Sinclair (1994), and Ulph and Ulph (1994). Our chapter also complements work by Nordhaus (1994), Nordhaus and Yang (1996), and Peck and Teisberg (1992, 1994), in which numerical methods are used to obtain the optimal carbon abatement and carbon tax profiles under different exogenous technological specifications.[3] Another related paper is by Kolstad (1996), who solves numerically for optimal emissions trajectories in the presence of endogenous learning. Kolstad's paper differs from ours, however, in that it focuses on learning that reduces uncertainty about CO_2-related damages, rather than on learning that improves abatement technologies and thus reduces abatement costs. Finally, our chapter is closely related to the previously mentioned studies by Wigley *et al.* (1996) and Ha-Duong *et al.* (1996), as well as to working papers by Grubb (1997), Goulder and Schneider (1999), and Nordhaus (1997) that analyze the implications of ITC for optimal climate policy.

The present investigation differs from each of these other studies in three ways. First, it derives *analytical* results revealing the impact of ITC on optimal time profiles for carbon taxes and carbon abatement. Second, it considers, in a unified framework, two channels for knowledge accumulation (R&D and learning by doing) and two policy criteria (cost-effectiveness and benefit–cost). In the model, policy makers (or the social planner) choose optimal paths of carbon abatement and carbon taxes, taking into account the impact of these taxes on technological progress and future abatement costs. Finally, it employs both analytical and numerical methods in an integrated, complementary way.

The analytical model reveals (contrary to what some analysts have suggested) that the presence of ITC generally implies a lower time profile of optimal carbon taxes.[4] The impact of ITC on the optimal abatement path varies. When knowledge is gained through R&D investments, the presence of ITC justifies shifting some abatement from the present to the future. However, when knowledge reflects learning by doing, the impact on the timing of abatement is analytically ambiguous.

When the government employs the benefit–cost policy criterion, the presence of ITC justifies greater overall (cumulative) abatement than would be warranted in its absence. This does not imply, however, that abatement rises in every period: when knowledge accumulation results from R&D expenditure, the presence of ITC implies a reduction of near-term abatement efforts, despite the overall increase in the scale of abatement over time.

Our numerical simulations reinforce the qualitative predictions of the analytical model. The quantitative impact on overall costs and optimal carbon taxes can be quite large in a cost-effectiveness setting but typically is much smaller under a benefit–cost policy criterion. The weak effect on the tax rate in the benefit–cost case reflects the relatively trivial impact of ITC on optimal CO_2 concentrations, associated marginal damages, and (hence) the optimal tax rate. As for the optimal abatement path, the impact of ITC on the timing of abatement is very weak, but the effect on overall abatement (which applies in the benefit–cost case) can be large, especially when knowledge is accumulated via learning by doing.

The rest of the chapter is organized as follows. Section 9.2 lays out the analytical model and applies it to the case in which the policy criterion is cost-effectiveness. Section 9.3 applies the model to the situation in which policy makers employ the broader benefit–cost criterion. Section 9.4 presents and interprets results from numerical simulations and includes a sensitivity analysis. The final section offers conclusions and indicates directions for future research.

9.2   Optimal Policy under the Cost-Effectiveness Criterion

In this section we consider optimal abatement when the policy criterion is cost-effectiveness. We assume that producers are competitive and minimize costs. Let $C(A_t, H_t)$ be the economy's (aggregate) abatement-cost function, where A_t is abatement at time t and H_t is the stock of knowledge—or alternatively, the level of technology—at time t. We assume $C_A(\cdot) > 0$, $C_{AA}(\cdot) > 0$, $C_H(\cdot) < 0$, and $C_{AH}(\cdot) < 0$. The last two properties imply that increased knowledge reduces, respectively, total and marginal costs of abatement. Later on, we consider the implications of alternative assumptions. We also allow for the possibility that costs may depend on the relative amount of abatement (A_t/E_t^0) rather than the absolute level (A_t). In this case baseline emissions become an argument of the cost function. For expositional simplicity, however, we usually suppress E_t^0 from the cost function in the main text.

9.2.1   Technological change via R&D

The Problem and Basic Characteristics of the Solution

Within our cost-effectiveness analysis, we consider two modes of knowledge accumulation. The first specification assumes that to accumulate knowledge, the economy must devote resources to R&D. We refer to this as the CE_R specification (where "R" indicates that the channel for knowledge accumulation is R&D). The planner's problem is to choose the time paths of abatement and R&D investment that minimize the costs of achieving the concentration target.[5] Formally, the optimization problem is

$$\min_{A_t, I_t} \int_0^\infty (C(A_t, H_t) + p(I_t)I_t)e^{-rt}dt \tag{9.1}$$

$$\text{s. t.} \quad \dot{S}_t = -\delta S_t + E_t^0 - A_t \,, \tag{9.2}$$

$$\dot{H}_t = \alpha_t H_t + k\Psi(I_t, H_t) \,, \tag{9.3}$$

$$\text{given } S_0, \ H_0$$

$$\text{and} \quad S_t \leq \bar{S} \ \ \forall t \geq T \,, \tag{9.4}$$

where A_t is abatement, I_t is investment in knowledge (i.e., R&D expenditure), S_t is the CO_2 concentration, H_t is the knowledge stock, $p(\cdot)$ is the real price of investment resources, r is the interest rate, δ is the natural rate of "removal" of atmospheric CO_2, E_t^0 is baseline emissions, α_t is the rate of *autonomous* technological progress, and k is a parameter that, as discussed below, indicates whether *induced* technological progress is present as well.

Expression (9.1) indicates that the objective is to minimize the discounted sum of abatement costs and expenditure on R&D into the infinite future. Expression (9.2) states that the change in the CO_2 concentration is equal to the contribution from current emissions $(E_t^0 - A_t)$ net of natural removal (δS_t).[6]

Expression (9.3) describes the evolution of the knowledge stock (H_t), that is, the process of technological change. In the case where $k > 0$, the planner will choose an optimal profile for investments in R&D consistent with meeting the concentration target at minimum cost. These R&D investments (I_t) serve to increase the stock of knowledge (H_t) through the knowledge-accumulation function ($\Psi(\cdot)$). This profile of investment can be interpreted as the *additional* R&D investment that the optimal carbon tax would induce on the part of competitive firms. Thus, the $k > 0$ case is the induced technological change, or "ITC" case. We also consider the situation where $k = 0$ and there is no possibility of ITC because the connection between additional R&D investments and the stock of knowledge is severed. We call this the no-ITC or "NITC" case. In much of this chapter, we will compare optimal abatement and carbon tax paths between the ITC and NITC cases. In addition to ITC, we also allow for *autonomous* technological change at the rate α_t: even if there were no climate policies in place, it seems reasonable to assume that some technological progress would still occur. There may be nonclimate reasons for such progress, such as a desire on the part of firms to economize on costly fuel inputs.

Expression (9.4) shows that the target CO_2 concentration, $\bar{S}$, must be met by time T and maintained after that point in time. We assume $p(\cdot)$ is nondecreasing in I_t; that is, the average cost of R&D investment increases with the level of R&D. This captures in reduced form the idea that there is an increasing opportunity cost (to other sectors of the economy) of employing scientists and engineers to devise new abatement technologies.[7] We also assume that the knowledge-accumulation function $\Psi(\cdot)$ has the following properties: $\Psi(\cdot) > 0$, $\Psi_I(\cdot) > 0$, and $\Psi_{II}(\cdot) < 0$.

The current-value Hamiltonian associated with the optimization problem for $t < T$ is[8]

$$
\begin{aligned}
\mathcal{H}_t &= -(C(A_t, H_t) + p(I_t)I_t) - \tau_t(-\delta S_t + E_t^0 - A_t) \\
&\quad + \mu_t(\alpha_t + k\Psi(I_t, H_t)),
\end{aligned}
\tag{9.5}
$$

where $-\tau_t$ and μ_t are the shadow values of S_t and H_t, respectively. For $t \geq T$, however, we must form the following Lagrangian:

$$
\mathcal{L}_t = \mathcal{H}_t + \eta_t(\bar{S} - S_t).
\tag{9.6}
$$

From the maximum principle, we obtain a set of first-order conditions, assuming an interior solution, as well as costate equations, state equations, and transversality conditions. Two key equations are

$$
C_A(\cdot) = \tau_t
\tag{9.7}
$$

and

$$
\dot{\tau}_t = \begin{cases} (r + \delta)\tau_t & \text{for } t < T \\ (r + \delta)\tau_t - \eta_t & \text{for } t \geq T. \end{cases}
\tag{9.8}
$$

In this problem, $-\tau_t$ is the shadow value of a small additional amount of CO_2 at time t. This shadow value is negative, since CO_2 is a "bad" from the policy maker's perspective. Thus τ_t represents the (positive) shadow *cost* of CO_2 or, equivalently, the benefit from an incremental amount of abatement (a small reduction in the CO_2 concentration). In a decentralized competitive economy in which all other market failures have been corrected, the optimal carbon tax is τ_t, the shadow cost of CO_2. By Equation (9.7), this is equal to the marginal abatement cost at the optimal level of abatement. Equation (9.7) states that abatement should be pursued to the point at which marginal cost equals marginal benefit, while Equation (9.8) states that the optimal carbon tax grows at the rate $(r + \delta)$ (at least for points in time up until T).[9] The two equations together imply that in an optimal program, the discounted marginal costs of abatement must be equal at all points in time (up to T), where the appropriate discount rate is $(r + \delta)$.[10] In Appendix 9.A we demonstrate that this corresponds to an optimal abatement profile that slopes upward over time (whether or not there is ITC) so long as baseline emissions are not declining "too rapidly."

Implications of ITC

We now examine the effect of ITC on abatement costs and on the optimal carbon tax and abatement profiles. We do this by considering the significance of a change in the parameter k. As mentioned above, the case of $k = 0$ corresponds to a scenario with no ITC (the NITC scenario), while positive values of k imply the presence of induced technological change (the ITC scenario). Our analysis will focus on incremental increases in k from the point $k = 0$.[11]

If (as is assumed) $C_H(\cdot) < 0$, then additional knowledge is clearly valuable (i.e., the multiplier μ is positive). When $k = 0$, all of the growth in knowledge is due to the autonomous term, and knowledge grows at the rate α_t. In contrast, for strictly positive values of k, the planner will find it optimal for society to accumulate at least some additional knowledge, assuming an interior solution.[12] This additional knowledge causes a decrease[13] in optimized costs to a degree dictated by μ_t. Thus, as would be expected, the introduction of the ITC option lowers the costs of achieving the given concentration target.

Next we examine the impact of introducing ITC on the optimal time profiles of abatement and carbon taxes. Differentiating Equation (9.7) with respect to k and rearranging, we obtain

$$\frac{dA_t}{dk} = \frac{d\tau_t/dk - C_{AH}(\cdot)dH_t/dk}{C_{AA}(\cdot)} . \tag{9.9}$$

For the moment, assume that the first term in the numerator is zero; that is, that ITC has no impact on the shadow cost of CO_2. Under this assumption, we are left only with what we shall refer to as the *knowledge-growth effect*: to the extent that knowledge has increased as a result of ITC $(dH_t/dk > 0)$ and has thus reduced marginal abatement costs $(-C_{AH}(\cdot) > 0)$, abatement tends to rise.[14]

The knowledge-growth effect is represented in Figure 9.1 by the upward pivot of the abatement profile from the initial path 1 to path 2. At time 0 path 2 coincides with the initial path because knowledge is initially fixed at H_0: there can be no knowledge-growth effect at time 0.[15] The distance between paths 1 and 2 grows over time, representing the fact that the knowledge-growth effect becomes larger over time. This follows from the fact that there is no depreciation of knowledge in our model: whatever additional knowledge was induced by ITC at time t remains

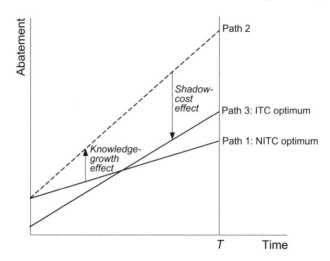

Figure 9.1. Knowledge-Growth and Shadow-Cost Effects (Drawn for CE_R Model).

at time $t' > t$, and there might have been a further increment to knowledge at this later time.

Note that path 2 involves more abatement in every period than does the first path. Given that the same $\bar{S}$ constraint holds and that the initial path satisfied this constraint, path 2 clearly cannot be optimal. Path 2 was obtained under the assumption that the introduction of ITC had no impact on the shadow cost of CO_2. In fact, however (as shown in Appendix 9.A), under the maintained assumption that $C_{AH}(\cdot) < 0$, the shadow cost of CO_2 at all points in time decreases in magnitude in the presence of ITC: $d\tau_t/dk \leq 0 \; \forall t$. The basic explanation for this *shadow-cost effect* is as follows. If we are armed with the potential to develop new technologies rapidly through ITC, the prospect of being given an additional amount of CO_2 at time t and still being expected to meet the $\bar{S}$ constraint by time T is less worrisome than it would be if we had only autonomously advancing abatement technologies at our disposal.[16] Note that since the optimal carbon tax is the shadow cost of CO_2, it follows that the presence of ITC lowers carbon taxes.

This result contradicts the notion that the induced-innovation benefit from carbon taxes justifies a higher carbon tax rate. Figure 9.2 demonstrates our result heuristically by offering a static representation of this dynamic problem.[17] Cost-effective abatement (depicted in the upper panel) is achieved by a carbon tax set equal to the marginal abatement cost (MC) at the desired level of abatement. Technological progress causes the MC curve to pivot down, thus implying a lower optimal tax: it now takes a lower tax to yield the same amount of abatement. Note that this result depends on the assumption that *marginal* abatement costs are lowered by technological progress; that is, $C_{AH} < 0$. It is possible to conceive of new technologies that involve higher marginal abatement costs

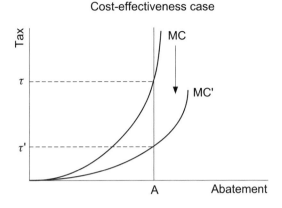

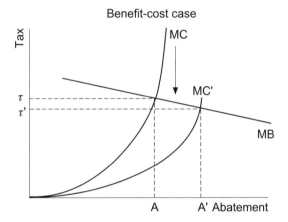

Figure 9.2. Optimal Climate Policy in a Static Setting.

but that are nonetheless attractive because of lower fixed (and overall) abatement costs; however, this seems to be an unusual case.

Now we return to our analysis of the impact of ITC on abatement. The shadow-cost effect, reflected in the first term of the numerator in Equation (9.9), shows up in Figure 9.1 as the downward shift from path 2 to path 3. The shift is not parallel: as shown in Appendix 9.A, tax rates at later points in time fall by greater absolute amounts than do early taxes, in such a way as to preserve the carbon tax growth rate at $(r + \delta)$. The downward shift is of a magnitude such that path 3 lies neither completely above nor completely below path 1: if it did, it would imply either overshooting or undershooting the constraint $\bar{S}$, which is likely to be suboptimal.[18] Together, the knowledge-growth and shadow-cost effects imply a new optimal abatement path that is steeper than the initial one: abatement is postponed from the present into the future.[19] Intuitively, ITC reduces the cost of future abatement relative to current abatement and thus makes postponing (some) current abatement more attractive. Thus, in a cost-effectiveness setting with R&D-based technological change, our analysis supports the claim of Wigley *et al.* (1996) that future technological developments justify a more gradual approach to abatement.

At any given time t, we cannot be sure whether abatement rises or falls— this depends on whether the knowledge-growth effect or the shadow-cost effect dominates at that particular moment. But we can say something definite about abatement at time 0. Because knowledge is initially fixed at H_0, only the shadow-cost effect comes into play at time 0:

$$\frac{dA_0}{dk} = \frac{d\tau_0/dk}{C_{AA}(\cdot)} \leq 0 . \tag{9.10}$$

Thus, *initial* abatement weakly declines as a result of ITC.

These results depend on our assumption that $C_{AH}(\cdot) < 0$—that knowledge lowers marginal abatement costs. However, the possibility that $C_{AH}(\cdot) > 0$ cannot be ruled out. In this case (which we find somewhat implausible), ITC raises marginal costs, but presumably lowers total costs through greatly reduced sunk costs. Under these circumstances, the shadow-cost effect is positive and the presence of ITC raises the optimal carbon tax. The net effect of an increase in k on abatement at any arbitrary time t is (again) ambiguous, but initial abatement unambiguously rises.

Summary

Our results to this point are as follows. First, the solution to the cost-minimization problem (for any value of k) involves carbon taxes that rise over time at the rate $(r + \delta)$ for $t < T$, and that grow more slowly, and perhaps even decline, afterward. Second, the optimal abatement profile is upward sloping for $t < T$, as long as baseline emissions are not too steeply declining. Finally, assuming that ITC reduces marginal (and total) abatement costs, opening the ITC option causes optimized costs to fall, makes the entire carbon tax path fall (and by an equal proportion at all t), and causes initial abatement to fall and later abatement to rise. Table 9.1 summarizes the results regarding the implications of ITC.

9.2.2 Technological change via learning by doing

The Problem and Basic Characteristics of the Solution

Here we analyze a variant of the model presented above; now abatement itself yields improvements in technology. This is the "CE_L" model, where the "L" refers to learning by doing. The optimization problem is now

$$\min_{A_t} \int_0^\infty C(A_t, H_t)e^{-rt}dt \tag{9.11}$$

$$\text{s. t. } \dot{S}_t = -\delta S_t + E_t^0 - A_t , \tag{9.12}$$

$$\dot{H}_t = \alpha_t H_t + k\Psi(A_t, H_t) , \tag{9.13}$$

Table 9.1. Summary of Analytical Results.

Policy criterion	Channel for technological change	Impacts of induced technological change on optimal solution	
		Tax path*	Abatement path
Cost-effectiveness	R&D	Falls by an equal proportion at all t	A_0 falls and later A rises; "steepening" of path
	Learning by doing	Falls by an equal proportion at all t	Ambiguous effect on A_0 and on slope of path
Benefit–cost	R&D	Falls	A_0 falls; cumulative abatement rises; "steepening" of path
	Learning by doing	Falls	Ambiguous effect on A_0; cumulative abatement increases

*Assuming damages are a convex function of the CO_2 concentration.

given S_0, H_0

and $S_t \leq \bar{S} \;\; \forall t \geq T$. (9.14)

This problem is virtually the same as the CE_R model of the previous section, except for a change in the $\Psi(\cdot)$ function: induced knowledge growth is now a function of the current level of abatement rather than R&D investment. Equivalently, current knowledge depends on cumulative abatement, which is regarded as a measure of experience. The first-order condition for abatement is now given by

$$C_A(\cdot) - \mu_t k \Psi_A(\cdot) = \tau_t . \tag{9.15}$$

Equation (9.15) states that the marginal benefit of abatement (τ_t, the value of the implied reduction in the CO_2 concentration) should equal the gross marginal cost of abatement ($C_A(\cdot)$) adjusted for the cost reduction associated with the learning by doing stemming from that abatement ($\mu_t k \Psi_A(\cdot)$).

As in the CE_R model, the optimal carbon tax here is equal to τ_t.[20] Since the costate equation for τ_t is unchanged from before, we can refer to earlier results and conclude that the carbon tax grows at the rate $(r + \delta)$ for $t < T$, and that it grows more slowly, and perhaps declines, thereafter.

Although the CE_R and CE_L models are similar as regards the carbon tax path, they differ with respect to the characteristics of the optimal abatement path. In particular, it is no longer unambiguously true that the abatement path is positively sloped for $t < T$, even in the case in which baseline emissions are growing over time, This is demonstrated in Appendix 9.A; the basic reason is that the cost reduction due to learning by doing does not necessarily grow with time.[21]

Implications of ITC

Now consider what happens to the optimal tax and abatement paths when we introduce ITC, that is, increase k from the point $k = 0$. As before, our assumption that $C_H(\cdot) < 0$ directly implies that the presence of ITC causes optimized costs to fall. Perhaps more substantively, under the assumption that $C_{AH}(\cdot) < 0$, we again find that the presence of ITC causes the shadow cost of the CO_2 concentration, and thus the optimal carbon tax, to decline (and increasingly so for higher t).

To analyze the impact of ITC on the abatement path, we differentiate Equation (9.15) with respect to k. Evaluating this at $k = 0$ yields

$$\frac{dA_t}{dk} = \frac{d\tau_t/dk + \mu_t \Psi_A(\cdot) - C_{AH}(\cdot)dH_t/dk}{C_{AA}(\cdot)}.$$ (9.16)

As in the CE_R model, we observe the negative shadow-cost effect $(d\tau_t/dk)$ and the positive knowledge-growth effect $(-C_{AH}(\cdot)dH_t/dk)$. In our learning-by-doing specification, however, the presence of ITC has an additional, positive effect on abatement which we term the *learning-by-doing effect* $(\mu_t \Psi_A(\cdot))$. This effect reflects the fact that in the learning-by-doing specification, there is an additional marginal benefit (the learning) from abatement. Other things being equal, this further marginal benefit justifies additional abatement. Thus, under this specification the presence of ITC has three effects on abatement, one negative (the shadow-cost effect), and two positive (the knowledge-growth and learning-by-doing effects).[22] The net effect is ambiguous. Even at time 0, when the knowledge-growth effect does not come into play, we are still left with the opposing shadow-cost and learning-by-doing effects:

$$\frac{dA_o}{dk} = \frac{d\tau_o/dk + \mu_0 \Psi_A(\cdot)}{C_{AA}(\cdot)}.$$ (9.17)

Thus, in contrast to the CE_R model, the presence of ITC no longer implies unambiguously that initial abatement will fall. If the learning-by-doing effect is strong enough, initial abatement rises. (This in fact happens in most of the numerical simulations presented in Section 9.4.)[23] These results offer partial support for Ha-Duong *et al.*'s (1996) claim that because of learning by doing, ITC justifies higher initial abatement. Higher initial abatement may be justified, but this is not always the case.

Summary

We can summarize our results for the CE_L case as follows. The optimal carbon tax grows at the rate $(r + \delta)$ for $t < T$, but will grow more slowly, and perhaps even decline, after that. The slope of the optimal abatement path is of ambiguous sign throughout (unless we are in an NITC scenario, in which case abatement unambiguously rises over time, at least for $t < T$, if baseline emissions do not decline too rapidly). The presence of ITC lowers optimized costs and makes the entire carbon tax path fall by an equal proportion at all $t < T$. The impact on initial abatement is analytically ambiguous. These effects of ITC are noted in Table 9.1.

9.3 Optimal Policy under the Benefit–Cost Criterion

We now analyze optimal tax and abatement profiles in a benefit–cost framework. No longer is there an exogenously given concentration target; rather, the object is to minimize the sum of abatement costs, investment costs (in the R&D model), *and damages from CO_2* over an infinite horizon.

9.3.1 Technological change via R&D

The Problem and Basic Characteristics of the Solution

In the R&D-based specification (hereafter referred to as the BC_R model), the problem is

$$\min_{A_t, I_t} \int_0^\infty (C(A_t, H_t) + p(I_t)I_t + D(S_t))e^{-rt} dt \tag{9.18}$$

$$\text{s. t. } \dot{S}_t = -\delta S_t + E_t^0 - A_t, \tag{9.19}$$

$$\dot{H}_t = \alpha_t H_t + k\Psi(I_t, H_t), \tag{9.20}$$

given S_0, H_0,

where $D(S_t)$ is the damage function, assumed to have the following properties: $D'(\cdot) > 0$ and $D''(\cdot) > 0$. This is not completely uncontroversial. Although most would accept that damages are a convex function of climate change, it is also widely felt [see, e.g., Dickinson and Cicerone (1986)] that climate-change forcing is a *concave* function of changes in the atmospheric CO_2 concentration. Thus our $D(\cdot)$ function—relating damages to concentrations—could be concave. The shape of the damage function is critical in predicting the impacts of ITC.

The current-value Hamiltonian associated with the optimization problem is

$$\begin{aligned} \mathcal{H}_t &= -(C(A_t, H_t) + p(I_t)I_t + D(S_t)) - \tau_t(-\delta\dot{S}_t + E_t^0 - A_t) \\ &+ \mu_t(\alpha_t + k\Psi(I_t, H_t)). \end{aligned} \tag{9.21}$$

From the maximum principle, assuming an interior solution, we obtain a set of necessary conditions, of which the most important to us are

$$C_A(\cdot) = \tau_t \tag{9.22}$$

and

$$\dot{\tau}_t = (r + \delta)\tau_t - D'(\cdot). \tag{9.23}$$

As before, $-\tau_t$ is the negative shadow value of a small additional amount of CO_2. Hence τ_t again represents the marginal benefit of abatement, Equation (9.22) thus states that abatement should be pursued up to the point at which marginal cost equals marginal benefit. Equation (9.23) can be integrated, using the relevant transversality condition as a boundary condition, to obtain

$$\tau_t = \int_t^\infty D'(S_s)e^{-(r+\delta)(s-t)}ds \, . \tag{9.24}$$

Equation (9.24) states that the shadow cost of an increment to the CO_2 concentration equals the discounted sum of marginal damages that this increment would inflict over all future time. Alternatively, the marginal benefit from incremental CO_2 abatement equals the discounted sum of the avoided damages attributable to such abatement.

As in the CE_R model, the optimal carbon tax is equal to τ_t, and thus, by Equation (9.22), to the marginal abatement cost at the optimum. Using Equation (9.24), we demonstrate in Appendix 9.A that in the BC_R model the optimal carbon tax may either rise or fall over time. This contrasts with the results from the cost-effectiveness models, in which the optimal carbon tax rose at the rate $(r + \delta)$ (at least for $t < T$). The reason for the ambiguity is that although there is a tendency for the BC_R shadow cost to grow at the rate $(r + \delta)$, there is also a tendency for it to decline over time because an extra amount of CO_2 later on would inflict marginal damages over a shorter time horizon. Appendix 9.A shows that given the convex damage function which we think reasonable, a sufficient condition ensuring that the tax path slopes upward is that the optimized path of CO_2 also slopes upward.

Given rising taxes and a baseline emissions path that rises (or at least does not fall too rapidly), we can also demonstrate that optimal abatement rises; otherwise, the slope of the abatement path is ambiguous. (See Appendix 9.A for details.)

Implications of ITC

As before, the presence of ITC leads to lower optimized total costs (where these now include CO_2-related damages as well as abatement and investment costs). Just as before (and as proved in Appendix 9.A), if we assume that knowledge reduces the marginal costs of abatement, the shadow cost of CO_2 declines in the presence of ITC: $d\tau_t/dk \leq 0$. The intuition is similar to what it was in both the CE_R and CE_L models. Technological progress makes marginal abatement cheaper. Thus, when R&D investments are capable of yielding advanced technologies ($k > 0$), the prospect of being given an additional amount of CO_2 is less worrisome than it would be if we knew only more primitive abatement technologies would be available ($k = 0$). Since the optimal carbon tax is the shadow cost of CO_2, the presence of ITC lowers carbon taxes (the shadow-cost effect).[24] In this benefit–cost setting, we can also appeal to another piece of intuition. When ITC gives us the prospect of having more advanced technologies at our disposal, it makes sense that we would aim for more ambitious CO_2 concentration targets. Given a convex damage function, this would imply that marginal damages would be lower in the ITC world, and thus, by Equation (9.24), optimal carbon taxes would be lower as well.

The result that ITC lowers optimal carbon taxes is perhaps surprising. Earlier, in a cost-effectiveness setting, we dismissed the claim that the presence of ITC should increase optimal taxes by appealing to a simple static graph; this graph showed that with ITC, it took a lower tax to achieve the same required level of abatement. But one might still have expected that in the broader, benefit–cost setting, if technology progressed sufficiently, it would make sense to *increase* the amount of abatement, and thus the optimal tax would increase.

The lower panel of Figure 9.2 heuristically indicates that this notion is incorrect, at least under the assumption that the damage function is convex in the CO_2 concentration. The optimal amount of abatement and the optimal carbon tax are given by the intersection of the upward-sloping MC curve and the downward-sloping marginal abatement benefit (MB) curve.[25] If the MC curve were to pivot downward as a result of technological progress, the optimal amount of abatement would increase, but the optimal carbon tax would fall because we move to a lower point on the marginal benefit (marginal damage) curve.

If the damage function were linear, implying a flat marginal damage schedule, then the MC pivot would increase the optimal amount of abatement while leaving the optimal carbon tax unchanged. On the other hand, if damages were concave in the CO_2 concentration, then the MB curve would be upward sloping, and it is possible to envision a scenario in which a technology-driven fall in the MC schedule could actually increase the optimal carbon tax.[26]

Next we examine the implications of increasing k. Using the same approach as in the CE_R model, we obtain

$$\frac{dA_t}{dk} = \frac{d\tau_t/dk - C_{AH}(\cdot)dH_t/dk}{C_{AA}(\cdot)} . \tag{9.25}$$

Once again, the impact of ITC on abatement at time t is ambiguous because the shadow-cost effect and the knowledge-growth effect oppose each other. At time 0, however, the stock of knowledge is fixed at H_0, and thus only the shadow-cost effect comes into play:

$$\frac{dA_0}{dk} = \frac{d\tau_0/dk}{C_{AA}(\cdot)} \leq 0 . \tag{9.26}$$

Thus initial abatement declines as a result of ITC (although this result is reversed if $C_{AH}(\cdot) < 0$).

In the cost-effectiveness analyses, where we had a fixed terminal constraint, $\bar{S}$, we knew that over the entire time horizon, cumulative abatement would be approximately the same under both ITC and NITC scenarios.[27] This implied that the shadow-cost and knowledge-growth effects would approximately balance each other out over the entire horizon; in terms of Figure 9.1, the area under path 1 would roughly approximate the area under path 3.

In the benefit–cost framework, however, this is not the case. As demonstrated in Appendix 9.A, the overall scale of abatement over the entire infinite horizon increases; that is to say, the knowledge-growth effect dominates the shadow-cost effect on average. Since CO_2 inflicts environmental damages, it seems reasonable that in the presence of ITC, which makes emissions abatement cheaper, the optimal balance of benefits and costs of emissions abatement would be struck at a higher level of abatement (on average) than would be optimal in the NITC scenario. This result is perhaps not very surprising.[28] Perhaps a more unexpected result is that *initial* abatement still falls, no matter how "large" or powerful the ITC option. Equation (9.25) indicates that this occurs because there is no separate analytical term representing an upward shift of abatement at all points in time. Rather, the increased scale of abatement is reflected completely in the steepening of the abatement path resulting from the interaction between the knowledge-growth and shadow-cost effects.

Summary

We have obtained the following main results for the BC_R case. First, the optimal carbon tax may either rise or fall over time, but if concentrations of CO_2 are increasing through time, then (given a convex damage function) the optimal carbon tax rises as well. Optimal abatement may either rise or fall over time, but, as long as baseline emissions are not falling too rapidly over time, it will rise if the carbon tax is rising. Second, as summarized in Table 9.1, introducing the ITC option lowers optimized net costs and causes the entire carbon tax path to fall. Initial abatement also falls, but cumulative abatement over the entire horizon rises; hence ITC implies a "steeper" abatement path.

9.3.2 Technological change via learning by doing

Finally, we examine a learning-by-doing specification in a benefit–cost framework (the BC_L model).

The Problem and Basic Characteristics of the Solution

The optimization problem is now

$$\min_{A_t} \int_0^\infty (C(A_t, H_t) + D(S_t))e^{-rt}dt \qquad (9.27)$$

$$\text{s. t. } \dot{S}_t = -\delta S_t + E_t^0 - A_t \,, \qquad (9.28)$$

$$\dot{H}_t = \alpha_t H_t + k\Psi(A_t, H_t) \,, \qquad (9.29)$$

given S_0, H_0 .

Thus, CO_2-related damages are part of the minimand, and abatement effort contributes to the change in the knowledge stock. The optimality conditions are the same as in the BC_R model, with one major change: the first-order condition for abatement is now

$$C_A(\cdot) - \mu_t k\Psi_A(\cdot) = \tau_t \,, \qquad (9.30)$$

which is just as it was in the CE_L model [Equation (9.15)].

As in the BC_R model, the slope of the carbon tax path is ambiguous (though it will be positive if the optimized CO_2 concentration rises over time, given convex damages). Thus the slope of the abatement path is ambiguous as well.

Implications of ITC

As always, the presence of ITC lowers overall optimized costs as well as the profile of optimal carbon taxes (assuming $C_H(\cdot) < 0$ and $C_{AH}(\cdot) < 0$). The impact of ITC on abatement is given by [29]

$$\frac{dA_t}{dk} = \frac{d\tau_t/dk + \mu_t \Psi_A(\cdot) - C_{AH}(\cdot)dH_t/dk}{C_{AA}(\cdot)}. \tag{9.31}$$

As in the CE_L model, ITC has three effects on abatement: the negative shadow-cost effect $(d\tau_t/dk)$, the positive learning-by-doing effect $(\mu_t \Psi_A(\cdot))$, and the positive knowledge-growth effect $(-C_{AH}(\cdot)dH_t/dk)$. The net effect on abatement at an arbitrary point in time t (including $t = 0$) is clearly ambiguous. At $t = 0$, in particular, the knowledge-growth effect drops out, leaving the negative shadow-cost and positive learning-by-doing effects

$$\frac{dA_0}{dk} = \frac{d\tau_0/dk + \mu_0 \Psi_A(\cdot)}{C_{AA}(\cdot)} \tag{9.32}$$

and we cannot even claim that *initial* abatement declines unambiguously.

Although the components of the analysis here are the same as in the corresponding cost-effectiveness case, their overall impact is different. In the CE_L model, since the overall scale of abatement was approximately the same in both ITC and NITC scenarios, all three effects roughly balanced out over the entire horizon. In contrast, in this benefit–cost case, cumulative abatement increases.[30] Thus, on average the learning-by-doing and knowledge-growth effects dominate the shadow-cost effect.

Summary

The key results are as follows. The slope of the optimal carbon tax path is ambiguous. However, if the optimized CO_2 concentration rises (given a convex damage function), the tax rises as well. These results are similar to those in the CE_L model. Moreover, the slope of the optimal abatement path is of ambiguous sign throughout (unless we are in an NITC world with rising taxes and baseline emissions that are not declining too rapidly). As noted in Table 9.1, although introducing the ITC option makes overall costs and the entire carbon tax path fall, it could lead to an increase in initial abatement. Furthermore, cumulative abatement over the entire time horizon increases.

9.4 Numerical Simulations

Here, we perform numerical simulations to gauge the quantitative significance of our results. We postulate functional forms and parameter values and solve for optimal paths. We then conduct a sensitivity analysis to assess the robustness of our results. The numerical simulations reinforce our analytical findings and also point up several striking empirical regularities, as discussed below. We begin this section by describing the choice of functional forms and the methods used to calibrate the various parameters of the model. We then present and discuss the numerical results.

9.4.1   Functional forms and parameter values

The numerical model is solved at 10-year intervals, with the year 2000 as the initial year. Although the planner's time horizon is infinite, we actually simulate over 41 periods (400 years) and impose steady-state conditions in the last simulated period. This enables us to project forward the values of this last period and thereby determine benefits and costs into the infinite future.[31]

The CO_2 concentration in 2000 is taken to be 360 parts per million by volume (ppmv), following the projections of the Intergovernmental Panel on Climate Change (IPCC, 1995). Baseline emissions for the period 2000 to 2100 roughly follow the IPCC's IS92(a) central scenario. After that time, we adopt a hump-shaped profile that peaks at 26 gigatons of carbon (GtC) in 2125 and flattens out to 18 GtC by 2200.[32]

In the analytical section, we assumed for expositional clarity that CO_2 in the atmosphere is naturally "removed" at a constant exponential rate. In the numerical simulations, we adopt Nordhaus' (1994) slightly more complex and realistic model, which applies short-term and long-term removal rates to the flow and "stock" of emissions, respectively:[33]

$$\dot{S}_t = \beta(E_t^0 - A_t) - \delta(S_t - PIL) , \tag{9.33}$$

where $\beta = 0.64$,

and $\delta = 0.008$.

Thus, only 64 percent of current emissions actually contribute to the augmentation of atmospheric CO_2, and the portion of the current CO_2 concentration in excess of the preindustrial level (PIL = 278 ppmv) is removed naturally at a rate of 0.8 percent per annum.

For our benefit–cost simulations, we need to specify a CO_2 damage function. We assume this function to be quadratic and, following Nordhaus (1994), who reviewed damage estimates from a number of studies, calibrate the remaining scale parameter so that a doubling of the atmospheric CO_2 concentration implies a loss of 1.33 percent of world output each year. Thus we have

$$D(S_t) = M_D S_t^{\alpha_D} , \tag{9.34}$$

where $M_D = 0.0012$,

and $\alpha_D = 2$.

The functional form assumed for the abatement-cost function is

$$C(A_t, H_t) = M_C \frac{A_t^{\alpha_{C1}}}{(E_t^0 - A_t)^{\alpha_{C2}}} \frac{1}{H_t} \,. \tag{9.35}$$

This form has the properties assumed in the analytical model, including the feature that knowledge lowers marginal abatement costs ($C_{AH}(\cdot) < 0$). It also has the property that marginal costs tend to infinity as abatement approaches 100 percent of baseline emissions.[34] We choose the parameters M_C, α_{C1}, and α_{C2} to meet the requirements that (1) a 25 percent emissions reduction in 2020 should cost between 0.5 and 4 percent of global gross domestic product (GDP)[35] and (2) the present value (at a 5 percent discount rate) of global abatement costs for reaching $S_t = 550$ ppmv by 2200 (in an NITC world) should be roughly US$600 billion (Manne and Richels 1997). The parameter values that best meet these requirements are $M_C = 83$, $\alpha_{C1} = 3$, and $\alpha_{C2} = 2$, but calibration of the cost function remains an area of considerable uncertainty, and sensitivity analyses in this respect are particularly important.

Following estimates common in the literature,[36] we take the rate of autonomous technological progress to be 0.5 percent per annum: $\alpha_t = 0.005$. The knowledge accumulation function exhibits the properties discussed in the analytical section and is given, in the R&D simulations, by

$$\Psi(I_t, H_t) = M_\Psi I_t^\gamma H_t^\phi \,, \tag{9.36}$$

where $M_\Psi = 0.0022$,

$\gamma = 0.5$,

and $\phi = 0.5$.

H_0, the initial knowledge stock, is normalized to unity. In the learning-by-doing simulations, the knowledge accumulation function is the same, with A_t replacing I_t. The function we use is fairly standard in the endogenous growth literature.[37] γ is chosen to be 0.5 to indicate diminishing returns to R&D investment,[38] while ϕ, which dictates the intertemporal knowledge spillover, is set to 0.5, a central value of the range typically seen in the literature. As ϕ is positive, it indicates that knowledge accumulation today makes future accumulation easier. This is the "standing-on-shoulders" case which has been used, for example, by Nordhaus (1997). It contrasts with the case where $\phi < 0$, which implies a limited pool of ideas which are slowly "fished out"—current knowledge accumulation makes future accumulation more difficult. M_Ψ is calibrated so that the cost savings from ITC are approximately 30 percent in the CE_R model. This is consistent with Manne and Richels (1992), who compare the costs of carbon abatement under different assumptions about technological progress.[39]

We assume that the price of investment funds is

$$p(I_t) = I_t \,. \tag{9.37}$$

Thus the average cost of R&D investment increases with scale; as mentioned earlier, this captures the idea that drawing scientists away from R&D in other sectors involves increasing costs. Following Manne and Richels (1997), we take the discount rate to be 5 percent.[40] Finally, we model the NITC cases by setting $k = 0$ and the ITC cases by setting $k = 1$.

9.4.2   Central cases

CE_R Simulation

In the cost-effectiveness cases (CE_R and CE_L), the concentration target ($\bar{S}$) is 550 ppmv, which must be reached by 2200. This scenario has received considerable attention in policy discussions. We first consider results for the CE_R case, both with and without ITC. The upper-left panels of Figures 9.3 and 9.4 depict, respectively, the optimal abatement and carbon tax paths in this case.

Abatement. As predicted by the analytical model, the optimal abatement paths slope upward for most of the horizon until 2200, the year in which the constraint is first imposed.[41] Figure 9.3 shows that the presence of ITC leads to a slightly "steeper" abatement profile, with less abatement during the first 125 years and more abatement after that. However, the effect of ITC on abatement is almost imperceptible. The minuteness of this "abatement-timing effect" is noteworthy, particularly in light of the fact that ITC lowers discounted average costs of abatement by 30 percent. The sensitivity analysis below will show that the weakness of ITC's abatement-timing effect is robust to different parameter specifications.

Carbon Tax. The upper-left panel of Figure 9.4 shows that the optimal carbon tax starts at a few dollars per ton and grows exponentially. Although not evident from the figure alone, the tax grows at the rate $(r + \delta)$, just as predicted by the analytical model. While ITC's impact on abatement was extremely small, its effect on the optimal tax is pronounced. The presence of ITC lowers the optimal carbon tax path at all points in time up to 2200 by about 35 percent, roughly in line with the 30 percent cost savings mentioned earlier.

CE_L Simulation

The upper-right panels of Figures 9.3 and 9.4 depict the abatement and tax paths, respectively, for the CE_L case. The results here are broadly similar to those in the CE_R case just discussed. Again the optimal abatement paths slope upward,[42] the optimal carbon tax rises at the rate $(r + \delta)$, and the presence of ITC causes a slight steepening of the abatement path and a sizable downward shift in the tax path. Here ITC implies a reduction in total costs of about 39 percent and a comparable (41 percent) lowering of the optimal carbon tax path.

Some differences between the CE_R and CE_L cases deserve mention. First, under learning by doing, the presence of ITC has an even smaller effect on optimal abatement timing than it does under R&D. This makes sense because the basic tendency toward postponing some abatement from the present to the future is offset in the CE_L case by the learning-by-doing effect, which prompts more

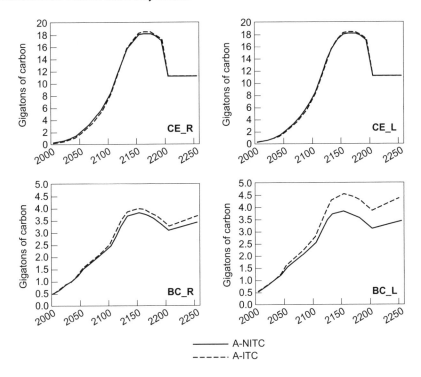

Figure 9.3. Optimal Abatement Paths.

abatement now in order to accumulate experience-based knowledge. (In fact, this learning- by-doing effect is large enough to cause initial abatement to rise in the CE_L simulation.) A second difference is that ITC has a larger impact on taxes and costs in the CE_L case than it does in the CE_R case. This reflects the fact that under ITC based on learning by doing, technological progress comes about as a "free" by-product of abatement, rather than as a result of costly expenditures on R&D.

BC_R Simulation

We now turn to the benefit–cost cases. The lower-left panels of Figures 9.3 and 9.4 depict the optimal abatement and tax paths, respectively, in the BC_R model.

Abatement. The analytical model indicated that as long as taxes were rising and baseline emissions were not declining "too rapidly," the abatement path would rise. In our simulations, abatement rises over the 2000–2150 interval and falls after that, largely matching the pattern of baseline emissions. As shown in Figure 9.3, in the presence or absence of ITC, there is much less abatement here than in the CE cases (note the different scales used on the vertical axes). Correspondingly, the CO_2 concentration in 2200 that results from the optimal abatement path is above 800 ppmv, considerably higher than the 550 ppmv imposed

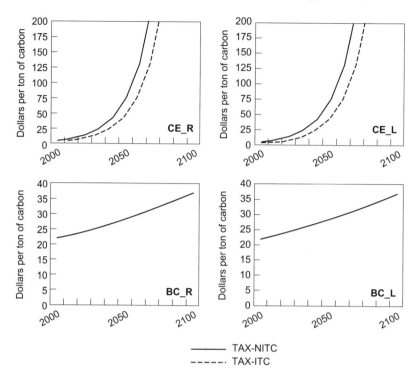

Figure 9.4. Optimal Carbon Tax Paths.

in the CE simulations. These differences imply that the 550 ppmv target in the cost-effective analysis—a target given much attention in policy discussion—is too stringent from an efficiency point of view (given the parameters used here for the cost and damage functions). The presence of ITC implies a slight increase in the overall scale of abatement and steepening of the abatement path. Nonetheless, initial abatement falls (though only slightly). These outcomes all square with the predictions of the analytical model.

Carbon Tax. The lower left-panel of Figure 9.4 shows that the optimal carbon tax profile is roughly linear in this simulation. This contrasts with the exponential shape in the CE simulations and conforms to the analysis of Section 9.3. Recall that the shadow cost of the CO_2 concentration (i.e., the carbon tax) is given by the sum of marginal damages that a small additional amount of CO_2 would cause into the infinite future, discounted at the rate $(r + \delta)$. Although the shadow cost tends to rise at the rate $(r + \delta)$, this is offset by the fact that as time goes on, less time remains over which the incremental amount of CO_2 can inflict marginal damages. The combination of these two effects produces a linear carbon tax profile.

Again in striking contrast to the CE simulations, the impact of ITC on the optimal carbon tax path is virtually imperceptible in the BC_R central case. There are two reasons for the difference. First, as suggested by Figure 9.2, the adjustment due to ITC is in both the quantity (that is, abatement) and price (tax) dimensions; in the CE cases, in contrast, adjustment can only occur in the price dimension because of the constraint on the terminal CO_2 concentration. In our central case, the marginal damage curve is very flat over the relevant range. As a result, nearly all of the adjustment to ITC in the BC cases comes via changes in the level of abatement.[43]

The second reason is more subtle and relates to the fact that the $\bar{S}$ constraint imposed in the cost-effectiveness scenarios is too stringent, as indicated by the fact that $\bar{S}$ is significantly lower than the optimal concentration for the year 2200 that emerges from the benefit–cost simulation. Consequently, levels of abatement are generally much higher in the cost-effectiveness cases, which implies that the potential gains from improved technology are also higher. Thus the benefits of accumulating knowledge are higher in the cost-effectiveness cases than in the benefit–cost settings. This implies that the downward pivot of the MC curve and associated impact on the carbon tax are larger in these cases.

Finally, the presence of ITC has an extremely small (2 percent) impact on average costs of abatement in the benefit–cost cases; as before, this contrasts with the result in the cost-effectiveness cases. Given that ITC in the benefit–cost cases has a small impact on the carbon tax, and in fact on the entire marginal abatement cost schedule, it should not be surprising that it has a correspondingly small impact on average costs. Similarly, the ITC-induced percentage increase in the net benefits of climate policy is very small.[44]

In a working paper circulated contemporaneously with early drafts of this chapter, Nordhaus (1997) independently obtained the result that, in a benefit–cost context, the presence of ITC has an imperceptible impact on the optimal carbon tax and on the net benefits from carbon abatement policy. Our BC results conform to Nordhaus', although, as discussed below, we obtain different results under alternative parameterizations.[45]

BC_L Simulation

Finally, we consider the BC_L case. The results for this case are displayed in the lower-right panels of Figures 9.3 and 9.4. The effect of ITC on abatement (Figure 9.3) is similar to the effect in the BC_R case, although the impact is somewhat more pronounced.[46] Initial abatement rises, indicating that the learning-by-doing effect outweighs the shadow-cost effect. Once again, the presence of ITC has a virtually imperceptible impact on the optimal carbon tax path, average costs per unit of abatement, and net benefits. The explanation is the same as that given in the BC_R case.

9.4.3   Sensitivity analysis

Here we examine the sensitivity of the numerical results to changes in key parameters. For each of the four models, we examine six sets of variants of the central case. Table 9.2 presents summary statistics describing, in each variant, the percentage impact that ITC has on the abatement profile, the cumulative amount of abatement over the 2000–2200 period, the terminal CO_2 concentration, the tax profile, and overall costs per unit of abatement. We report the impacts on abatement and taxes in the years 2000, 2050, and 2200 (or 2190).[47] For the benefit–cost cases, we also report the percentage impact of ITC on the net benefits of optimal climate policy relative to a zero-abatement baseline.

A higher discount rate (case 2a) reduces the importance of future benefits or costs relative to current ones. Since the costs of ITC are borne today, whereas the benefits are spread more uniformly through time, a higher discount rate tends to reduce the net benefits from ITC. This means that there will be less knowledge accumulation, which implies that the abatement-timing effect (the pivoting of the abatement path) is smaller. The reduced attractiveness of ITC implies, in the benefit–cost cases, that there will be a smaller impact on the overall scale of abatement as well. The opposite results hold under a lower discount rate (case 2b).

The next variant involves changes either in the constraint on year-2000 concentrations (in the cost-effectiveness cases) or in the parameters of the damage function (in the benefit–cost cases). In a cost-effectiveness setting, a tighter concentration constraint (case 3a) enforces greater overall abatement and therefore entails higher marginal costs of abatement. This confers higher value to ITC in terms of greater cost savings. The reverse applies when the constraint is more lax (case 3b).[48] In the benefit–cost simulations, case 3a imposes higher curvature on the damage function.[49] Thus the marginal damage function is steeper in the relevant range. As a result, ITC has a larger impact on the optimal carbon tax and there is less impact on quantity (abatement). Case 3b imposes a linear damage function, so that the marginal damage schedule is perfectly flat. In this case, there is no impact on the optimal carbon tax profile. All of the adjustment occurs in quantity (abatement). Even in this case, however, the effect on the abatement levels is quite small because the marginal cost curve is quite steep. Finally, in case 3c (applicable only in the EC simulations), we introduce a concave damage function, so that the marginal damage curve is upward sloping in abatement (but still flatter than the MC curve). As expected, taxes *rise* in this case.

In cases 4a, 4b, and 4c we alter the curvature of the cost function such that the *marginal* cost curve is, respectively, more convex than in the central case, strictly linear (less convex), and concave (much less convex). For the BE models, M_C in cases 4a, 4b, and 4c is calibrated such that the optimal tax path in the NITC world coincides with what it was in the central case.[50] For the BC models, M_C is calibrated such that the total amount of NITC abatement over the 2000–2200 period stays constant. Changes in the curvature of the cost function are most important to the results of the BC simulations. In case 4a, the marginal cost function is convex and steep in the relevant range. As a result, the downward pivot of this function caused by ITC does not greatly alter the optimal levels of abatement. In contrast, when the marginal cost function is linear or concave (cases 4b and 4c) and much flatter in the relevant range, ITC has pronounced effects on optimal abatement. Indeed, in the concave case, ITC implies a 26 percent increase in cumulative abatement in the BC_R model and a 93 percent increase in the BC_L

Table 9.2. Sensitivity Analysis (Percentage Changes, ITC Case Relative to NITC Case).

	Abatement in year			Cumulative abatement	CO_2 concn. in 2200	Tax in year			Cost per unit of abatement
	2000	2050	2190			2000	2050	2190	
1. Central Case	−18.47	−12.39	1.46	−0.89	0.00	−34.83	−34.83	−34.83	−29.83
2a. $r = 0.075$	−14.92	−12.02	1.04	−0.60	0.00	−27.88	−27.88	−27.88	−24.14
2b. $r = 0.025$	−21.63	−11.18	2.09	−1.32	0.00	−44.49	−44.49	−44.49	−37.87
3a. $\bar{S} = 350$	−19.69	−4.16	0.52	−0.78	0.00	−57.14	−57.14	−57.14	−52.85
3b. $\bar{S} = 836.39$	−4.63	−3.98	0.97	−0.26	0.00	−9.06	−9.06	−9.06	−6.76
4a. $C = 12.14 * A^4/((E - A)^2 * H)$	−12.04	−8.35	1.52	−0.90	0.00	−34.74	−34.74	−34.74	−29.39
4b. $C = 105.7 * A^2/H$	−39.30	−31.89	0.00	−0.28	0.00	−39.30	−39.30	−39.30	−36.05
4c. $C = 438.4 * A^{1.5}/H$	0.00	−51.21	0.00	−0.06	0.00	−37.75	−37.75	−37.75	−36.19
5a. $M_\Psi = 0.0058$	−31.35	−21.16	2.31	−1.48	0.00	−54.43	−54.43	−54.43	−48.83
5b. $M_\Psi = 0.0015$	−9.74	−6.51	0.81	−0.47	0.00	−19.37	−19.37	−19.37	−15.97
6a. $\phi = 0.75$	−21.72	−15.27	2.00	−1.14	0.00	−40.14	−40.14	−40.14	−34.53
6a. $\phi = -0.5$	−10.96	−6.08	0.53	−0.39	0.00	−21.64	−21.64	−21.64	−18.20
7a. $\alpha_t = 0.01$	−12.71	−8.15	0.76	−0.50	0.00	−24.56	−24.56	−24.56	−20.43
7b. $\alpha_t = 0.00$	−25.60	−17.86	2.59	−1.47	0.00	−46.68	−46.68	−46.68	−41.28

Policy Criterion: Cost-Effectiveness CE_R Model

Table 9.2. Continued.

	Abatement in year			Cumulative abatement	CO$_2$ concn. in 2200	Tax in year			Cost per unit of abatement
	2000	2050	2190			2000	2050	2190	
1. Central Case	19.63	−6.01	1.32	−0.53	0.00	−41.15	−41.15	−41.15	−39.06
2a. $r = 0.075$	47.40	−2.00	0.86	−0.35	0.00	−36.36	−36.36	−36.36	−34.70
2b. $r = 0.025$	1.37	−7.25	2.06	−0.84	0.00	−46.25	−46.25	−46.25	−43.65
3a. $\bar{S} = 350$	−8.88	−2.48	0.31	−0.42	0.00	−42.44	−42.44	−42.44	−42.57
3b. $\bar{S} = 836.39$	79.70	15.92	1.77	−0.05	0.00	−24.88	−24.88	−24.88	−22.08
4a. $C = 12.14 * A^4/((E - A)^2 * H)$	0.34	−6.02	1.37	−0.65	0.00	−44.21	−44.21	−44.21	−42.40
4b. $C = 105.7 * A^2/H$	613.85	122.37	0.00	0.29	0.00	−27.10	−27.10	−27.10	−25.74
4c. $C = 438.4 * A^{1.5}/H$	15,109.73	3,003.29	0.00	0.18	0.00	−20.12	−20.12	−20.12	−18.73
5a. $M_\Psi = 0.0058$	18.04	−11.05	2.08	−0.95	0.00	−60.02	−60.02	−60.02	−57.79
5b. $M_\Psi = 0.0015$	16.30	−3.18	0.79	−0.29	0.00	−25.85	−25.85	−25.85	−24.26
6a. $\phi = 0.75$	18.24	−7.66	1.77	−0.71	0.00	−46.17	−46.17	−46.17	−43.95
6a. $\phi = -0.5$	24.25	−2.10	0.50	−0.16	0.00	−27.81	−27.81	−27.81	−26.05
7a. $\alpha_t = 0.01$	29.68	−3.07	0.83	−0.30	0.00	−33.94	−33.94	−33.94	−31.88
7b. $\alpha_t = 0.00$	9.11	−9.56	2.03	−0.88	0.00	−49.00	−49.00	−49.00	−46.96

Policy Criterion: Cost-Effectiveness
CE.L Model

Table 9.2. Continued.

Policy Criterion: Benefit–Cost
BC-R Model

	Abatement in year			Cumulative abatement	CO_2 concn. in 2200	Tax in year			Cost per unit of abatement	Net benefits
	2000	2050	2200			2000	2050	2200		
1. Central Case	-0.00	1.59	5.77	3.73	-0.45	-0.01	-0.06	-0.52	-2.29	0.74
2a. $r = 0.075$	-0.00	1.04	4.08	2.57	-0.25	-0.00	-0.03	-0.27	-2.04	0.25
2b. $r = 0.025$	-0.03	3.01	9.63	6.41	-1.04	-0.07	-0.25	-1.24	-1.06	3.16
3a. $D = 1.14E^{-6} * S^3$	-0.01	1.42	5.43	3.56	-0.44	-0.02	-0.11	-1.02	-2.01	0.84
3b. $D = 1.64 * S$	0.00	1.80	6.14	3.92	-0.46	0.00	0.00	0.00	-2.55	0.67
3c. $D = 85.3 * S^{0.5}$	0.00	1.93	6.35	4.03	-0.47	0.00	0.04	0.26	-2.67	0.65
4a. $C = 47.62 * A^4/((E - A)^2 * H)$	-0.00	1.06	3.90	2.39	-0.28	-0.01	-0.05	-0.33	-1.43	0.41
4b. $C = 2.441 * A^2/H$	-0.02	4.29	17.58	11.00	-1.39	-0.02	-0.16	-1.89	-7.59	1.62
4c. $C = 5.608 * A^{1.5}/H$	-0.04	7.90	34.46	25.71	-3.52	-0.02	-0.17	-5.15	-15.35	4.59
5a. $M_\Psi = 0.0058$	-0.01	3.97	13.97	9.15	-1.10	-0.02	-0.15	-1.26	-5.32	1.83
5b. $M_\Psi = 0.0015$	-0.00	0.66	2.43	1.57	-0.19	-0.00	-0.03	-0.22	-0.99	0.31
6a. $\phi = 0.75$	-0.00	1.66	6.94	4.26	-0.52	-0.01	-0.07	-0.61	-2.71	0.78
6b. $\phi = -0.5$	-0.00	1.37	3.15	2.41	-0.28	-0.01	-0.05	-0.31	-1.25	0.61
7a. $\alpha_t = 0.01$	-0.00	1.42	3.75	2.83	-0.44	-0.01	-0.06	-0.51	-1.56	0.68
7b. $\alpha_t = 0.00$	-0.00	1.79	8.74	4.82	-0.45	-0.01	-0.06	-0.51	-3.19	0.79

Table 9.2. Continued.

Policy Criterion: Cost-Effectiveness
BC_L Model

	Abatement in year			Cumulative abatement	CO_2 concn. in 2200	Tax in year			Cost per unit of abatement	Net benefits
	2000	2050	2190			2000	2050	2190		
1. Central Case	0.47	5.61	23.64	14.45	-1.76	-0.03	-0.22	-2.06	-8.98	3.23
2a. $r = 0.075$	0.22	5.00	22.54	13.64	-1.34	-0.01	-0.13	-1.53	-10.21	1.50
2b. $r = 0.025$	1.69	6.91	24.67	15.57	-2.54	-0.19	-0.61	-3.12	-5.03	8.95
3a. $D = 1.114E^{-6} * S^3$	0.57	5.22	21.80	13.73	-1.69	-0.06	-0.41	-3.95	-7.95	3.78
3b. $D = 1.64 * S$	0.41	6.03	25.65	15.14	-1.81	0.00	0.00	-0.00	-9.91	2.80
3c. $D = 85.3 * S^{0.5}$	0.38	6.26	26.74	15.48	-1.84	0.02	0.13	1.07	-10.32	2.63
4a. $C = 47.62 * A/((E - A)^2 * H)$	0.23	4.38	17.41	10.35	-1.23	-0.03	-0.20	-1.44	-6.78	2.15
4b. $C = 2.441 * A^2/H$	1.55	15.47	74.51	43.18	-5.53	-0.10	-0.59	-7.75	-23.15	15.02
4c. $C = 5.608 * A^{1.5}/H$	3.86	23.16	141.78	93.45	-13.03	-0.07	-0.52	-18.99	-37.16	15.02
5a. $M_\Psi = 0.0058$	0.93	11.17	45.82	28.43	-3.45	-0.06	-0.45	-4.02	-15.74	6.43
5b. $M_\Psi = 0.0015$	0.25	2.91	12.39	7.52	-0.92	-0.02	-0.12	-1.07	-4.98	1.67
6a. $\phi = 0.75$	0.48	5.87	29.17	16.74	-2.06	-0.03	-0.24	-2.46	-10.55	3.42
6a. $\phi = -0.5$	0.45	4.76	12.63	9.16	-1.08	-0.03	-0.18	-1.21	-5.24	2.67
7a. $\alpha_t = 0.01$	0.48	5.08	16.66	11.66	-1.85	-0.03	-0.23	-2.16	-6.99	3.07
7b. $\alpha_t = 0.00$	0.46	6.18	32.65	17.44	-1.62	-0.03	-0.22	-1.88	-11.05	3.37

model! These larger impacts on abatement are associated with significant effects on average costs (costs per unit of abatement) and on the net benefits from optimal abatement. Thus, even if ITC's impact on the tax profile is small (a result attributable to the flatness of the marginal damage schedule), it may have a significant impact on abatement levels, abatement costs, and net benefits if the marginal cost function is concave and flat in the relevant range. Further research regarding the shape of the abatement cost schedule would seem necessary before one could confidently accept the Nordhaus (1997) conclusion that ITC has only negligible effects.

In variants 5a and 5b we change the ease of accumulating knowledge when ITC is present by altering the multiplicative parameter M_Ψ in the $\Psi(\cdot)$ function. As expected, when the ITC option is made more powerful (case 5a), the effects of ITC are magnified. The reverse occurs when the ITC option is made weaker (case 5b).

Next, in variants 6a and 6b we consider alternative values for ϕ, which governs the intertemporal knowledge spillover. The central value is 0.5, indicating some degree of "standing on shoulders." Case 6a involves a value of 0.75 (a stronger positive intertemporal spillover); as expected, the effects of ITC are magnified, though only by a small amount. In case 6b, we set ϕ to -0.5 (which represents "fishing out"); here, the opposite holds, and the effects of ITC are (slightly) diminished.

Finally, in variants 7a and 7b we consider alternative rates of autonomous technological progress. The higher the rate of autonomous technological change, the more muted the effects of ITC. This is highly sensible, given the idea of diminishing returns to R&D investments or learning-by-doing efforts.

9.5   Conclusions

This chapter has employed analytical and numerical models to examine the implications of induced technological change for the optimal design of CO_2 abatement policy. We obtain optimal time profiles for carbon taxes and CO_2 abatement under two channels for knowledge accumulation—R&D-based and learning-by-doing-based technological progress—and under both a cost-effectiveness and a benefit–cost policy criterion.

The analytical model reveals that, in contrast with some recent claims, the presence of ITC generally lowers the time profile of optimal carbon taxes. The impact of ITC on the optimal abatement path varies: when knowledge is gained through R&D investments, some abatement is shifted from the present to the future, but if the channel for knowledge growth is learning by doing, the impact on the timing of abatement is analytically ambiguous.

When the government employs the benefit–cost policy criterion, the presence of ITC justifies greater overall (cumulative) abatement than would be warranted in its absence. However, ITC does not always promote greater abatement in all periods. When knowledge accumulation results from R&D expenditure, the presence of ITC implies a reduction in near-term abatement, despite the increase in overall abatement.

The numerical simulations reinforce the qualitative predictions of the analytical model. The quantitative impacts depend critically on whether the government is adopting the cost-effectiveness criterion or the benefit–cost criterion. ITC's effect on overall costs and optimal carbon taxes can be quite large in a cost-effectiveness setting: thus, policy-evaluation models that neglect ITC can seriously overstate both the costs of reaching stipulated concentration targets and the carbon taxes needed to elicit the desired abatement. On the other hand, the impact on costs and taxes is typically much smaller under a benefit–cost policy criterion. The weak effect on the tax rate in the benefit–cost case reflects the relatively trivial impact of ITC on CO_2 concentrations, associated marginal damages, and (hence) the optimal tax rate. As for the optimal abatement path, the impact of ITC on the timing of abatement is very weak, but the effect (present in the benefit–cost case) on total abatement over time can be large, especially when knowledge is accumulated via learning by doing.

Our work abstracts from some important issues. One is uncertainty. We have assumed both that knowledge accumulation is a deterministic process and that the cost of damage functions are perfectly known. In doing so, we have avoided difficult issues of abatement timing relating to irreversibilities and the associated need to trade off the "sunk costs and sunk benefits" of abatement policy.[51]

In addition, in this model the sole policy instrument available to the decision maker (social planner) is a tax on CO_2 emissions. It would be useful to extend the model to include two instruments: namely, a carbon tax and a subsidy to R&D. This would allow explorations of public policies that simultaneously consider two market failures—one attributable to the external costs from emissions of CO_2, and one attributable to knowledge spillovers, which force a wedge between the social and private returns to R&D. In this broader model, one could investigate optimal combinations of carbon taxes and subsidies to R&D. It would also permit investigations of second-best policies: for example, optimal R&D subsidies in a situation in which the government is not able to levy a carbon tax. This approximates the situation implied by recent policy proposals of the Clinton administration.

Appendix 9.A

9.A.1. The cost-effectiveness criterion

Technological Change via R&D

We first demonstrate the basic characteristics of the slope of the optimal abatement path. We then go on to establish the implications of ITC. To determine how abatement changes over time, we differentiate the first-order condition governing abatement with respect to t. Note that the abatement-cost function is not necessarily time stationary, because costs may depend on baseline emissions, which usually vary through time. Differentiating Equation (9.5) with respect to t yields

$$C_{AA}(\cdot)\dot{A}_t + C_{AH}(\cdot)\dot{H}_t + \frac{\delta C_A(\cdot)}{\delta t} = \dot{\tau}_t$$

$$\Leftrightarrow \dot{A}_t = \frac{\dot{\tau}_t - C_{AH}(\cdot)\dot{H}_t - C_{AE}(\cdot)\dot{E}_t^0}{C_{AA}(\cdot)} . \tag{9.38}$$

We have established that for $t < T$, $\dot{\tau}_t > 0$ [see Equation (9.8)], and we know that $\dot{H}_t \geq 0$. Previously we had assumed that $C_{AH}(\cdot) < 0$ and $C_{AA}(\cdot) > 0$. If costs do not depend on the level of emissions, then $C_{AE}(\cdot) = 0$ and Equation (9.39) implies that abatement increases over time ($\dot{A}_t \geq 0$).

It is plausible that $C_{AE}(\cdot) < 0$, namely, that the lower the level of baseline emissions, the greater the marginal cost of a fixed amount of abatement. This is consistent with the idea that abatement costs depend on relative, rather than absolute, levels of abatement. In this circumstance, $\dot{A}_t \geq 0$ so long as baseline emissions are not declining "too rapidly."

Next we move to the ITC/NITC comparison. Under the assumption that $C_{AH}(\cdot) < 0$, we prove the claim that $d\tau_0/dk \leq 0$. Suppose the opposite; that is, suppose that

$$\frac{d\tau_0}{dk} > 0 . \tag{9.39}$$

Equation (9.8) in the main text can be integrated, using the relevant transversality condition as a boundary condition, to obtain the following expression:

$$\tau_t = \int_{\max[t,T]}^{\infty} \eta_s e^{-(r+\delta)(s-t)} ds . \tag{9.40}$$

η_t, the multiplier on the $\bar{S}$ constraint, is zero if the constraint does not bind, and is typically positive, representing the shadow value of relaxing the constraint, if the constraint does bind. Thus, Equation (9.40) states that the shadow cost of having a small additional amount of CO_2 at time t is dictated by how binding the $\bar{S}$ constraints are into the infinite future.[52] Combining Equation (9.40) with our supposition (9.39) yields (assuming the proper regularity conditions hold)

$$\int_T^{\infty} \frac{d\eta_s}{dk} e^{-(r+\delta)s} ds > 0 , \tag{9.41}$$

which states that overall, the $\bar{S}$ constraints from T onward become more binding, or costly. The supposition that τ_0 rises implies, from Equation (9.10), that A_0 also rises. Noting from Equation (9.8) that $\tau_t = \tau_0 e^{(r+\delta)t}$ for $t < T$, we see that our supposition implies that τ_t rises for all $t < T$. This in turn implies, from Equation (9.9) (since $-C_{AH}(\cdot)dH_t/dk$ is clearly nonnegative), that A_t strictly rises for all $t < T$. In fact, as we shall now show, A_t strictly rises for all t, even beyond T. If abatement has strictly risen at every point in time up until T, then we know that S_T is now strictly less than it used to be in the NITC scenario, and thus certainly strictly less than $\bar{S}$. This itself is acceptable: it is easy to imagine situations in which, given a convex abatement-cost function and an emissions baseline that rises sharply after time T, an optimal program involves undershooting the constraint at the first point in time when it is imposed. However, the fact that S_T is now strictly less than $\bar{S}$ implies, by complementary slackness, that $\eta_T = 0$, and thus, since η_T is always nonnegative, that η_T is less than or equal to its value before the increase in k. In other words, $d\eta_T/dk \leq 0$. But we know from Equation (9.41) that the constraints from T onward are, on the whole, more binding, and thus we can now conclude, for sufficiently small ϵ' and all $\epsilon \in (0, \epsilon')$ as well, that

$$\int_{T+\epsilon}^{\infty} \frac{d\eta_s}{dk} e^{-(r+\delta)s} ds > 0 \Leftrightarrow \frac{d\tau_{T+\epsilon}}{dk} > 0 \Leftrightarrow \frac{dA_{T+\epsilon}}{dk} > 0 . \tag{9.42}$$

Now we know that abatement has strictly risen for all $t < T + \epsilon$. The above argument can be repeated, in the style of a proof by induction, to show that $dS_{T+\epsilon}/dk < 0$, implying that $d\eta_{T+\epsilon}/dk \leq 0$, and that, in turn, $dA_t/dk > 0 \; \forall t$. Our supposition that $d\tau_0/dk > 0$ has led us to the conclusion that abatement rises at all points in time. Given that the initial program satisfied the constraints, a new program in which abatement is higher at every point clearly cannot be optimal. Thus we have a contradiction. We may conclude that $d\tau_0/dk \leq 0$, and thus that $dA_0/dk \leq 0$. Since the multiplier simply grows at the constant rate $(r + \delta)$ until time T, we have also shown that $d\tau_t/dk \leq 0 \; \forall t < T$, and in fact, that the absolute fall in the multiplier increases with t over this time range, but in such a way as to preserve the growth rate as $(r + \delta)$.

Note that if we had assumed $C_{AH}(\cdot) > 0$, then the above proof could be reversed to show that initial abatement and the entire tax path weakly rise. We would find that, in contrast to the normal case, ITC would cause a "flattening" rather than a "steepening" of the optimal abatement profile.

Technological Change via Learning By Doing

We start by establishing the slope of the optimal abatement path. It is necessary, however, first to examine the profile of μ_t, the shadow value of knowledge. The costate equation for μ_t, which is the same in both the CE_R and CE_L models, states that

$$\dot{\mu}_t = \mu_t(r - \alpha_t - k\Psi_H(\cdot)) + C_H(\cdot) . \tag{9.43}$$

The shadow value grows at r because it is a *current-value* multiplier. The value of knowledge falls at α_t because new knowledge is being generated autonomously at that rate. Next, depending on the sign of $\Psi_H(\cdot)$—that is, depending on whether knowledge accumulation is characterized by "standing on shoulders" or "fishing out"—there is a third tendency for the shadow value either to fall or to rise over time. For example, when $\Psi_H(\cdot) < 0$, the "fishing out" case where further knowledge accumulation becomes more difficult the larger the current stock of knowledge, it is preferable to suffer this disadvantage over as short a time interval as possible. Thus in this case, the shadow value tends to rise over time. The opposite holds in the "standing on shoulders" case where $\Psi_H(\cdot) > 0$. Finally, since $C_H(\cdot) < 0$, there is a tendency for the shadow value of knowledge to fall over time because we have a shorter time range over which the knowledge will serve to reduce abatement costs. These four effects combine to make the slope of the μ_t path ambiguous in sign.

We now focus on the slope of the optimal abatement path in the CE_L model. Differentiating Equation (9.15) and rearranging, we obtain

$$\dot{A}_t = \frac{\dot{\tau}_t + \mu_t k \Psi_A(\cdot) + (\Psi_{AH}(\cdot)\mu_t k - C_{AH}(\cdot))\dot{H}_t - C_{AE}(\cdot)\dot{E}_t^0}{C_{AA}(\cdot) - \mu_t k \Psi_{AA}(\cdot)} . \qquad (9.44)$$

The denominator is positive, but the numerator is of ambiguous sign because of the second and third terms. If we consider an NITC scenario in which $k = 0$, then we obtain

$$\dot{A}_t = \frac{\dot{\tau}_t - C_{AH}(\cdot)\dot{H}_t}{C_{AA}(\cdot)} - C_{AE}(\cdot)\dot{E}_t^0 , \qquad (9.45)$$

which, at least for $t < T$, is clearly positive, as discussed above, as long as $\dot{E}_t^0$ is not too negative. In the general learning-by-doing case with ITC, however, the optimal abatement path may very well slope downward (even if the emissions baseline is growing over time), in contrast to the R&D case.[53]

Now we examine the implications of ITC. The proof that $d\tau_0/dk \leq 0$ proceeds along the same lines as in the CE_R appendix. We suppose that τ_0 strictly rises, and this implies that abatement rises for all t, which cannot be optimal. [The extra learning-by-doing effect in Equation (9.16) is positive and thus only strengthens the link between τ_t's rising and A_t's rising.[54]] We conclude, as in the CE_R model, that the entire path of carbon taxes falls, and increasingly so for higher t (up to T). As noted in the text, however, this finding is not enough to assure us that initial abatement also falls.

9.A.2. The benefit–cost criterion

Technological Change via R&D

First let us analyze the slope of the carbon tax path. We rearrange Equation (9.23) to see that

$$\dot{\tau}_t = (r + \delta)\tau_t - D'(\cdot) . \qquad (9.46)$$

The first term on the right-hand side contributes to growth in τ_t, while the second contributes to its decline over time (an additional amount of CO_2 later on inflicts marginal damages over a shorter horizon). It is thus possible for the optimal carbon tax to decline over time.

Let us now consider the conditions under which the carbon tax will necessarily rise. Substituting Equation (9.24) into Equation (9.46) yields

$$\dot{\tau}_t = (r + \delta)e^{(r+\delta)t} \int_t^\infty D'(S_s)e^{-(r+\delta)s}ds - D'(S_t) . \qquad (9.47)$$

If we had a linear damage function, such that $D'(S_s)$ were constant and equal to $D'(S_t)$ for all $s > t$, then the first term in Equation (9.47) would reduce to $D'(S_t)$, and we would conclude that $\dot{\tau}_t = 0$; that is, the optimal tax path would be flat. If, however, $D'(S_s) > D'(S_t) \ \forall_s > t$, the first term in Equation (9.47) would be larger than $D'(S_t)$, and the tax path would be upward sloping. Given the convex damage function which we (and others, typically) assume, having an optimized S_t path that slopes upward ensures $D'(S_s) > D'(S_t) \ \forall S > t$, and is thus a sufficient condition for having an upward-sloping tax path. Given that many other authors' simulations involve a steadily increasing optimized CO_2 concentration, it is easy to see why the literature frequently obtains optimal carbon taxes that forever rise.

What can we say about the slope of the abatement path in the BC_R model? Differentiating Equation (9.22) with respect to t and rearranging, we obtain, as in the CE_R model,

$$\dot{A}_t = \frac{\dot{\tau}_t - C_{AH}(\cdot)\dot{H}_t - C_{AE}(\cdot)\dot{E}_t^0}{C_{AA}(\cdot)} . \tag{9.48}$$

The denominator and the second term in the numerator[55] are clearly positive, but the ambiguous slope of the optimal carbon tax path prevents us from concluding that optimized abatement must always rise over time. Once again, if the optimized S_t path is rising and the damage function is convex, then taxes rise, and thus so does abatement, as long as the emission baseline is not declining too rapidly. As we see in our numerical simulations of the BC_R model, even though taxes are always rising, the optimal abatement profile actually slopes down during the time when baseline emissions are steeply decreasing.

Now we turn to the analysis of the implications of ITC; that is, the effects of increasing k. We shall prove that $d\tau_0/dk \leq 0$, that $dA_0/dk \leq 0$, that the overall scale of abatement increases when we raise k, that the abatement path thus becomes steeper, and that $d\tau_t/dk \leq 0 \ \forall t$. If we were to assume that knowledge *raises* marginal abatement costs ($C_{AH}(\cdot) \geq 0$), then the entire proof could be reversed to demonstrate that taxes rise, initial abatement rises, the overall scale of abatement falls, and the abatement path thus becomes flatter.

Suppose that τ_0 rises, and that thus, by Equation (9.26), A_0 rises as well. We have, using Equations (9.24) and (9.22),

$$\frac{d\tau_0}{dk} > 0 \quad \Leftrightarrow \quad \int_0^\infty \frac{dD'(S_s)}{dk} e^{-(r+\delta)s} ds > 0 \tag{9.49}$$

$$\Leftrightarrow \quad \int_0^\infty D''(S_s) \frac{dS_s}{dk} e^{-(r+\delta)s} ds > 0$$

$$\Leftrightarrow \quad \int_0^\infty D''(S_s) e^{-(r+\delta)s} \int_0^s \frac{dA_m}{dk} e^{-\delta(s-m)} \tag{9.50}$$

$$dm \ ds < 0 .$$

Given our convex damage function, Equation (9.50) means that the overall scale of abatement becomes less ambitious when k rises.

For $t > 0$, we can use similar steps to obtain

$$\frac{d\tau_t}{dk} = -\int_t^\infty D''(S_s)e^{-(r+\delta)(s-t)} \int_0^s \frac{dA_m}{dk} e^{-\delta(s-m)} dm\, ds$$

$$\Leftrightarrow \frac{d}{dt}\left(\frac{d\tau_t}{dk}\right) = -(r+\delta)\int_t^\infty D''(S_s)e^{-(r+\delta)(s-t)}$$

$$\int_0^s \frac{dA_m}{dk} e^{-\delta(s-m)} dm\, ds$$

$$+ D''(S_t)\int_0^t \frac{dA_m}{dk} e^{-\delta(t-m)} dm\, . \qquad (9.51)$$

Note that $d(d\tau_t/dk)/dt$ is clearly positive if each of the two terms on the right-hand side of Equation (9.51) is positive. Given Equation (9.50), the first term is definitely nonnegative if

$$\int_0^t D''(S_s)e^{-(r+\delta)s} \int_0^s \frac{dA_m}{dk} e^{-\delta(s-m)} dm\, ds \geq 0\, . \qquad (9.52)$$

The second term is clearly positive, assuming a convex damage function, if

$$\int_0^t \frac{dA_m}{dk} e^{-\delta(t-m)} dm \geq 0\, . \qquad (9.53)$$

We thus are led to the following lemma. Given our assumption that the initial tax rises [Equation (9.49)], and assuming a convex damage function, then

$$\int_0^t D''(S_s)e^{-(r+\delta)s} \int_0^s \frac{dA_m}{dk} e^{-\delta(s-m)} dm\, ds \geq 0 \qquad (9.54)$$

and

$$\int_0^t \frac{dA_m}{dk} e^{-\delta(t-m)} dm \geq 0\, . \qquad (9.55)$$

These together imply

$$\frac{d}{dt}\left(\frac{d\tau_t}{dk}\right) \geq 0\, . \qquad (9.56)$$

Equations (9.49) and (9.26) together tell us that $dA_0/dk > 0$. This means (using inductive reasoning much like that used in the CE_R appendix),[56] that sufficiency conditions (9.54) and (9.55) hold for $t = \epsilon$ sufficiently close to 0, and also for $t = \epsilon'\ \forall\ \epsilon' \in (0, \epsilon)$. Thus we conclude that

$$\frac{d}{dt}\left(\frac{d\tau_\epsilon}{dk}\right) \geq 0 \Leftrightarrow \frac{d\tau_\epsilon}{dk} \geq \frac{d\tau_0}{dk} > 0 \Leftrightarrow \frac{dA_\epsilon}{dk} > 0\, , \qquad (9.57)$$

where this last implication is only strengthened by the $-C_{AH}(\cdot)dH_\epsilon/dk$ effect in Equation (9.26). The whole chain of reasoning can be repeated inductively to imply that abatement strictly rises at every point in time. This, however, contradicts

244 Optimal CO$_2$ Abatement

Equation (9.50), which says that the overall scale of abatement is less ambitious. Thus, our supposition must be wrong. Thus, we conclude that τ_0 (weakly) falls, A_0 falls, a more ambitious overall scale of abatement is adopted, and the abatement path becomes "steeper," all as a result of the increase in k. That is to say,

$$\frac{d\tau_0}{dk} \leq 0 \,,$$

$$\frac{dA_0}{dk} \leq 0 \,,$$

and

$$\int_0^\infty D''(S_s)e^{-(r+\delta)s} \int_0^s \frac{dA_m}{dk} e^{-\delta(s-m)} \, dm \, ds \geq 0 \,. \tag{9.58}$$

Using arguments similar to those used above, it is also possible to demonstrate that the entire path of carbon taxes must weakly fall: $d\tau_t/dk \leq 0 \; \forall t$. This does not mean, however, that *abatement* always weakly falls; the growth in H_t as a result of k counters the effect of the weakly falling carbon taxes, and we know, in fact, that overall we end up with a weakly more ambitious abatement path.

Technological Change via Learning By Doing

Using methods virtually identical to those in the previous sections, we prove that (1) the carbon tax falls at all points in time, including time 0, (2) the impact on A_0 is ambiguous, and (3) the overall scale of abatement increases. Please refer to earlier sections of this appendix corresponding to the CE_L and BC_R models; the proofs here are not substantively different.

Acknowledgments

We have benefited from very helpful comments from Michael Dalton, Michael Grubb, Chad Jones, Alan Manne, Robert Mendelsohn, William Nordhaus, Richard Richels, Stephen Schneider, Sjak Smulders, David Wheeler, and an anonymous referee. Financial support from Department of Energy Grant DE-FG03-95ER62104 and National Science Foundation Grant SBR9310362 is gratefully acknowledged.

Notes

1. Some have suggested that the innovation-related benefits from a carbon tax might be as large as the direct abatement costs associated with such a tax. If this were the case, then the overall cost (ignoring environmental benefits) of a carbon tax would be zero. Porter and van der Linde (1995) advance a general argument consistent with this view, maintaining that environmental regulation often stimulates substantial technological progress and leads to significant long-run cost savings that make the overall costs of regulation trivial or even negative.

2. This is equivalent to minimizing the sum of abatement costs and CO_2-related damages to the environment.

3. The present chapter also complements the work of Manne and Richels (1992), who employ a multiregion computable general equilibrium model to solve for Pareto-efficient paths of carbon abatement and taxes.

4. However, in a benefit–cost setting, the opposite could be true if damages were concave in the atmospheric CO_2 concentration.

5. Our analysis focuses on the social planner's problem. We disregard the market failure associated with knowledge spillovers, that is, with the inability of firms to appropriate the full social returns on their investments in knowledge. Our model implicitly assumes that any market failures associated with this appropriability problem have already been addressed through public policies.

6. For analytical convenience, we postulate a simple stock-flow relationship here. A more complicated equation of motion, such as the one introduced in the numerical simulations, would not alter the qualitative analytical results obtained here.

7. This issue is discussed in greater detail by Goulder and Schneider (1999).

8. This Hamiltonian actually corresponds to the problem of maximizing the negative costs.

9. After T, matters are complicated by the η_t term in Equation (9.6).

10. The appropriate discount rate is not simply r. Consider an arbitrary path of emissions leading to a given concentration S_T at time T. Since CO_2 is removed naturally, altering this path by increasing emissions slightly at time t and reducing emissions slightly at a later time t' leads to greater overall removal and thus leads to a CO_2 concentration at time T that is less than S_T. Equivalently (as seen in the sensitivity analysis in Section 9.3), S_T can be achieved with less cumulative emissions abatement if the path of abatement is oriented more toward the future. Hence there is a value to postponing abatement beyond that implied by interest rate r; this additional value is captured in the appearance of δ in the discount rate.

11. The focus here on differential changes does not limit the generality of the analysis. Our analytical results are independent of the initial value of k. Given the smooth nature of our problem, results that hold for small changes in k around any initial value will carry over qualitatively for large changes around the point 0. This is confirmed in the numerical simulations.

12. A corner solution arises if even the first increment of knowledge has marginal returns smaller than marginal costs. In this case, the social planner does not invest in additional knowledge; even here, though, we know that knowledge at least will not decrease from the baseline path.

13. Throughout, when we use the words "increase" and "decrease" we mean nonstrict increases and decreases, thus including the possibility that the variable stays constant.

14. Note that the denominator of Equation (9.7) is positive by assumption.

15. We are stating that $dH_0/dk = 0$. This simply expresses the notion that the initial value for H_t (that is, H_0) is not affected by different values for k. It remains true, however, that dH_t/dt is positive at all points in time. Even at time 0, the time-derivative of H_t is positive as a result of autonomous knowledge growth and induced investment in knowledge.

16. The decline in the shadow cost reflects the maximum potential of the ITC option over the entire time horizon; that is, the fall in the shadow cost corresponds to optimal R&D, as is sensible in this model of an optimizing planner.

17.   The figure is not meant to represent a single year in the program, but rather an independent one-period analogue to our abatement problem.

18.   If baseline emissions were to rise sharply after T, then given the convexity of the abatement cost function, it might be optimal to more than meet the $\bar{S}$ requirement at time T to reduce the amount of abatement required afterward. However, it is the case nevertheless that one curve cannot lie above or below another over the entire infinite horizon: perpetual over- or undershooting of the constraint cannot be optimal.

19.   In characterizing the path as "steeper" we do not mean that the slope of the new path is everywhere greater than that of the old path. In fact, in the numerical simulations we will see that this is often not the case. We simply mean that, loosely speaking, less abatement is undertaken early on, and more later on.

20.   We have assumed no spillovers in the model (or at least none that have not been fully addressed by other government policies); the cost-reduction from learning by doing is fully appropriated by agents.

21.   In an NITC scenario, the abatement path will unambiguously slope upward for $t < T$, given that baseline emissions do not decline too rapidly. See Appendix 9.A for details.

22.   Evaluating at an arbitrary nonzero initial value of k adds extra terms which are difficult to sign. Unlike in the R&D specification, here we cannot be fully confident that our differential analysis around the point $k = 0$ carries over to the case of large increases in k from 0. However, the numerical simulations below indicate that the qualitative results obtained here carry through even for large changes in k.

23.   The learning-by-doing effect can be quite large, as abatement, by increasing the stock of knowledge, lowers the cost of future abatement over the entire remaining time horizon.

24.   Unlike in the cost-effectiveness models, however, it is not necessarily true that taxes later on fall by greater amounts than do early taxes. See Appendix 9.A.

25.   This MB curve conveys the same information as the schedule of marginal damages from additions to the stock of CO_2.

26.   See Repetto (1987) for a discussion of nonconvex damages. Also note that, as before, if technological progress were to raise the MC schedule, then even with convex damages, the optimal carbon tax would rise (and the optimal scale of abatement would fall). This is confirmed in Appendix 9.A.

27.   We say "approximately" because natural removal implies that two abatement paths leading to $\bar{S}$ need not involve exactly the same cumulative abatement. In fact, as will be seen in the sensitivity analysis in Section 9.4, paths which concentrate relatively more abatement in the future need less cumulative abatement to reach the same $\bar{S}$ constraint because they take better advantage of natural removal than do more heavily "front-loaded" abatement paths.

28.   What may be surprising, however, is that the result depends on the convexity of the damage function. With concave damages and a marginal damage schedule steeper than marginal cost (unlikely, given that we have a stock pollutant), a downward pivot in the marginal cost schedule could lead to less abatement.

29.   As in the CE_L analysis, we restrict our attention to the neighborhood around $k = 0$.

30.   See Appendix 9.A for details.

31.   Specifically, we impose the requirement that the CO_2 concentration remain constant after the last period. For this to occur, abatement must also remain constant (given that baseline emissions are constant at that point in time). In the R&D simulations, we

also impose the steady-state constraint that investment go to zero. Even when these constraints are imposed, our model does not yield a steady state with constant abatement costs because the stock of knowledge continues to grow; this continued growth is due both to autonomous technological change and (in the learning-by-doing simulations) to the increments to knowledge stemming from continued experience with abatement. In solving the numerical model, we assume that abatement costs beyond the last simulation period are constant, even though the analytical structure of the model implies that abatement costs would fall as knowledge continued to accumulate. Thus our approach overestimates the true future costs. We have verified that this inconsistency has no numerical significance by comparing numerical results under this approach with those from simulations that assume costs after year 2400 are zero and thus underestimate future costs. These two alternative specifications bound the truth about future costs. The numerical results under these two very different approaches are indistinguishable: discounting over a 400-year horizon makes the terminal conditions unimportant in practice.

32.   This profile is patterned after a scenario used by Manne and Richels (1997).

33.   Some scholars endorse more sophisticated formulations, such as the five-box model of Maier-Reimer and Hasselmann (1987).

34.   There is no backstop technology in the model.

35.   These calculations are based on results of a literature review in EPRI (1994) and are extrapolated to the global economy.

36.   See Manne and Richels (1992, 1997).

37.   See, for example, Romer (1990), Jones (1996), or Jones and Williams (1996). We are grateful to William Nordhaus and Chad Jones for recommending this function and alerting us to its usefulness.

38.   Jones and Williams (1996) dub this the "stepping-on-toes effect," for "an increase in R&D effort induces duplication that reduces the average productivity of R&D."

39.   In the work by Manne and Richels (1992:64), GDP costs of abatement policy are approximately 90 percent lower in an optimistic technology scenario than in the central-case technology scenario. This difference in GDP costs does not account for the costs of developing the improved technologies that distinguish the optimistic scenario from the central-case scenario. We assume that R&D investments have a social rate of return of 50 percent [as in Nordhaus (1997)] and then calculate the net cost savings from technological progress to be roughly 30 percent. [The R&D costs that generate 0.90 of abatement-cost savings amount to $(1/1.5)0.90$. Thus, the net cost savings from technological progress are given by $0.90 - (1/1.5)0.90 = 0.30$.] We assume that this figure is relevant to the induced technological change which we study in our chapter, and we then choose M_{Ψ} to generate this level of savings.

40.   The discount rate represents, in this context, the marginal product of capital, rather than the pure rate of time preference.

41.   A slight decline begins around 2170, as baseline emissions are declining rapidly at this point; this is fully consonant with the analytical model. In both the ITC and NITC cases, the level of abatement drops discontinuously in the year 2200 and stays constant thereafter, maintaining the CO_2 concentration at the level $\bar{S}$. The constraint on the year-2200 concentration forces this discontinuity.

42.   Recall that the analytical model was unable to guarantee this result for the ITC scenario.

43.   In the central case, when the MC curve is relatively steep, the effect on abatement is not very large either.

44. In other words, the net benefits of optimal abatement relative to a baseline of no abatement whatsoever are scarcely bigger in an ITC scenario than they are in an NITC scenario.

45. Nordhaus obtains his result in a dynamic optimization model in which technological change is driven by R&D expenditure. Some differences between the Nordhaus study and the present study are worth noting. His analysis explicitly models a production function, while ours represents production (or ease of substitution) in reduced form through the abatement-cost function. This enables us to obtain analytical results where Nordhaus relies solely on numerical simulations. Another difference is that the Nordhaus study considers only the BC_R case. In contrast, the present study considers both the benefit–cost and cost-effectiveness cases, and considers learning by doing as well as R&D-based technological change. The present chapter's attention to alternative policy specifications and knowledge-generation channels, along with the broad sensitivity analysis below, enable it to map out more broadly the conditions under which ITC has (or does not have) a significant impact on economic outcomes.

46. As when we compared the CE_R and CE_L models, this difference is due to the fact that ITC is "free" under a learning-by-doing specification but costly under R&D.

47. In the cost-effectiveness simulations, we report these results for the year 2190; the year-2200 statistic is uninformative as both abatement and taxes in this first year of the constraint are identical across ITC and NITC runs.

48. In case 3b, the value 836.39 is chosen to match the optimized value of the year-2200 CO_2 concentration in the NITC scenario of the BC_R simulation.

49. The multiplicative parameter in the damage function is recalibrated so that the cumulative amount of abatement in the NITC run remains constant.

50. Thus we are ensuring, for comparability across the cases, that the MC curve always intersects the vertical constraint (in the upper panel of Figure 9.2) at the same point.

51. See Pindyck (1993) and Ulph and Ulph (1997).

52. As noted in the main text, this is in contrast to the benefit–cost cases, in which shadow cost is given by the discounted sum of the marginal damages that a small additional amount of CO_2 would cause into the infinite future.

53. Our numerical solutions confirm, however, that the abatement path typically does slope upward, even in the ITC learning-by-doing case.

54. The learning-by-doing effect, however, prevents us from reversing the proof for the $C_{AH} > 0$ case. In that case, under a learning-by-doing specification, we cannot conclude anything about the impact of ITC on taxes or abatement.

55. Assuming $C_{AH}(\cdot) < 0$.

56. Where we used a small ϵ and appealed to continuity to justify an inductive proof in a continuous-time problem.

References

Dickinson, R.E., and Cicerone, R.J., 1986, Future global warming from atmospheric trace gases, *Nature*, **319**(January):109–115.

EPRI (Electric Power Research Institute), 1994, Economic Impacts of Carbon Taxes: Detailed Results, EPRI TR-104430-V2, November, Palo Alto, CA, USA.

Farzin, Y.H., 1996, Optimal pricing of environmental and natural resource use with stock externalities, *Journal of Public Economy*, **62**:31–57.

Farzin, Y.H., and Tahvonnen, 0., 1996, Global carbon cycle and the optimal time path of a carbon tax, *Oxford Economic Papers*, **48**:515–536.

Goulder, L.H., and Schneider, S., 1999, Induced technological change, crowding out, and the attractiveness of CO_2 emissions abatement, *Resource & Energy Economics*, **21**:211–253.

Grubb, M., 1997, Technologies, energy systems, and the timing of CO_2 abatement: An overview of economic issues, *Energy Policy*, Butterworths, **25**(2):159.

Ha-Duong, M., Grubb, M., and J.-C Hourcade, 1996, Optimal Emission Paths Towards CO_2 Stabilization and the Cost of Deferring Abatement: The Influence of Inertia and Uncertainty, Working Paper, CIRED, Montrouge, France.

IPCC (Intergovernmental Panel on Climate Change), 1995, *Climate Change 1994: Radiative Forcing of Climate Change and an Evaluation of the IPCC 1S92 Emission Scenarios*, J.T. Houghton, L.G. Meira Filho, J. Bruce, H. Lee, B.A. Callander, E. Haites, N. Harris, and K. Maskell, eds, Cambridge University Press, Cambridge, UK.

Jones, C.I., 1996, Human Capital, Ideas, and Economic Growth, Working Paper, VIII Villa Mondragone International Economic Seminar, 25–27 June, Rome, Italy.

Jones, C.I., and Williams, J., 1996, Too Much of a Good Thing?, Working Paper, Stanford University, Stanford, CA, USA.

Kolstad, C.D., 1996, Learning and stock effects in environmental regulation: The case of greenhouse gas emissions, *Journal of Environmental Economics and Management*, **31**:1–18.

Maier-Reimer, E., and Hasselmann, K., 1987, Transport and storage in the ocean: An unorganic ocean-circulation carbon cycle model, *Climate Dynamics*, **2**:63–90.

Manne, A., and Richels, R., 1992, *Buying Greenhouse Insurance: The Economic Costs of CO_2 Emission Limits*, MIT Press, Cambridge, MA, USA.

Manne, A., and Richels, R., 1997, Toward the Stabilization of CO_2 Concentrations: Cost-Effective Emission Reduction Strategies, Working Paper, Electric Power Research Institute, Palo Alto, CA, USA.

Nordhaus, W.D., 1980a, Thinking about Carbon Dioxide: Theoretical and Empirical Aspects of Optimal Growth Strategies, Cowles Foundation Discussion Paper No. 565, Yale University, October, New Haven, CT, USA.

Nordhaus, W.D., 1980b, How fast should we graze the global commons?, *American Economic Review*, **172**:242–246.

Nordhaus, W.D., 1994, *Managing the Global Commons: The Economics of Climate Change*, MIT Press, Cambridge, MA, USA.

Nordhaus, W.D., 1997, Modeling Induced Innovation in Climate-Change Policy, Working Paper, NBER Summer Institute, PEG Workshop, National Bureau of Economic Research, Cambridge, MA, USA.

Nordhaus, W.D., and Yang, Z., 1996, A regional dynamic general-equilibrium model of alternative climate change strategies, *American Economic Review*, **86**:741–765.

Peck, S.C., and Teisberg, T.J., 1992, CETA: A model for carbon emissions trajectory assessment, *Energy Journal*, **13**:71–91.

Peck, S.C., and Teisberg, T.J., 1994, Optimal carbon emissions trajectories when damages depend on the rate or level of global warning, *Climatic Change*, **28**:289–314.

Peck, S.C., and Wan, Y.S., 1996, Analytical Solutions of Simple Optimal Greenhouse Gas Emission Models, Working Paper, Electric Power Research Institute, Palo Alto, CA, USA.

Pindyck, R.S., 1993, Sunk Costs and Sunk Benefits in Environmental Policy, Working Paper, MIT, Cambridge, MA, USA.

250   *Optimal CO₂ Abatement*

Porter, M.E., and van der Linde, C., 1995, Toward a new conception of the environment-competitiveness relationship, *Journal of Economic Perspectives*, **9**:97–118.

Repetto, R., 1987, The policy implications of non-convex environmental damages: A smog control case study, *Journal of Environmental Management*, **14**:13–29.

Romer, P., 1990, Endogenous technological change, *Journal of Political Economy*, **98**:S71–S102.

Sinclair, P.J.N., 1994, On the optimum trend of fossil fuel taxation, *Oxford Economic Papers*, **46**:869–877.

Ulph, A., and Ulph, D., 1994, The optimal time path of a carbon tax, *Oxford Economic Papers*, **46**:857–868.

Ulph, A., and Ulph, D., 1997, Global warming, irreversibility and learning, *The Economic Journal*, **107**(442):636–650.

Wigley, T.M.L., Richels, R., and Edmonds, J., 1996, Economic and environmental choices in the stabilization of atmospheric CO_2 concentrations, *Nature*, **379**:240–243.

Chapter 10

Modeling Uncertainty of Induced Technological Change

Andrii Gritsevskyi and Nebojsa Nakicenovic

10.1 Introduction

This chapter presents a new method for modeling induced technological learning and uncertainty in energy systems. Three related features are introduced simultaneously: (1) increasing returns to scale for the costs of new technologies; (2) clusters of linked technologies that induce learning depending on their technological "proximity" and the technology relations through the structure of the energy system; and (3) uncertain costs of all technologies and energy sources.

The energy systems-engineering model MESSAGE developed at the International Institute for Applied Systems Analysis (IIASA) was modified to include these three new features. MESSAGE is a linear programming optimization model. The starting point for this new approach was a global (single-region) energy systems version of the MESSAGE model that includes more than 100 different energy extraction, conversion, transport, distribution, and end-use technologies. A new feature is that the future costs of all technologies are uncertain and assumed to be distributed according to the log-normal distribution. These are stylized distribution functions that indirectly reflect the cost distributions of energy technologies in the future based on the analysis of the IIASA energy technology inventory. In addition, the expected value of these cost distributions is assumed to decrease and variance to narrow with the increasing application of new technologies. This means that the process of technological learning is uncertain even as cumulative experience increases. New technologies include, for example, fuel cells, photovoltaics (PVs), and wind energy conversion technologies.

The technologies are related through the structure of the energy system in MESSAGE. For example, cheaper wind energy has direct and indirect effects on other technologies that produce electricity upstream and on electric end-use technologies downstream. In addition, technologies are grouped into clusters that depend on technological "proximity." For example, the costs of all fuel cells for mobile applications are a function of their combined installed capacity weighted

Reprinted from *Energy Policy*, Volume 28, Gritsevskyi, A., and Nakicenovic, N., Modeling uncertainty of induced technological change, pp. 907–921, ©2000, with permission from Elsevier Science.

according to their expected unit sizes. This relationship depends on how closely the technologies are related. This varying degree of "collective" technological learning for technologies belonging to the same cluster is also uncertain.

Each scenario of alternative future developments for a deterministic version of the global energy systems model MESSAGE requires approximately 10 minutes of run-time on a PC. Therefore, it is simply infeasible to generate alternative future developments under uncertainty based on a simple Monte Carlo type of analysis where one sequentially draws observations from the more than 200,000 cost distributions (100 technologies, 11 time steps, 10 technological clusters with 22 technologies included) assumed here for modeling technological learning and uncertainty. Instead, the new approach proposed here starts with a large but finite number of alternative energy system "technology dynamics" and generates "in parallel" another large but finite number of deterministic scenarios by sampling from the distributions simultaneously for each of these technology dynamics. In this application, about 130,000 scenarios were generated. There were 520 alternative technology dynamics, each with about 250 alternative deterministic scenarios resulting from the simultaneous stochastic samplings. Both numbers were initially varied before deciding that about 500 is a sufficient number of different technology dynamics required for a wide spectrum of alternative technological learning possibilities and that about 250 different deterministic scenarios are sufficient to generate most of the interesting future energy system structures for each of the technology dynamics based on the analysis that in total produced roughly one million different scenarios. These large numbers of scenarios represent a very small subset of the basically infinite number of all possible scenarios. They were not chosen randomly, but are a result of applying adaptive global search techniques to the formulated nonconvex, non-smooth stochastic problem.

From the 520 alternative technology dynamics, about 53 resulted in scenarios with very similar overall energy systems costs. They have fundamentally different technological dynamics and produce a wide range of different emergent energy systems, but can be considered to be approximately equivalent with respect to "optimality" criteria (in this case, simultaneous cost and risk minimization). Thus, one result of the analysis is that different energy system structures emerge with similar overall costs; in other words, there is a large diversity across alternative energy technology strategies. The strategies are path dependent and it is not possible to choose *a priori* an "optimal" direction of energy systems development.

Another result of the analysis is that the endogenous technology learning with uncertainty and spillover effects will have the greatest impact on the emerging energy system structures during the first few decades of the twenty-first century. Over these "intermediate" periods of time, these two processes create endogenous lock-in effects and increasing returns to adoption. In the very long run, however, none of these effects is of great importance. The reason is that over such long periods many doublings of capacity of all technologies with inherent learning occur, so that little relative cost advantage results from large investments in only a few technologies and clusters. Therefore, the main finding is that, under uncertainty, the near-term investment decisions in new technologies are more important in determining the direction of long-term development of the energy system than are decisions that are made later, toward the end of the time horizon. Thus, the most dynamic phase in the development of future energy systems will occur during the next few decades. It is during this period that there is a high degree of freedom of choice across future technologies, and many of these choices lead to high spillover learning effects for related technologies.

One policy implication that can be made based on the emerging dynamics and different directions of energy systems development in this analysis is that future research, development, and demonstration (RD&D) efforts and investments in new technologies should be distributed across "related" technologies rather than directed at only one technology from the cluster, even if that technology appears to be a "winner." Another implication is that it is better not to spread RD&D efforts and technology investments across a large portfolio of future technologies. Rather, it is better to focus on (related) technologies that might form technology clusters. Finally, the results imply that fundamentally different future energy system structures might be reachable with similar overall costs. Thus, future energy systems with low carbon dioxide (CO_2) emissions need not be associated with costs higher than those of systems with high emissions.

10.2　The Modeling Approach

Fundamental changes in global energy systems tend to occur slowly. The replacement of traditional energy sources—such as the substitution of coal for fuelwood with the advent of steam, steel, and railways—took most of the nineteenth century. The subsequent replacement of coal with oil and gas and associated technologies lasted the better part of the twentieth century. In contrast to these very slow processes of change in the global energy system, in some parts of the energy system change can be more dynamic—especially in the evolution of end-use technologies. However, the fact that fundamental changes occur over many decades rather than a few years means that technological changes that have inherently shorter time constants need to be consistent with the overall slower processes of change in the energy system. Thus, the many generations of individual technologies that are replaced through the normal rate of capital turnover are a part of the overall slow change from older to newer sources of energy and other related structural changes in energy systems. This means that many generations of new technologies are likely to come and go before the possible transition to the post-fossil era or to new fossil systems is achieved. The directions of these future transitions are uncertain. Future energy systems could rely on renewable energy sources, on clean coal, on less carbon-intensive fossils such as natural gas, or on nuclear power. Therefore, there is an infinite number of alternative scenarios that lead to all possible future energy systems.

As mentioned, the replacements of primary energy sources have in each case required the better part of a century, and similar changes are conceivable during this century. Climate change is characterized by long time constants, just as energy systems are. It might take a few decades to resolve the uncertainty surrounding the influence of human intervention in the climate system resulting from emissions of greenhouse gases and aerosols. The main sources of emissions for most of these gases are associated with energy activities. This and other environmental concerns are another reason why the direction of technological changes in the energy system is important. Changes in the energy system that lead to radically lower future emissions would need to be implemented before this uncertainty about possible climate change is resolved because of the long time constants of change in both the energy and climate systems. This is especially true for the introduction of new energy technologies if sufficient cumulative experience with these technologies is to be achieved in time to facilitate rapid technological learning and their widespread diffusion.

An important motivation for developing this new approach for endogenizing technological learning and uncertainty in energy system scenarios was the desire to capture the different directions of possible future technological change resulting from the many technology replacements and incremental improvements that may occur during this century. Our basic assumption is that endogenous learning is a function of cumulative experience, measured by cumulative installed capacity, and that this process is uncertain. Clearly, this is a strong oversimplification. Although there are many other indicators of technological learning, we chose this one because it is relatively easy to measure. Nevertheless, we feel that the oversimplification is justifiable in a tool for analyzing the cumulative effect that incremental investments in new technologies have on the direction of alternative energy systems development.

Energy services are expected to increase dramatically during the twenty-first century, especially in today's developing countries. This means that the installed capacities of energy extraction, conversion, transport, distribution, and end-use technologies will increase accordingly, perhaps at a somewhat lower rate owing to the overall improvements of efficiencies throughout the energy system as older technologies are replaced by newer ones. Here again, the alternative directions of energy systems development are important. To a large extent, they will determine the eventual energy requirements needed to satisfy the increasing demand for energy services. The actual energy requirements for a given provision of energy services can range from very high to extremely low compared with current standards. Similarly, the future environmental impacts of energy systems will vary accordingly, as well. For example, CO_2 emissions range from 10 times the current levels to virtually no net emissions by 2100 for scenarios in the literature. Figure 10.1 shows the range of future CO_2 emissions derived by the new modeling approach for the set of 520 technology dynamics (some 130,000 scenarios) versus the set of 53 "optimal" dynamics (more than 13,000 scenarios). In comparison, Figure 10.2 shows the range of emissions for some 400 scenarios from the published literature collected for the new Intergovernmental Panel on Climate Change (IPCC) Special Report on Emissions Scenarios (Morita and Lee 1998; Nakicenovic *et al.* 1998b). The emissions range from 7 to 41 gigatons of carbon (GtC) by 2100, compared with about 6 GtC in 1990. As these figures illustrate,

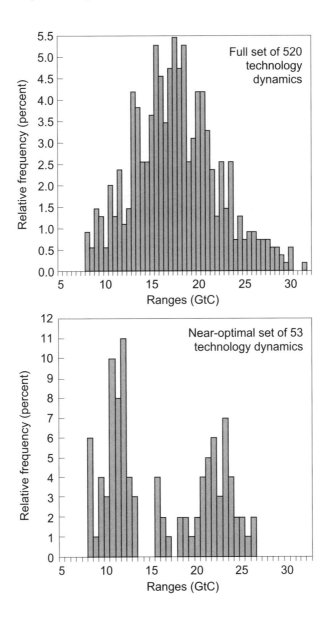

Figure 10.1. Global Carbon Dioxide Emissions Ranges for (*Top*) the Full Set of 130,000 Scenarios with Endogenous Technological Change (Comprising 520 Different Technology Dynamics) versus (*Bottom*) the Ranges of More than 13,000 "Optimal" Scenarios from 53 Different Technology Dynamics.

All scenarios share a given useful energy trajectory; emissions ranges in gigatons of carbon (GtC).

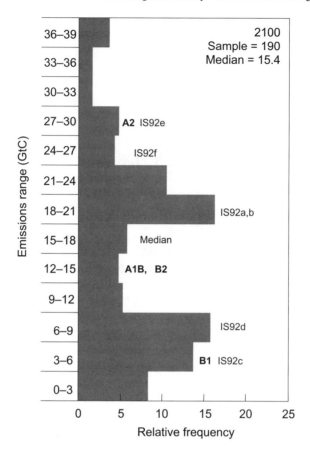

Figure 10.2. Global Carbon Dioxide Emissions for the Range of Some 400 Scenarios from the Literature.

Emissions ranges in gigatons of carbon (GtC) (Morita and Lee 1998; Nakicenovic *et al.* 1998b). Some of the IPCC IS92 and SRES scenarios are indicated within the appropriate emissions intervals shown in the histogram.

the set of scenarios developed for capturing endogenous technological learning and uncertainty covers most of this range. The scenarios from the literature span this range owing to the variation of the driving forces of future emissions, such as energy demand. In contrast, the set of scenarios with endogenous learning spans the range owing to different technological dynamics alone. It is interesting to note that the "optimal" scenarios match the distribution of the scenarios from the literature quite closely, but with a somewhat narrower range (they leave the extreme tails of the distribution uncovered). In contrast, the frequency distribution of the full set of 520 technology dynamics is different from the other two, with many more scenarios in the mid-range of the distribution. This means that the optimal or most "cost-effective" development paths correspond quite closely

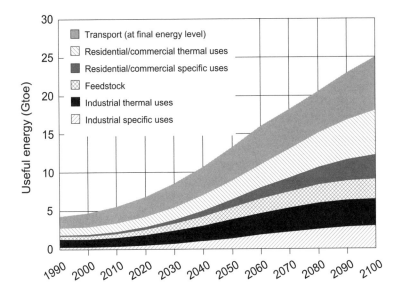

Figure 10.3. Global Useful Energy Demand Trajectory Common to All 520 Different Technology Dynamics (130,000 Scenarios), in Gigatons Oil Equivalent (Gtoe).

to the scenario distribution from the literature. The "median" or "central" futures are underrepresented both in the literature and among the scenarios, indicating a kind of "crowding-out" effect surrounding balanced and median scenarios. In any case, technological learning as specified in our approach leads to future energy systems that are marked by either high or low emissions ranges with a single useful energy demand trajectory, demonstrating a kind of implicit bifurcation across the range of possible emissions.

To simplify matters, we have assumed a single trajectory of global useful (end-use) energy requirements as an input assumption for all 130,000 scenarios considered in this analysis. What is varied endogenously are the technologies that make up the energy system and their costs. Figure 10.3 shows the single useful energy trajectory that is common to all scenarios. It represents relatively high useful energy demand compared with the scenarios in the literature. However, it is associated with considerable variations of final and primary energy demand trajectories across the scenarios. The figure shows that a very wide portfolio of future energy systems characteristics is consistent with a single end-use demand trajectory. The scenarios map the higher part of the range of future primary energy requirements found in the scenario literature, but leave uncovered the lower part of the range, which is associated with very low demand scenarios in the literature. As mentioned, the scenarios cover most of the emissions range.

10.3 Increasing Returns and Uncertainty

Time horizons of a century or longer are frequently adopted in energy studies. Modeling energy systems developments over such long time horizons imposes a number of methodological challenges. Over longer horizons, technological change becomes fluid and fundamental changes in the energy system are possible. It has been especially difficult to devise an appropriate representation of endogenous technological change and its associated uncertainties. In general, induced technological change and uncertainties are interconnected. It is widely recognized that together they play a decisive role in shaping future energy systems. Many approaches to modeling these processes have included elements of increasing returns to scale and decreasing uncertainty to scale. This basically means that technologies improve with cumulative experience, as expressed by the scale of their application. Costs and uncertainty are assumed to decline with increasing scale of application. Such processes are frequently represented by learning or experience curves.

In contrast, the "standard" modeling approaches with diminishing returns do not allow for such consequences of technological learning processes. Despite this deficiency, diminishing returns dominate standard economic theory, perhaps because of the very elegant and simple concept of equilibrium that can be achieved under those conditions. Diminishing returns to scale generate negative feedbacks, which tend to stabilize the system by offsetting major changes and inevitably produce a unique equilibrium independent of the initial state of the economy. In mathematical terms, the models are convex and generally lead to unique solutions.

Increasing returns, on the other hand, lead to disequilibrium tendencies by providing positive feedbacks. After (generally large) initial investments in RD&D and early market introduction, the incremental costs of further applications become cheaper and cheaper per unit capacity (or as assumed here, per unit output). Thus, the more widely adopted a technology becomes, the cheaper it becomes (with lower uncertainties, leading to lower risks to adoption). There are many incarnations of this basic principle. One of the better known is the concept of "lock-in." As a technology becomes more widely adopted it tends to increasingly eliminate other possibilities, thus the lock-in. Another concept frequently used in empirical analysis is the so-called learning or experience curve. At the core of all of these processes is technological learning: the more experience that is gained with a particular technology, the greater the improvements in performance, costs, and other important technology characteristics.

Despite the fundamental importance of technological learning, modeling of these processes has received inadequate attention in the literature. Several reasons may explain the apparent lack of systematic approaches. Among them, the complexity of appropriate modeling approaches is perhaps the most critical. Increasing returns to scale lead to nonconvexities. Thus, the standard optimization techniques cannot be applied. In conjunction with the treatment of uncertainties, modeling of technological learning becomes methodologically and computationally very demanding. It requires the development of so-called global

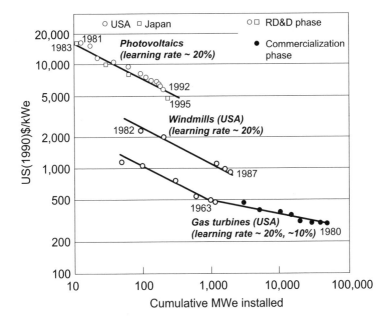

Figure 10.4. Technology Learning Curves for Three Different Electricity-Generation Technologies: Gas Turbines, Windmills, and Photovoltaics.

Cost improvements per unit installed capacity, in US(1990)$ per kilowatt electric (kWe), versus cumulative installed capacity, in megawatts electric (MWe), on a logarithmic scale. *Sources*: MacGregor *et al.* (1991); Christiansson (1995); Nakicenovic *et al.* (1998a).

nonsmooth stochastic optimization techniques, which are only now under development (Ermoliev and Norkin 1995, 1998; Horst and Pardalos 1995).

Figure 10.4 gives learning or experience curves for three electricity-generating technologies. Costs per unit installed capacity are shown versus cumulative installed capacity. The lowest curve shows cost improvements of gas turbines. Today, gas turbines are the most cost-effective technology for electricity generation. This was certainly not the case three decades ago, when the costs were high and it was by no means certain that the great technology improvements suggested by the curve would be achieved. The technology can be characterized as "precommercial" until the early 1960s: the costs were very high and the improvement rates were particularly rapid, about a 20 percent reduction in unit costs per doubling of cumulative capacity. After about 1963 the improvement rate declined, and it has since averaged less than 10 percent per doubling. This development phase was no doubt also associated with a significant reduction in uncertainties. In the early development phases, the investments in this technology were indeed risky, as many accounts indicate.

Figure 10.4 also shows two relatively new electricity-generating technologies. Wind power is becoming a "commercial" technology in many parts of the world, especially where wind is abundant. A typical example is wind electricity generation in Denmark. The cost reductions for this technology are impressive at about 20 percent per doubling of cumulative capacity. However, as a source of electricity, wind is on average significantly costlier than gas turbines and the risk is higher. PVs show equally impressive performance improvements of about 20 percent unit cost reductions per doubling, but from a very high level of costs. They are about an order of magnitude more expensive than gas turbines per unit capacity. The prospects for this technology are thus very promising, but they are also associated with great risks for potential investors.

The learning curves have been used in a stylized form in a number of energy modeling approaches to capture elements of endogenous technological change. At IIASA, Messner (1995) and Messner *et al.* (1996) incorporated learning curves for six electricity-generation technologies in the simplified version of the (deterministic) energy systems engineering model MESSAGE. As this is a linear programming framework, integer programming was needed to deal with emerging nonconvexities in the problem formulation. It was assumed that "new" energy technologies have a certain cost reduction per doubling of cumulative installed capacity. The approach was very innovative and led to a number of important insights for further modeling of endogenous technological change (Grübler and Messner 1996; Nakicenovic 1996, 1997). However, the principal drawbacks were the significantly greater complexity and the very high computational demands. Another important deficiency of the approach was that the learning rates were deterministic. MESSAGE is a model with perfect foresight, so that early investments in new, costly technologies were always rewarded with increasing returns. While it is clear that such reductions are possible on average, they are associated with considerable uncertainty.

The next step at IIASA was to introduce uncertainties into the distributions of future costs. The basis for this approach was the IIASA technology inventory, which now contains information on the costs and technical and environmental characteristics of some 1,600 energy technologies (Messner and Strubegger 1991). Figure 10.5 shows future cost distributions for three energy technologies from the inventory (Nakicenovic *et al.* 1998a). The figure illustrates that the distributions are not symmetric and that they have very pronounced tails with both very "pessimistic" and very "optimistic" views on future costs per unit capacity. Such cost distributions were introduced explicitly into a simple, stochastic version of MESSAGE and have led to spontaneous "hedging" against this uncertainty as an emerging property of the model (Golodnikov *et al.* 1995; Messner *et al.* 1996).

Finally, both approaches to modeling endogenous learning and uncertainty were combined for a very highly stylized stochastic version of MESSAGE with increasing returns for "just" three "technologies" (see Chapter 11 in this volume). One technology is characterized by no learning whatsoever; another displays moderate learning of about 10 percent per doubling; and the third shows a much more rapid 20 percent per doubling.[1] The last two learning rates are associated with uncertainties based on the future cost distribution functions discussed above. In this much more complicated approach, the diffusion of new technologies occurs spontaneously, displaying the S-shaped patterns so characteristic of

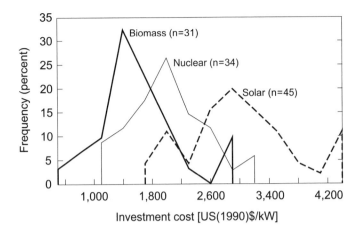

Figure 10.5. Range of Future Investment Cost Distributions from the IIASA Technology Inventory for Biomass, Nuclear, and Solar Electricity-Generation Technologies, in US(1990)$ per kilowatt (kW).

Sources: Messner and Strubegger (1991); Nakicenovic *et al.* (1998a).

technological diffusion. This occurs without any explicit technology inducement mechanisms other than uncertain learning and hedging. The disadvantage of the approach was that it is very computationally demanding and basically infeasible for application with a large number of technologies, as is required for development of long-term energy scenarios.

Here, we retain this basic approach and combine technological learning with uncertain outcomes while significantly extending the application to about 100 technologies. This is possible with the use of new global nonsmooth stochastic optimization techniques in conjunction with "parallel" problem structure and computing techniques. Cost reductions are assumed to be uncertain and thus are not specified by a given deterministic learning rate value. The learning rates are uncertain and are captured by assumed distribution functions. We assume that the generic cost reduction function has the following form:

$$CI_t = \left(2^{-\beta}\right)^{ND_t} , \qquad (10.1)$$

where CI_t is the *cost reduction index*, or the ratio between the technology unit costs (or, more precisely, the annual levelized costs) at time t and the initial cost in the base year; ND_t is the number of doublings of cumulative output achieved by time t compared with the initial output; and β is the progress ratio indicating the cost reduction rate per doubling of output. β is a random variable with a known distribution function. We have assumed that β is normally distributed with known mean and variance. It is important to note that the suggested algorithmic approach is not limited to the type of distribution assumed here and that it does not require any prior knowledge about the type of distribution function.[2]

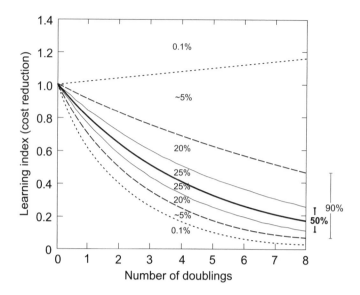

Figure 10.6. Uncertain Cost Reductions Represented by the Learning Index as a Function of Number of Doublings of Cumulative Output.

The expected value of β, the mean learning index (rate), corresponds to a 20 percent cost reduction per doubling of cumulative output. The numbers between the isolines of different learning indices indicate probability ranges. There is a small probability of no learning at all between any given doubling.

Figure 10.6 illustrates the uncertain learning index as a function of each doubling of cumulative output. In the example shown, the expected value for the cost reduction rate is 20 percent per doubling. The numbers between the isolines indicate the probability ranges of the occurrence of different learning rates. For example, there is a 50 percent chance that the cost reduction rate will fall between 14 percent and 25 percent per doubling. Note that there is a small chance ($\sim$5 percent) that the cost reductions will range from very small to an actual cost increase, and that there is a very small probability (0.1 percent) of a significant cost *increase* per doubling. This indicates a real possibility of negative learning or "induced forgetting" rather than learning. Such representation of uncertain learning illustrates the true risk of investing in new technologies. There is a high chance that technology will improve with accumulated experience, but there is also a small chance that it will be a failure and an even smaller chance that it will be a genuine disaster.

Here, we extend the application of uncertain learning to many new technologies, ranging from wind and PVs to fuel cells and nuclear energy. In keeping with IIASA's earlier approaches to capturing learning, we assume that traditional, "mature" technologies do not benefit from learning (another interpretation is that cost reductions as a result of learning are insignificant compared with other uncertainties that affect costs). Altogether there are 10 clusters of new technologies that benefit from induced learning.

As mentioned, we also assume that all technologies—both traditional and new ones—have stochastic costs with known distributions in any given period (similar to the distributions of electricity-generation technologies used by Golodnikov *et al.* 1995). The difference in the treatment of new and traditional technologies is that we assume that the cost distributions of traditional technologies are *static* over time and the costs in different time periods are independent random values. Owing to possible cost reductions resulting from technological learning (as described above), the costs for new technologies are *dynamic*. They are specified by conditional probabilities that result from the realization of a particular value for the uncertain learning rate. Again for simplicity we assume that all initial cost distributions are *log-normal* with different means and variances based on empirical analysis of technological characteristics using the IIASA technology inventory (see Strubegger and Reitgruber 1995).

We assume that the cost distribution function for each new technology at any given moment in time t, under the condition that N doublings of cumulative output have been achieved and the realized value for a random learning rate β is equal to b, is defined by the following expression:

$$F_t(\zeta | ND_t = N, \beta = b) = F_0(m_t, s_t) \,,$$

$$m_t = m_0(2^{-b})^N, s_t = Km_t \,, \tag{10.2}$$

where $F_0(\cdot, \cdot)$ is the initial log-normal distribution function with parameters m_0 and s_0, and K is the ratio between the standard deviation and the expected mean value and defines the compactness of the distribution. K is assumed to be a function of typical unit size (K increases with the unit size). We decided to keep K constant over time because of the lack of empirical data; therefore, it can be obtained simply by solving the following equation:

$$s_0 = Km_0 \,, \tag{10.3}$$

where m_0 and s_0 are derived empirically from statistical analysis.[3]

In addition to the uncertain learning rates, another new feature of our approach is that the future costs of all technologies are uncertain and assumed to be distributed according to the log-normal distribution. These are stylized distribution functions that indirectly reflect the cost distributions of energy technologies in the future based on analysis of the IIASA energy technology inventory. In addition, the mean value of these cost distributions is assumed to decrease and variance is assumed to narrow with increasing application of new technologies according to the generic cost reduction function (specified above) with the normally distributed progress ratio. This means that the process of technological learning is uncertain even as cumulative experience increases. The uncertainty of new technologies is characterized by the joint distribution of cost uncertainty and learning uncertainty. In summary, we assume uncertain future costs for all technologies and uncertain learning for new technologies.

Another uncertainty considered here is associated with the magnitudes and costs of energy reserves, resources, and renewable potential, and their extraction and production costs. Based on estimates by Rogner (1997), Nakicenovic *et al.* (1996), and others, we assume a very large global fossil resource base corresponding to some 5,000 Gtoe and correspondingly large renewable potentials. We also assume that the energy extraction and productions costs are uncertain, varying by a factor of more than five. Following the approach proposed by Rogner (1997), we formulated aggregate, global, upward-sloping supply curves with uncertain costs. Thus, the supply of fossil and non-fossil energy sources is characterized by expected increasing marginal costs and is one of the few areas where we have not assumed increasing returns, although we do assume uncertain costs.

10.4   Technological Spillovers

Technologies are related to one another. For example, jet engines and gas turbines for electricity generation are related technologies—in fact, the latter were derived from the former. These kinds of relationships among technologies are common. They imply that improvement in some technologies may be transferable to other, related technologies. For example, improvements in automotive diesel engines might lead to better diesel-electric generators, because the technologies are closely related to each other. Improvements in one area that lead to benefits in other areas are often referred to as spillover effects. In the case of related technologies, this is a real possibility. For example, we consider the different applications of fuel cells, such as for stationary electricity generation and for vehicle propulsion. We also consider fuel cells that have the same end-use application but different fuels, for example, hydrogen and methanol mobile fuel cells. These fuel cells are different but they are related in the technological sense, so that improvements in one technology may lead to improvements in the other. In this new approach to modeling technological learning and uncertainty, we explicitly consider the possibility of such spillover effects among energy technologies.

However, operational implementation of spillovers is not trivial. One important barrier is the lack of a technology "taxonomy." Presumably, the possibility of positive spillovers from technological learning is greater for technologies that are "more" similar than for those that are "less" similar. Thus, some kind of measure or metric of technological "proximity" or "distance" is required, even though a genuine taxonomy does not exist. A number of proposals have been made that could conceivably lead to the development of a taxonomy in the future (Foray and Grübler 1990). Instead of venturing into more complex representations of technology relationships, we simply assume that there are basically two explicit types of spillover effects. One is indirect through the connections among energy technologies within the energy system. For example, cheaper gas turbines mean cheaper electricity, so that *ceteris paribus* this could favor electricity end-use technologies for providing a particular energy service compared with other alternatives. The other effect is more direct. Some technologies are related through their "proximity" from the technological point of view, as was suggested by the example of hydrogen and methanol mobile fuel cells. We explicitly define "clusters" of technologies, where learning in one technology may spill over into another technology. The spillover effects are assumed to be strong within clusters and weaker across clusters.

Technology clusters were explicitly prespecified. Table 10.1 shows the groupings of technologies into 10 clusters. Each cluster consists of technologies that are related either because they are technologically "close" (i.e., are similar) or because they enable and support one another through the connections among them within the energy system.

The nature of the spillover effects is assumed to be different within and across clusters. Technologies from the same cluster share total cumulative output and are assumed to have the same learning rate, but their actual costs are drawn independently from their respective distributions.

Figure 10.7 illustrates the spillover effects within one cluster of technologies. The example shown gives two density functions of technology costs in 2030 for decentralized fuel cells. The density function with the lower overall costs is for the case of spillover effects within the technology cluster; that with the higher overall costs is for the case without spillover effects. The costs are given in US(1990) cents per kilowatt hour (kWh) of electricity generation *without* the fuel costs. Both the expected costs and their variance are substantially higher without the spillover effects. Thus, the costs, as well as the uncertainty, are expected to be lower with spillover effects. Therefore, the probability of lower costs is overall much higher with spillovers. However, the high tail of the density distribution is proportionally more pronounced in the case of spillover effects. This is an interesting feature of these density functions: the expected costs are generally lower with spillovers, but at the same time the possibility of realizations of very high costs compared with the mean is higher. Thus, spillovers also amplify somewhat the small chance of induced "forgetting."

Spillover rates between clusters are proportional (weighted) to the technological "proximity," for instance, how closely the technologies are related to one another. One example is additive learning from all kinds of fuel cells. For instance, stationary fuel cells can contribute significantly to learning for mobile ones because of the large capacity (size); conversely, experimenting with small-scale mobile units could be an important factor in the early development of stationary units.

Figure 10.8 is a schematic diagram of the 10 technology clusters indicating how they are related to one another with respect to the assumed learning spillover effects within the energy system structure. Two of the technology clusters, the nuclear high-temperature reactors (HTRs) and hydrogen infrastructure clusters (also shown in Table 10.1), are characterized by generally large "unit size" compared with other technologies. Consequently, very large cumulative output is required for achieving a doubling compared with the other clusters. This leads to correspondingly high risks associated with induced learning. The expected learning rates are indicated for each cluster. The modular (smaller "unit size") technologies generally have higher mean learning compared with the other technologies.

Table 10.1. Ten Technology "Clusters" of New Energy Technologies That Are Assumed to Benefit from Spillovers Associated with Technological Learning from One Technology to Another within the Cluster.

Technology cluster	Expected learning rate, %	Cumulative output at base year, GWyr^a	Typical lifetime, years	Technologies
Fuel cells (FCs) in transportation^b	20	0.01	10	Hydrogen-, liquid-hydrogen-, and methanol-based FCs in transportation, three technologies
Decentralized FCs (industrial and residential & commercial sectors)^c	20	0.1	20	Hydrogen-based FCs in industrial and R&C sectors, two technologies
Centralized FCs in energy sector^d	20	1	30	Natural-gas- and coal-based large-scale FCs, two technologies
Solar photovoltaic (PV)	25	0.2	20	Solar panels in industrial and R&C sectors and in energy sector, three technologies
Hydrogen infrastructure	10	1	40	Hydrogen and liquid hydrogen transportation and distribution infrastructure, two technologies
Solar to hydrogen	10	1	20	Solar to hydrogen production, one technology
Nuclear high-temperature reactors (HTRs)	10	10	40	Nuclear HTRs with possible hydrogen output, one technology
Wind	15	2	30	Wind power generators, one technology
Synthetic fuels and hydrogen production	20	3	30	Synthetic fuels and hydrogen production from biomass, gas, and coal, six technologies
Liquid hydrogen production	10	0.5	30	Hydrogen liquefaction, one technology

^{a}Part of model assumptions; in many cases, there are no reliable statistics for global cumulative output.
^{b}Contribute to other fuel cell clusters with weight 0.5 to decentralized and centralized units; accelerated by input from stationary units with weight 0.1 for decentralized and 0.01 for centralized installation.
^{c}Contribute to other fuel cell clusters with weight 0.5 to centralized units and 0.1 to transportation; accelerated by input from centralized units with weight 0.1 and from transportation with weight 0.5.
^{d}Contribute to other fuel cell clusters with weight 0.1 to decentralized units and 0.01 to transportation; accelerated by input from decentralized and transportation units with weight 0.5.

The highest mean learning rate is indicated for the solar photovoltaic cluster; the lowest rates are for the solar thermal to hydrogen, the nuclear HTRs, and the hydrogen infrastructure clusters.

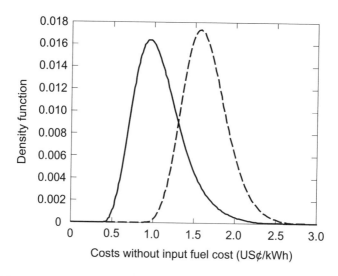

Figure 10.7. Spillover Effects within the Cluster of Decentralized Fuel Cell Technologies: Two Density Functions of Fuel Cell Costs in 2030, in US(1990) Cents per Kilowatt Hour (kWh) without Fuel Costs.

The density function with lower overall costs is for the case of spillover effects within the technology cluster; that with higher overall costs is for the case without spillover effects.

10.5 Model Structure and Implementation

In the presence of uncertainties any realistic policy bears risks, particularly the risk of under- or overestimating future technology costs. Explicit introduction of these risks into the model structure creates a driving force for the development of new technologies needed to make the energy system flexible enough to withstand possible instabilities and surprises. Thus uncertainty concerning future technology costs and characteristics in itself induces technological change. When this uncertainty is broadened to include technological learning and spillovers, the complex interplay between all three mechanisms leads to the same patterns of technological change that are encountered in deterministic modeling approaches. But whereas in deterministic modeling these patterns emerge under conditions of exogenous constraints, here this behavior is the result of induced technological change that occurs "spontaneously" owing to the stochastic nature of technological learning within the energy system.

The conventional approaches of control theory are applicable only in cases with a small number of variables (e.g., for simple energy systems), since such approaches deal with unrealistically detailed long-term strategies attempting to provide the best choice for every combination of uncertainties and designs that may occur before the given moment in time. This "chess game" solution concept is essential for the application of standard dynamic programming equations.

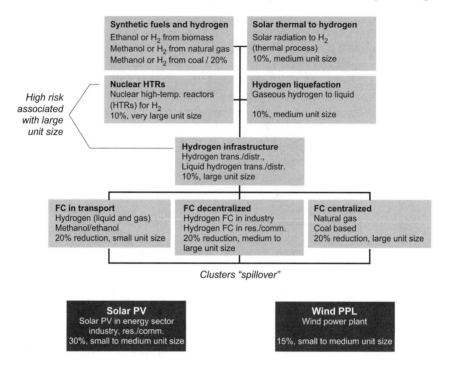

Figure 10.8. Schematic Diagram of the 10 Technology Clusters and Their Relationships to One Another with Respect to the Assumed Learning Spillover Effects within the Energy System Structure.

Technologies in each cluster are listed along with their assumed expected mean learning rates.

The same type of solution concept is used in multistage stochastic optimization models. Although in such cases large-scale optimization techniques are used instead of recurrent equations, the actual size of solvable problems is again small. The actual size of the problem is essentially connected with the solution concept, which requires the expansion of the original finite-dimensional model to a model with an infinite number of variables. Both approaches seem to be meaningful only for "online" or short-term energy planning problems. They are unrealistic for the analysis of long-term energy policies.

As it is impossible to explore all the details of long-term energy developments, our approach is based on the so-called two-stage dynamic stochastic optimization model with a rolling horizon. The solution concept in this case depicts the *ex ante* path of developments, which is flexible enough to be adjusted to possible *ex post* revealed uncertainties ("surprises"). The concept of a rolling horizon requires adjustments of *ex ante* strategies each time essential new information is revealed. A particular type of this model was proposed by Ermoliev and Norkin (1995) for the analysis of global change issues and is ideally suited for energy system engineering analyses as represented in some applications of the MESSAGE model.

The stochastic version of MESSAGE (see Golodnikov *et al.* 1995) is also a two-stage dynamic stochastic optimization model. This model explicitly incorporates risks of underestimating costs, which leads to a convex, generally nonsmooth, stochastic optimization problem.

The overall approach is based on the idea of representing energy systems development as a dynamic network where flows from one energy form to another and transformations of one energy form into another correspond to energy technologies such as electricity generation from coal or gas power plants. Figure 10.9 illustrates the assumed reference energy system as one composed of about 100 different technologies. Four types of energy flows are shown: (1) energy extraction from energy resources; (2) conversion of primary energy into secondary energy forms; (3) transport and distribution of energy to the point of end use, resulting in the delivery of final energy; and (4) the conversion at the point of end use into useful energy forms that fulfill the specified demands (as discussed above). All possible connections between the individual energy technologies are also specified in Figure 10.9. Various demands for useful energy are shown for different sectors of the economy. Each technology in the system is characterized by levelized costs, unit size, efficiency, lifetime, emissions, etc. In addition to various balance constraints, there are limitations imposed by the resource availability as a function of (uncertain) costs. The overall objective is to fulfill various demands using technologies and resources with minimal total discounted system costs.[4]

When future costs, demands, and other parameter values are known, it is possible to find a unique "optimal" solution for the evolution of the reference system shown in Figure 10.9. It is obtained by solving the following deterministic, linear optimization problem:

$$\min \sum_{t=0}^{T} ds^t < C^t, x^t >$$

$$B_t x^t \geq d^t, t = 0, 1, ..., T,$$

$$\sum_{k=0}^{t} R_k x^k = r, t = 0, 1, ..., T,$$

$$\sum_{k=0}^{T} P_k x^k \leq e^t, t = 0, 1, ..., T,$$

$$0 \leq x^t \leq \bar{x}^t, t = 0, 1, ..., T, \tag{10.4}$$

where

- $x^t = (x_1^t, ... x_n^t)$ are activity levels of technologies and resources at time t;

- B_t is the matrix of input and output relations among the technologies and d^t is the demand vector;

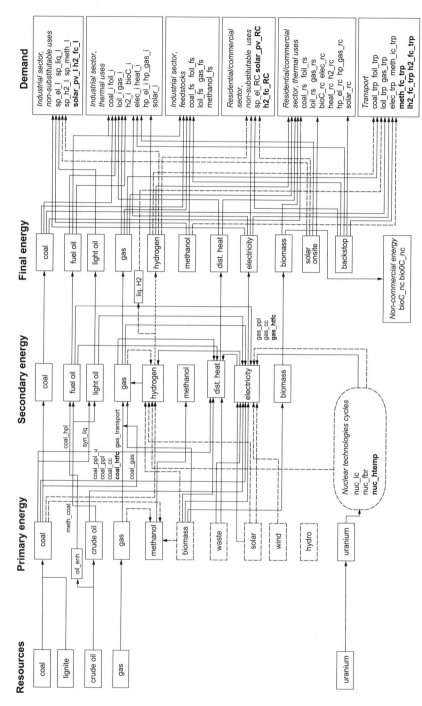

Figure 10.9. Schematic Diagram of the Reference Energy System Showing Some 100 Individual Technologies.

- R_t is the matrix for approximating the quadratic costs of resources and balances for resource use and r are corresponding quantities;

- P_k is the matrix of systems constraints, like market penetration constraints and maximum shares of specific resource and technology activities, and e^t are corresponding limits; and

- $\bar{x}^t$ is the upper limit on technological activities.

Such deterministic formulations of future energy systems development result in highly restrained possibilities. In addition, the dynamics of future developments are prescribed by the system of assumed constraints. In contrast, there is a wide range of possible alternative future developments of the energy system, especially in the long run (at the scale of a century). This is amply demonstrated by the enormous range of future energy requirements and CO_2 emissions in energy scenarios in the literature (see Figure 10.1).

In contrast, the alternative formulation of the problem proposed here is highly unrestrained and "open." We assume that there is a priori "freedom of choice" among fundamentally different future structures of the energy system and possible future dynamics. The uncertainty is resolved through a simultaneous drawing from all distributions from each particular technology dynamics (see the box on Terminology). To make a rational choice among alternative energy system structures, technology dynamics are compared on the basis of expected systems costs and risks associated with each particular technology dynamics. Risks or benefits are defined here as functions of the difference between the expected and realized costs of each technology dynamics. There are a number of ways to quantify risk (see, e.g., Markowitz 1959). We adopt a technique whereby the risk is represented by piecewise linear functions of the following form:

$$\min \mathrm{E} \sum_{t=0}^{T} ds^t < C^t(\omega), x^t > + \mathrm{ER}(C^t(\omega), x|_0^t)$$

$$= \sum_{t=0}^{T} ds^t < \mathrm{E}C^t(\omega), x^t > + r_i \sum_{i=o}^{l} \mathrm{E} \max(0, < C^t(\omega) - \mathrm{E}C^t(\omega)$$

$$-\Delta \hat{c}_i, x|_0^t) . \tag{10.5}$$

This clearly asymmetric form of the risk function has an obvious advantage over a more standard approach based on variance minimization. Splitting the risk function into two parts representing risk associated with cost underestimation[5] and the benefit associated with cost overestimation is a natural reflection of the highly asymmetric risk perception of "losses" and "gains." Moreover, different "actors" (energy agents) may have quite different levels of risk aversion. In principle, this approach allows for the representation of different risks perceived by different decision actors or agents.

This asymmetric treatment of "risk" and "benefit" significantly increases the complexity of the problem—risk cannot be expressed simply in terms of a functional relation of expected values and corresponding variances, as is done in Markowitz's formulation (Markowitz 1959). Formally speaking, the objective function specified above is a nonsmooth and obviously highly complicated nonconvex function defined on probabilistic space. In its general form, it is an analytically intractable problem, even in the case of a relatively small system where just a few technologies are considered. The problem could be solved using a stochastic approximation technique (see Ermoliev and Norkin 1995). This stochastic approximation approach is based on the idea of estimating the solution of the original problem by solving another stochastic problem where the original probabilistic space is replaced with a finite (sufficiently large) number of simultaneously generated "samples" according to the distribution function for uncertain parameters (see Grübler and Gritsevskyi, 1998). This approach differs significantly from the conventional Monte Carlo approach. All drawings are performed *simultaneously* and the resulting policy conclusion is formulated against the background of all considered outcomes. There are strong methodological similarities between so-called *exploratory modeling* (see Bankes 1993; Lempert *et al.* 1996; Robalino and Lempert 2000) and the approach used here, although we use a very different implementation and analysis technique.

A systematic approach to aggregating scenario-specific solutions into a robust solution is examined in Ermoliev and Wets (1988). These techniques require explicit characterization of scenario-specific solutions, which may lead to extremely large optimization problems. Different stochastic optimization techniques deal with the design of robust solutions from a set of previously or sequentially simulated scenarios. In the latter case, the stochastic optimization procedure can be viewed as a sequential adaptation of a given initial energy policy by learning from the simulated history of its implementation.

In our test runs we initially used between 100 and 500 simultaneously drawn scenarios from each technology dynamics specification as an approximation of a theoretically infinite number of possible realizations. These ranges for the appropriate number of scenarios were obtained as a result of practical experiments and represent optimal trade-offs between exponentially growing computational complexity and reasonable accuracy of obtained solutions. Eventually, we decided that 250 simultaneously drawn scenarios are sufficient for a given technology dynamics. It is important to emphasize that it is not necessary to maintain high accuracy by using many drawings during the initial calculation steps, when the value of the objective function is far from "optimal," as the difference between a solution value and the value of the new draw is much larger than the errors resulting from "rough" stochastic approximation. However, at the final stage the number of drawings needs to be increased. At that stage, we use an alternative drawing technique to obtain better estimates for error bounds.

Terminology

A *scenario* is a particular deterministic realization of a future energy system. Here it specifies unique values for all activity levels, such as energy flows, increases in capacities, total systems costs, energy extraction, etc. *Technological dynamics* denotes a more generic characterization of future developments with inherent uncertainties surrounding, for instance, future costs. Each resolution of these uncertainties inherent in technological dynamics results in a given scenario. There is an infinite number of possible scenarios that share exactly the same technological dynamics. Thus, technology dynamics specifies a set of uncertain, generic relations. In particular, it specifies the set of uncertain cost reductions as a function of doublings of output, the cost distributions in any given period, and possible spillover effects within and across the 10 technology clusters within the reference energy system. Our approach to analyzing and comparing alternative technological dynamics is to assume specific distribution functions for uncertain parameters and relations. The uncertainty is resolved through a simultaneous drawing from all distribution functions for a given technology dynamics that then results in a deterministic scenario. After many such drawings, expected costs and other characteristics of the scenario sample for a particular technology dynamics can be estimated. The expected costs and other sample statistics can then be used to obtain risk estimates associated with each technology dynamics. Each scenario within the set belonging to one specific technology dynamics can be characterized by a conditional probability relative to the other scenarios in that set.[6] *Feasible* technology dynamics are those that satisfy given energy demands and other systems constraints. A *run* of scenarios refers to all scenarios generated from a given set of technology dynamics through simultaneous drawings from all uncertain distributions. In this application we have analyzed 520 alternative technology dynamics and have drawn some 250 scenarios for each one, resulting in a run of about 130,000 scenarios.

We call a given technology dynamics *optimal (suboptimal)* for a given run if it is optimal (suboptimal) compared with all other technology dynamics in the run of scenarios with respect to the weighted sum of its expected systems costs, and risk functions based on these costs, for all drawn scenarios.

More formally, the problem is given by the following:

$$\min \mathrm{E} \sum_{t=0}^{T} ds^t < C^t(x|_0^t, \omega), x^t > + \mathrm{ER}(C^t(x|_0^t, \omega), x^t)$$

$$= \sum_{t=0}^{T} ds^t < \mathrm{E}C^t(x|_0^t, \omega), x^t > + r_i \sum_{i=0}^{l} \mathrm{E}\max(0, < C^t(x|_0^t, \omega)$$

$$-\text{E}C^t(x|_0^t, \omega) - \Delta\hat{c}_i, x^t) \underset{(N\to\infty)}{\approx}$$

$$\sum_{t=0}^{T} ds^t < \text{E}C^t(x|_0^t, \omega), x^t > + \frac{1}{N}\sum_{s=1}^{N}$$

$$\left(r_i \sum_{i=0}^{l} \text{E}\max(0, < C^t(x|_0^t, \omega_s) - \text{E}C^t(x|_0^t, \omega) - \Delta\hat{c}_i, x^t) \right)$$

$$B_t x^t \geq d^t, t = 0, 1, ..., T, \quad \sum_{k=0}^{t} R_k(\omega_s)x^k = r(\omega_s), t = 0, 1, ..., T$$

$$\sum_{k=0}^{T} P_k x^k \leq e^t, t = 0, 1, ..., T, \quad 0 \leq x^t \leq \bar{x}^t, t = 0, 1, ..., T, \quad (10.6)$$

where

- $x|_0^t$ is $(x^0, x^1, ..., x^t)$;

- $C^t(x|_0^t, \omega)$ are stochastic costs under the condition that technology dynamics $x|_0^t$ is chosen and such that

- $CI_t = \frac{C^t(x|_0^t, \omega)}{C^0(\omega)}$, the cost reduction index, has a distribution function as described before (with the number of doubling ND_t calculated from $x|_0^t$) and the initial distribution function for $C^0(\omega)$ is equal to $F_0(\cdot)$;

- $\Delta\hat{c}_i$ are given "threshold" values for total cost deviations; and

- $R_k(\omega)$ and $r(\omega)$ reflect uncertain quantity-to-cost relations.

10.6   Solution Technique and Computational Approach

As mentioned above, original stochastic global nonsmooth optimization is approximated by solving a sequence of large-scale linear optimization problems. This is done by applying a two-level nested structure. The global optimization part, which defines technological dynamics with respect to new unit installations for technologies with increasing returns to scale, is an implementation of an *adaptive global optimization random search algorithm* specifically tailored to network flows optimization problems. [For a description of such algorithms, see Horst and Pardalos (1995) and Pinter (1996).] The inner algorithm is the *interior-point method* for linear optimization. The PCx and $pPCx$ solvers used in this study were provided by the Argonne National Laboratory (Wright 1996a, 1996b; Czyzyk *et al.* 1997). These solvers are written in C code, modified to increase computational efficiency for our specific problem formulation and to link the solvers directly to the global optimization part of the structure.

A big advantage of the adaptive random search algorithm is that it does not require strict sequential updating of the approximated solution. Rather, it refines the approximated solution when new information becomes available. This allowed us to devise a "parallel" adaptation of this technique. The inner linear optimization problem is relatively large and difficult to solve. Finding a solution for a given technology dynamics (with fixed uncertainty distribution parameters) using the global optimization algorithm requires approximately 10–40 minutes of run time, depending on the number of simultaneous drawings from uncertain distributions, the number of parameters to be considered, and whether or not approximation of the starting point is available (the partial "hot" restart technique).

The original problem implementation was done using a CRAY T3E-900 supercomputer at the National Energy Research Scientific Computing Center (NERSC), in the United States. NERSC is funded by the US Department of Energy, Office of Science, and is part of the Computing Sciences Directorate at the Lawrence Berkeley National Laboratory. All initial feasibility runs and a number of experiments were performed using 32–64 processing units on the CRAY T3E-900. The problem was then ported to the IIASA computer network environment and re-implemented using the Message Passing Interface (MPI) standard. We used a public portable implementation of MPI–MPICH developed and supported by the Argonne National Laboratory and a special implementation for Windows NT network clusters (WMPI) provided by the University of Coimbra, Portugal. Currently it is operational on a network cluster that contains from 6–16 Intel Pentium II 233 MHz PCs (the number of PCs can be changed dynamically). Typical runs take from 22–46 wall clock hours. Owing to the extended logging procedure, calculations can be easily operated remotely, for example, stopping and re-activating at any time. This technique allows the computer to be used during "off-peak" and weekend hours.

10.7 Major Findings and Conclusions

From the 520 alternative technology dynamics, about 53 resulted in scenarios with very low overall energy system costs. They all fall within 1 percent of the best values achieved. We designate this set of 53 technology dynamics as "optimal" because they fulfill the "optimality" criteria. Most of the statistical and other analyses here focus on these 53 optimal technology dynamics.

These 53 optimal, but fundamentally different, technology dynamics produce a wide range of alternative emergent energy systems. They all share the same useful energy demand trajectory but cover most of the range of CO_2 emissions found in the literature and unfold into all possible future energy system structures. The underlying scenarios include futures that range from an increasing dependence on fossil energy sources to a complete transition to alternative energy sources and nuclear energy. Thus, one result of the analysis is that different energy system structures emerge with similar overall costs; in other words, there is great diversity across alternative energy technology strategies. The strategies are path dependent, and it is not possible to choose *a priori* "optimal" directions of energy systems development.

The scenarios from the literature span a wide range of future energy requirements and emissions owing to the variation of the driving forces of future emissions, such as energy demand. In contrast, the set of scenarios with endogenous learning spans the range as a result of different technological dynamics alone. It is interesting to note that the "optimal" scenarios quite closely match the distribution of the scenarios from the literature, but with a somewhat narrower range (they leave the extreme tails of the distribution uncovered). In contrast, the frequency distribution of the full set of 520 technology dynamics is different from the other two, with many more scenarios in the mid-range of the distribution. This means that the optimal or most "cost-effective" development paths correspond quite closely to the scenario distribution from the literature. The "median" or "central" futures are underrepresented in the literature and among the scenarios, indicating a kind of "crowding-out" effect surrounding balanced and median scenarios. In any case, technological learning as specified in our approach leads to future energy systems that are marked by either high or low emission ranges (with a single useful demand trajectory), demonstrating a kind of implicit bifurcation across the range of possible emissions.

Another finding from the analysis is that endogenous technological learning with uncertainty and spillover effects will have the greatest impact on the emerging energy system structures during the first few decades of the twenty-first century. Over these "intermediate" periods of time, these two processes create effective lock-in effects and increasing returns to adoption. In the very long run, however, none of these effects is of great importance. The reason is that over such long periods many doublings of capacity of all technologies with inherent learning occur, so little relative cost advantage results from large investments in only a few technologies and clusters. Therefore, the main finding is that, under uncertainty, the near-term investment decisions in new technologies are more important in deciding the direction of long-term development of the energy system than are decisions made toward the end of the time horizon. Thus, the most dynamic phase in the development of future energy systems will occur during the next few decades. It is during this period that there will be a high degree of freedom of choice among the future technologies, and many of these choices will lead to high spillover learning effects for related technologies.

Our analysis of the emerging dynamics and different directions of energy systems development suggests some policy implications, First, future RD&D efforts and investments in new technologies should be distributed across "related" technologies rather than directed at only one technology from the cluster, even if that technology appears to be a "winner." Second, RD&D efforts and technology investments should not be spread across a large portfolio of future technologies, but should focus on (related) technologies that might form technology clusters.

Acknowledgments

We would like to thank Sabine Messner, Gordon J. MacDonald, Yuri Ermoliev, and Manfred Strubegger, all from IIASA, for their help and advice. Sabine Messner and Gordon J. MacDonald worked with us on the grant from the National Energy Research Scientific Computing Center at Lawrence Berkeley National Laboratory that is funded by the US Department of Energy. This grant allowed the original problem implementation on a CRAY T3E-900 supercomputer, and we are grateful for the financial support provided by the US Department of Energy. The original MESSAGE model implementation and technology assumptions are based on the work of Sabine Messner. We are also grateful to Yuri Ermoliev, who helped with the development of the solution methods and continuously provided help and advice. Last, but not least, we thank Manfred Strubegger, who provided the fossil energy supply functions, implemented important changes in the problem solution processing module, and developed the new script that was used for storing the results of this analysis.

We would also like to thank colleagues from other institutions who have provided software and support for our research, including Michael Wagner from the Argonne National Laboratory, Steve Wright of Cornell University, and Francesca Verdier from the National Energy Research Scientific Computing Center at Lawrence Berkeley National Laboratory for their assistance.

Notes

1. Cost reduction and the *learning rate* may be quite different, depending on how "learning" is measured. The learning rate for photovoltaics in Figure 10.4 is about 20 percent per doubling of cumulative capacity. For example, Watanabe (1995) analyzed direct investment in photovoltaics in Japan and found that the unit costs decreased by about 50 percent per doubling of cumulative investment. Grübler (1998) estimated the learning rate at 30 percent per doubling of cumulative installed capacity based on the data set from Watanabe (1995).

2. To use our approach, we must be able to compute the mean value for the corresponding distribution and to generate random samples based on that distribution. Implementation in the form of a "black box" is perfectly suitable.

3. The suggested technique does not require or utilize the specific relation between F_t and the initial distribution F_0. It also does not require that K be kept constant over time. In the absence of a better understanding of the quite complex and nonlinear relationship and owing to the lack of empirical data, we decided to use the simplest assumptions possible— the type of distribution stays the same the (the distribution does not change its shape), the mean value follows the realized cost reduction curve, and variance decreases proportionally with the expected cost reduction.

4. As in many other models, a 5 percent discount rate was adopted.

5. More then one factor could lead to underestimation of realized cost. For all new technologies (especially in the early stages of development), even in cases where the cost reduction rate is as good or better then expected, there is significant cost fluctuation resulting from uncertainty associated with such cost distributions (high variance, heavy tails, and so on). Factors such as a high dependence on a particular resource form, a low level of technological diversification, and a strong linkage between system parts largely contribute

to the increasing probability of substantial cost underestimation. Such analysis would be nearly impossible to perform on the basis of simple cost-to-cost analysis for alternative energy supply chains.

6. Each scenario has exactly zero probability of realization. It makes sense to talk about scenario probability under some conditions. For example, under conditions that from the set of N scenarios *one* should happened, it is possible to introduce and compare relative probability defined on this set of N scenarios.

References

Bankes, S.C., 1993, Exploratory Modeling and Policy Analysis, RAND/RP-211, Santa Monica, CA, USA.

Christiansson, L., 1995, Diffusion and Learning Curves of Renewable Energy Technologies, WP-95-126, International Institute for Applied Systems Analysis, Laxenburg, Austria.

Czyzyk, J., Mehrotra, S., Wagner, M., and Wright, S.J., 1997, PCx User Guide (Version 1.1), Technical Report, OTC 96/01, http://www-unix.mcs.anl.gov/otc/Tools/PCx/doc/PCx-user.ps.

Ermoliev, Y.M., and Norkin, V., 1995, On Nonsmooth Problems of Stochastic Systems Optimization, WP-95-096, International Institute for Applied Systems Analysis, Laxenburg, Austria.

Ermoliev, Y.M., and Norkin, V., 1998, Monte Carlo Optimization and Path Dependent Nonstationary Laws of Large Numbers, IR-98-009, International Institute for Applied Systems Analysis, Laxenburg, Austria.

Ermoliev, Y.M., and Wets, R.J.-B., 1988, *Numerical Techniques for Stochastic Optimization*, Springer-Verlag, Berlin, Germany.

Foray, D., and Grübler, A., 1990, Morphological analysis, diffusion and lock-out of technologies: Ferrous casting in France and Germany, *Research Policy*, 19(6):535–550.

Golodnikov, A., Gritsevskyi, A., and Messner, S., 1995, A Stochastic Version of the Dynamic Linear Programming Model MESSAGE III, WP-95-094, International Institute for Applied Systems Analysis, Laxenburg, Austria.

Grübler, A., 1998, *Technology and Global Change*, Cambridge University Press, Cambridge, UK.

Grübler, A., and Gritsevskyi, A., 1998, A model of endogenized technological change through uncertain returns on learning, http://www.iiasa.ac.at/Research/TNT/WEB/Publications/

Grübler, A., and Messner, S., 1996. Technological uncertainty, in N. Nakicenovic, W.D. Nordhaus, R. Richels, and F.L. Toth, eds, *Climate Change: Integrating Science, Economics, and Policy*, CP-96-001, International Institute for Applied Systems Analysis, Laxenburg, Austria.

Horst, R., and Pardalos, P.M., eds, 1995, *Handbook of Global Optimization*, Kluwer, Dordrecht, Netherlands.

Lempert, R.J., Schlesinger, M.E., and Bankes, S.C., 1996, When we don't know the costs or the benefits: Adaptive strategies for abating climate change, *Climatic Change*, 33(2):235–274.

MacGregor, P.R., Maslak, C.E., and Stoll, H.G., 1991, *The Market Outlook for Integrated Gasification Combined Cycle Technology*, General Electric Company, New York, NY, USA.

Markowitz, H., 1959, *Portfolio Selection*, Wiley, New York, NY, USA.

Messner, S., 1995, Endogenized Technological Learning in an Energy Systems Model, WP-95-114, International Institute for Applied Systems Analysis, Laxenburg, Austria.

Messner, S., and Strubegger, M., 1991, Part A: User's Guide to CO2DB: The IIASA CO_2 Technology Data Bank–Version 1.0, WP-91-031, International Institute for Applied Systems Analysis, Laxenburg, Austria.

Messner, S., Golodnikov, A., and Gritsevskyi, A., 1996, A stochastic version of the dynamic linear programming model MESSAGE III, *Energy*, 21(9):775–784.

Morita, T., and Lee, H.-C., 1998, Appendix to emissions scenarios database and review of scenarios, *Mitigation and Adaptation Strategies for Global Change*, 3(2–4):121–131.

Nakicenovic, N., 1996, Technological change and learning, in N. Nakicenovic, W.D. Nordhaus, R. Richels, and F.L. Toth, eds, *Climate Change: Integrating Science, Economics, and Policy*, CP-96-001, International Institute for Applied Systems Analysis, Laxenburg, Austria.

Nakicenovic, N., 1997, Technological Change as a Learning Process, paper presented at the Technological Meeting '97, International Institute for Applied Systems Analysis, Laxenburg, Austria.

Nakicenovic, N., Grübler, A., Ishitani, H., Johansson, T., Marland, G., Moreira, J.R., and Rogner, H.-H., 1996, Energy primer, in *Climate Change 1995: Impacts, Adaptations and Mitigation of Climate Change: Scientific-Technical Analysis*, Contribution of Working Group II to the Second Assessment Report of the IPCC, Cambridge University Press, Cambridge, UK, pp. 75–92.

Nakicenovic, N., Grübler, A., and McDonald, A., eds, 1998a, *Global Energy Perspectives*, Cambridge University Press, Cambridge, UK.

Nakicenovic, N., Victor, N., and Morita, T., 1998b, Emissions scenarios database and review of scenarios, *Mitigation and Adaptation Strategies for Global Change*, 3(2–4): 95–120.

Pinter, J., 1996, *Global Optimization in Action*, Kluwer, Dordrecht, Netherlands.

Robalino, D., and Lempert, R.J., 2000, Carrots and sticks for new technology: Crafting greenhouse gas reduction policies for a heterogeneous and uncertain world, *Integrated Assessment*, 1(1):1–19.

Rogner, H.-H., 1997, An assessment of world hydrocarbon resources, *Annual Review of Energy and the Environment*, 22:217–262.

Strubegger, M., and Reitgruber, I., 1995, Statistical Analysis of Investment Costs for Power Generation Technologies, WP-95-109, International Institute for Applied Systems Analysis, Laxenburg, Austria.

Watanabe, C., 1995, Identification of the role of renewable energy, *Renewable Energy*, 6(3):237–274.

Wright, S.J., 1996a, Modified Cholesky Factorizations in Interior-Point Algorithms for Linear Programming, Preprint ANL/MCS-P600-0596, Argonne National Laboratory, Argonne, IL, USA.

Wright, S.J., 1996b, *Primal-Dual Interior-Point Methods*, SIAM, London, UK.

Chapter 11

A Model of Endogenous Technological Change through Uncertain Returns on Innovation

Arnulf Grübler and Andrii Gritsevskyi

11.1 Introduction

Changes in products, devices, processes, and practices—that is, changes in technology—largely determine the development and consequences of industrial society. Historical evidence (e.g., Freeman 1989; Mokyr 1990; Maddison 1991; Grübler 1998) and economic theory (Tinbergen 1942; Solow 1957; Denison 1962, 1985; Griliches 1996) confirm that advances in technological knowledge are the single most important contributing factor to long-term productivity and economic growth. Technology is also central to the long-term evolution of the environment and to development problems now on policy agendas worldwide under the general rubric, "global change." Although technology is central, technological change is typically the least satisfactory aspect of global change modeling. Each of the factors that determine the wide range of projected emissions of, say, carbon dioxide (CO_2)—such as the future level of economic activity (largely driven by advances in productivity), the energy required for each unit of economic output, and the carbon emitted for each unit of energy consumed—is a function of technology. This also applies to the technological linkages in any kind of macro or sectoral production function (Abramovitz 1993). Consequently, technology largely accounts for the wide range seen in published long-term carbon emission estimates. A recent review of the literature (Nakicenovic *et al.* 1998b) indicates an emissions range spanning from 2 GtC (gigatons, 10^{15} grams, of elemental carbon) to well above 40 GtC by 2100. This wide span is largely explained by differences in technology-related assumptions such as macroeconomic productivity growth, energy intensities, and availability and costs of low- and zero-carbon technological alternatives.

In this chapter we outline a model of endogenous technological change that is applied to the energy sector and the CO_2 emissions problem. Like any model, it is an abstraction of our understanding of how "the system works." Therefore, we begin by outlining a number of stylized abstractions from our review of the theoretical and empirical literature (Section 11.2). We emphasize in particular that, like all knowledge, improved technological knowledge can exhibit *increasing returns*. These are, however, highly uncertain, resulting in diverse innovation strategies reflecting different technological "expectations" (Rosenberg 1996) by a multitude of actors. Adopting the term "innovation" in the Schumpeterian sense means that all innovative activity is essentially *economic*, that is, technology arises from "within" (Schumpeter 1934) the economy and society at large. Innovation is costly, requiring up-front expenditures in improving technological knowledge in its disembodied form [typically research and development (R&D)] and in its embodied form (plants and equipment). This leads to the formulation of a *multi-actor, multiregion* model of *uncertain increasing returns* to technological innovation. In this model, innovation costs include both R&D and expenditures on physical plants and equipment that lead to improvements via learning by doing and learning by using. The basic elements of the model and its most salient parameterizations and input assumptions are outlined in Section 11.3 in a nonmathematical way. Section 11.4 presents some illustrative quantitative model results, exploring in particular changes in patterns of technology diffusion and carbon emissions under alternative assumptions concerning uncertainty about resource availability and costs, energy demand, technology characteristics, and the existence of uncertain environmental limits that are examined in both the absence and the presence of uncertain increasing returns phenomena. Conclusions are presented in Section 11.5, highlighting in particular the implications for analytical "next steps" toward the challenge of a deeper theoretical and empirical understanding of the mechanisms and incentives driving technological change.

11.2   Main Features of Technological Change

11.2.1   Introduction

A review of the literature and of empirical case studies (e.g., Freeman 1994; Grübler 1998) suggests that technological change can be summarized as being *dynamic, cumulative, systemic*, and *uncertain* in nature. Technological change is by definition *dynamic*, as it is characterized by the continuous introduction of new varieties ("species") and continuous improvements and modifications of existing ones. These new and improved technologies essentially represent new forms of technological knowledge that cannot be created *ex nihilo* (i.e., without prior knowledge), hence technological change is inherently *cumulative*. Change is also *systemic*: because of technological interdependencies and infrastructure needs, any change in a component of a larger technological system requires corresponding changes in other up- or downstream components of that system. Finally, technological change is inherently *uncertain*. The outcomes of the innovation process (the form and applicability of new technological knowledge) as well as of technology diffusion (market potentials, economics, etc.) cannot be known beforehand.

How should the above features be factored into a model of technological change? Basically, we argue that traditional representations of technological change need to be extended—first, by considering the types and *characteristics* of technological knowledge as well as *who* creates and uses it, and, second, by explicitly considering uncertainty and the systemic aspects of technological change.

11.2.2   Forms of technological knowledge

The dynamic nature of technological change poses an epistemological problem. The nature of the object under investigation, "technology," is very different at its conception, at its introduction into the market, and at the point when it is widely applied. Ayres (1987) reviews various conceptualizations and formulations of technology life-cycle models. Although Ayres is skeptical of the integrative capability of the model as a theory "of everything" in technological change, his review demonstrates the powers of the technology life-cycle model as a conceptual classification tool for understanding the different stages of technological change and the very different mechanisms at work transforming the technology itself as well as the market (and ultimately even the social) environment in which it is embedded.

Different technology life-cycle stages can be distinguished using a variety of criteria pertaining to characteristics such as technology performance, market volume and structure, etc. While of interest, they do not address the issue of sources and mechanisms of technological change proper. Distinctions based on the relative contributions of different forms of knowledge generation (Cowan and Foray 1995) or of disembodied versus embodied technological change (Grübler 1998) are more relevant for the discussion here.

A simple four-stage taxonomy of a technology's life cycle drawing on the work of Schumpeter (1911, 1934) distinguishes four stages in a technology's evolution: invention, innovation, niche markets, and diffusion. Whereas the invention stage (the first demonstration of the feasibility of a new solution) and the innovation stage (the first regular production of a new technology) are discrete time events, the niche market and (large-scale) diffusion stages overlap. The distinction between invention and diffusion in terms of mechanisms of technological change is straightforward. At the invention stage, change is by definition disembodied and new technological knowledge is generated exclusively via directed research activity (e.g., basic R&D). By contrast, during the diffusion phase, technological change is largely embodied (in the form of new plants, equipment, and products) and improved knowledge is generated primarily via learning by doing and learning by using—that is, in direct connection with embodied technological change (change in capital stock via investments). Knowledgewise, there is also an important contribution from applied R&D (involved, e.g., in process modifications or in scaling up the unit size of technological installations). The innovation and niche market phases combine both embodied and disembodied technological change—knowledge generation via R&D as well as actual "hands-on" experience.

It is important to emphasize that the technology life-cycle model is used here as a conceptual and terminological framework. It does not imply a linear causality chain or a strict temporal sequence, as the literature is unanimous in rejecting such models of technological change because of the multitude of interlinkages and feedbacks at work [for a review of the literature, see Freeman (1994)].

In a conceptual simplification, in the subsequent discussion we merge the above two dichotomies. Disembodied technological change takes the form of "blueprints" (e.g., a patent application) resulting from directed research activity (e.g., basic R&D, with an emphasis on research). Embodied technological change represents changes in hardware; technological knowledge improves primarily via hands-on experience in market applications, with a costwise relatively small but knowledgewise important contribution from directed research activities (applied R&D, with an emphasis on development work).[1] We recognize that no embodied technological change can proceed without disembodied change; in other words, investments in new plants and equipment usually also require changes in production organization, management, marketing strategies, etc. There are also important feedback mechanisms at work between embodied technological change and disembodied (technological and other) knowledge generation, for example, in the form of new instrumentation and measurement technologies that improve scientific analysis and understanding (Foray and Grübler 1996). For the analysis of technological change, however, the above distinctions, even if oversimplified, appear to be more useful than the traditional dichotomy between science and technology [conceptualized as applied science; see Toynbee (1962)]. This is because many of the improvements in technological knowledge have less to do with advances in science than with gaining hands-on experience, improving management, etc.—factors generally subsumed under the heading, "learning by doing."

An important, if evident, observation is that, in addition to the different forms of technological knowledge generation, there is also a *large variety of actors* that develop that knowledge. Heterogeneity of agents is therefore an important aspect emphasized throughout the economics and sociological literature dealing with the creation and diffusion of technological knowledge (or, in our terminology, the development and application of disembodied and embodied technological knowledge). Heterogeneity is a central theme, for instance, when discussing the influence of industry structure on invention and innovation (e.g., Scherer 1980); when explaining differences in the adoption rates of industrial process innovations among different firms (e.g,. Mansfield 1968, 1977); or for understanding differences in adoption rates and levels of consumer products (Rogers 1983; Mahajan *et al.* 1991).

While different technologies or technology life-cycle phases are characterized by different relative contributions of tangible and intangible elements of technological change, it is worthwhile to recall two main conclusions from the literature.

First, different forms of knowledge generation (basic applied R&D, learning by doing, and learning by using) are more complementary than substitutable (Rosenberg 1990; Pavitt 1993). Second, knowledge is acquired from both internal and external sources. The relative proportions vary widely across industries and technologies, and with firm characteristics [see the review by Freeman (1994) and the model of Silverberg (1991)], yet complementarity prevails over substitutability. New technological knowledge cannot be assimilated exclusively from external sources without corresponding internal knowledge. As Freeman (1994) notes, "all firms make use of external sources [of learning]." In other words, technological knowledge cannot be increased exclusively by internal efforts. Cohen and Levinthal (1989) have shown empirically the positive relation between the generation of internal technological knowledge (measured by R&D efforts) and the ability to assimilate external technological knowledge. Recognizing the importance of external knowledge sources for technological innovation and change raises an important externality problem: new knowledge is expensive to generate, but comparatively inexpensive to assimilate. "Leaky" (Mansfield 1985) technological knowledge implies that benefits do not necessarily accrue to those generating new knowledge, but rather to those who can most effectively *apply* new knowledge. This externality accrues first at the level of individual technology actors rather than at higher levels of aggregation (economic sectors, national economies, society at large).

Thus, at the highest level of abstraction, the following features of technology dynamics appear to be relevant for trying to model technological change as an endogenous process. First, the simple technology life-cycle model allows a separation of two distinct phases of technology genesis. At its earliest phase (invention), technological change is by definition "disembodied" (i.e., does not give rise to investments in new plants, equipment, and products), and knowledge generation relies exclusively on dedicated activity (typically R&D). Apart from the fact that such activities require resources and provide the essential basis for potential subsequent innovation, the act of invention is basically an act of human creativity that cannot be modeled. The second phase is characterized by complementary relationships between disembodied and embodied technological change as sources of improved knowledge. The relative proportions of different forms of knowledge generation (e.g., R&D, versus learning by doing) are highly variable over a technology's life cycle and across different technologies. As a generalization, we might conclude that the further a technology progresses in its technology life cycle, the more "embodied" technological change becomes. However, strict and neat separations in terms of cause-and-effect relationships and temporal sequences are not possible. Thus, second, both disembodied and embodied technological change, as well as different forms of knowledge generation, must be treated as interrelated and interdependent. Third, there is no social planner devising new technologies and improving existing ones. The existence of a multitude of different actors, as well as the fact that knowledge "leaks" and spillovers are central to technological evolution, gives rise to an important knowledge externality and results in different attitudes and behaviors of technology actors.

11.2.3 Characteristics of technological knowledge

New scientific and technological knowledge builds on previous knowledge and experience (de Solla Price 1965). Like all forms of knowledge, it is nonrival, complementary, and cumulative. A new technological artifact, like a new biological species, is seldom designed from scratch. Evolving designs (e.g., for successive generations of aircraft models or semiconductors) are perfect examples of evolutionary strategies of technological change. Moreover, they illustrate that the improvements in design, performance, economics, etc., that a new technology represents are deeply rooted in the experience and knowledge gained by designing (and using) its predecessors. From that perspective, most technological innovation is Usherian (incremental) rather than Schumpeterian (revolutionary), to paraphrase Ruttan's distinction (1959). Knowledge as applied in production also exhibits cumulativeness: initial defects are eliminated progressively as production volumes grow, production processes are scaled up, costs fall, model varieties and regional product differentiations are introduced, etc. Knowledge concerning the use of technologies (like knowing how to drive a car) is also cumulative. The need for extensive consumer learning in order to use novel artifacts has well-demonstrated effects of slowing down diffusion (see Rogers and Shoemaker 1971).

If technological knowledge is cumulative, then the development of a new design, the organization of a new production run, the use of an improved consumer product, etc., all benefit from the entirety of previous experiences and knowledge generated (subject to knowledge depreciation, discussed below). This can substantially improve the performance and costs of the latest technology "species" produced and speed its diffusion. In short, cumulativeness of knowledge implies the possibility of *increasing returns*. The most popular example of this in the technological literature is manufacturing "learning" or "experience" curves (Wright 1936; Arrow 1962; Alchian 1963; Argote and Epple 1990). Even comparatively "simple" technological learning processes, such as reductions in labor requirements in the mass production of a standardized technological artifact, build on whole series of earlier prototypes. In other words, a design and engineering competence needs to be accumulated before any large-scale industrial production can take place. A good example is the detailed case study of the B-17 Flying Fortress bomber by Michina (1992, 1999). That case study also demonstrates that learning by doing in manufacturing involves the accumulation of managerial and production organizational knowledge far more complex than simple scale economies or improved performance of repetitive production tasks by individuals and the workforce as a whole. Improved technological knowledge thus draws on a multitude of sources that cannot be reduced to single mechanisms or actors involved (Cantley and Sahal 1980).

Two important caveats are appropriate. First, technological innovation is a high-risk, high-uncertainty business. Few of the proposed solutions ultimately succeed, and anticipated future improvement rates for particular technologies are uncertain at best. While the payoffs from possible increasing returns on technological innovations are indeed enormous, they represent the extreme tails of distribution functions that cover the ultimate fate of *all* technological innovations, including those that never see widespread diffusion. Second, while technological knowledge is cumulative, it also depreciates if it is not applied (or is applied in a "stop-and-go" fashion). To paraphrase Rosegger (1996), the corollary of learning by doing is "forgetting by not doing." Numerous examples from the aircraft industry are given in the literature (Epple *et al.* 1996; Michina 1999). Watanabe (1995, 1999) identified a time lag of about three years in the translation between disembodied and embodied technological change for Japanese photovoltaic (PV) cells, and a depreciation of the technology-specific knowledge stock (measured by cumulative R&D expenditures) of about 20 percent per year (owing to the relatively short lifetime of about five years for a given PV technology, a result of rapid technological progress).

Again abstracting from our review of the literature, we arrive at the following generalizations. Technological improvements result from the accumulation of different sources of knowledge. The cumulative nature of technological knowledge can result in increasing returns. In other words, the more a technology is researched, experimented with, and tried out (including market applications), the better its performance, the lower its costs, etc. Sources of improved technological knowledge include both changes in design and experience gained in the production and utilization of new artifacts, and it is impossible to draw neat distinctions between the two without reverting to the now widely dismissed linear model of innovation. Yet, accumulation of technological knowledge does not automatically result in improved technology. Uncertainty is inherent: a few "big hits" can lead to significant jumps in productivity and efficiency [i.e., what Nordhaus (1997) considers to lie outside "routine innovations"], and most technological innovations are bound for oblivion, either as straightforward failures or as examples of innovations that never enjoyed widespread market application (i.e., diffusion).

Uncertainty is pervasive and persistent in technological evolution, representing both risks and opportunity for technological innovation. For those emphasizing the risk aspect of uncertainty, technological innovation can be seen as simple contingency planning. For those emphasizing the opportunity aspect of uncertainty, innovation corresponds to Schumpeter's (1934, 1942) conceptualization of the expectations of exceptional rates of profit to be reaped from temporary monopolies after the introduction of successful innovations. Both supply- and demand-side aspects of uncertainty need to be considered. Taking the example of energy technologies, uncertainties in energy demand and resource availability and costs (development of exploration and production technologies) need to be considered, along with uncertainties in future states of energy supply and end-use technologies (e.g., availability and costs) and the expenditures necessary to arrive at that future state—if it can be realized at all. Recognizing the importance of technological expectations points up the significance of heterogeneity of risk/opportunity perceptions among actors, which needs to be reflected in modeling. Technological expectations are also inherently *subjective*, defying any attempts to capture technological uncertainty under increasing returns with classical probability analysis.

11.2.4 Technology systems

Technological evolution is systemic. It cannot be treated as a discrete, isolated event that concerns only one artifact. A new technology not only needs to be invented and designed, but it needs to be produced as well. This requires a whole host of other technologies. Consequently, Kline (1985) refers to technology as "socio-technical systems of production and use" of artifacts. In most cases, technologies rely on infrastructures. A telephone needs a telephone network. A car needs both a road network and a gasoline distribution system, each of which consists of whole "bundles" of individual technologies. This interdependence of technologies frequently results in "lock-in" phenomena (Arthur 1983, 1989) that cause enormous difficulties in implementing large-scale changes, specifically in dealing with technological obsolescence in the sense of Frankel (1955), or what has more recently been referred to as technological "inertia" (e.g., Ha-Duong *et al.* 1997). But this interdependence is also what causes technological changes to have such pervasive and extensive impacts once they are implemented. The systemic aspects of technology are well recognized in the economic and technological literature in the form of "forward" and "backward" linkages (see, e.g., Fishlow 1965, von Tunzelmann 1982, and Freeman 1989) and "network externalities" (Katz and Shapiro 1985), as well as in conceptualizations of technological "families," "trajectories" (Dosi 1982), or "clusters" (Grübler 1998). In the absence of *a priori* knowledge of all possible systemic linkages between up- and downstream individual technological realizations, and with the daunting empirical problems of taxonomic classification methods such as morphological analysis (Foray and Grübler 1990; Godet 1997), simplified approaches that focus on key relationships (demand and supply of new technologies, their related infrastructures, and "upstream" technologies for the most important factor inputs) may not do justice to the complexity of large technical systems. Nonetheless, they represent an advance over conceptualizations of singular, "island" technologies, as typified by classical concepts advanced in the literature, such as that of "backstop" technologies (Nordhaus 1973).

Thus, as a basic conclusion, we retain the need for a (even if highly simplified) "bottom-up" representation of technological systems that aims at representing the most important technological/infrastructural interdependencies as well as possible synergies and spillovers within various "families" or clusters of technologies.

11.3 A Stylized Model

Below, we present the main features of a stylized model of endogenous technological change. It is a multiregion, multi-actor model of uncertain increasing returns on technological innovation that also considers most of the salient uncertainties in the market environment in which technology choice takes place (e.g., demand, resource availability, environmental limits). A distinguishing feature of the model is that the omniscient social planner is replaced by a set of actors, each of which optimizes its own part of a global system while remaining interdependent via negotiated energy and technology trade flows. In other words, the concept of a global optimum (viz. cost minimum) is replaced by a Pareto-optimal formulation. "Optimality" in this model is not based on finding a cost-minimum solution (which could be calculated only with perfect foresight) or on simply minimizing an expected value of corresponding energy systems costs (the most common approach using the so-called best guess technique). Instead, an optimal solution represents the best hedging strategy vis-à-vis a large number of persistent uncertainties; that is, uncertainties that are not assumed to be reduced at any arbitrary future date, as is done in classical stochastic optimization models. Some of these uncertainties cannot be estimated using classical statistical approaches because of the lack of reliable data. Others, as in the case of future performance of new technologies, are impossible to quantify because of their strong dependence on intervening policy actions (e.g., R&D expenditures, taxes, and other forms of market interventions). In such cases, subjective estimates based on "expert" judgments and model scenario analyses under a range of assumptions are unavoidable.

The model takes a global long-term perspective representing a stylized energy sector in which technology choice is studied under conditions of uncertainty. Model formulations draw on previous modeling work at the International Institute for Applied Systems Analysis (IIASA) (Golodnikov *et al.* 1995; Messner *et al.* 1996; Messner 1997; Grübler and Gritsevskyi 1998; Gritsevskyi and Nakicenovic 2000), with parameterizations largely derived from recent studies of long-term energy perspectives (Nakicenovic *et al.* 1998a; IPCC 2000) in which *inter alia* major long-term uncertainties for the energy sector were analyzed based on a scenario approach. The model overview presented here is short and nonmathematical. For a more formal exposition, see Grübler and Gritsevskyi (1998), and Chapter 10 in this volume.

11.3.1   The actors

The model presented here does not assume the existence of a global social planner with perfect foresight. Instead, different actors are distinguished, all of which operate under uncertainty. Figure 11.1 gives a schematic overview of the actors in the model and their main interdependencies. Endogeneously determined flows are denoted by straight lines (e.g., Pareto-optimal trade flows between regions); uncertain model variables, by zigzagged lines. Uncertain model variables are both exogenous (e.g., demand, resource availability, etc.) and endogenous (e.g., future costs as a function of cumulative demand, in the case of uncertain increasing returns).

First, we distinguish between developers/users of energy sector technologies and the energy sector proper; that is, the aggregation of actors that supply the exogenously specified, uncertain energy demand. Energy technology developers/users or agents are differentiated based on differences in their technological competences (knowledge) and other characteristics, most notably financing capability. At one end of the spectrum are truly "global" agents that have vast financing capabilities and that, through their size and financial resources, can diversify their

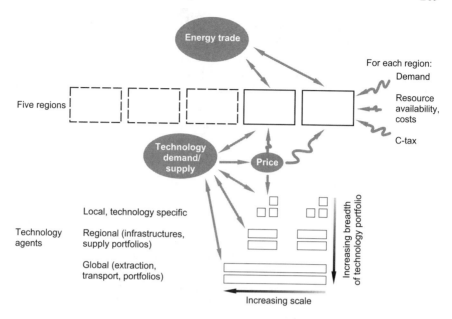

Figure 11.1. Schematic Overview of Actors and Their Main Interdependencies in the Model.

Flows denoted by straight arrows are endogenously determined variables. Zigzagged arrows indicate uncertain variables. Note that for clarity of exposition only two of five regions are illustrated in the figure. The three types of technology actors shown at the bottom are modeled via one representative agent each.

investments across all regions and technology portfolios. Such agents have or can acquire technological competences over the entire technology spectrum. A typical example is a company like Shell, a global player in the oil and gas business that recently has begun diversifying into a wide range of other energy technologies, including biomass plantations, independent power production, fuel cells, manufacturing of PVs, etc. At the other end of the spectrum are small, innovative actors with highly specialized technological knowledge and limited financial resources (e.g., the joint venture between DaimlerChrysler and the fuel cell manufacturer Ballard, or a wind turbine manufacturer cooperating closely with a "green utility" in Denmark). Intermediate actors are agents that essentially operate at a regional scale and at intermediate levels of technology specialization. A typical example is a large national or regional electric utility that builds and operates a diversified electricity generating park (coal, nuclear, hydropower, etc.) as well as transport and distribution infrastructures.

The second group of actors consists of five representative energy-producing/consuming regions. These five regions follow a geographical breakdown that is by now customary in global integrated assessment modeling. Countries are grouped on the basis of similarities of socioeconomic development status, economic and energy systems structure, export/import balances, and re-source availability. The model distinguishes two Organisation for Economic Co-operation and Development (OECD) regions (Canada, the United States, Australia, and New Zealand, on the one side; Western Europe and Japan, on the other), the reforming economies of Central and Eastern Europe and the former Soviet Union (EEFSU), and two developing regions (Asia; and Latin America, Africa, and the Middle East).

11.3.2   The objective function and interaction between actors

Components of the Objective Function

Figure 11.2 summarizes the various elements of the objective function. Moving from the top of the figure down, more and more elements of uncertainty are represented in the objective function. In the model runs reported in Section 11.4, we consider step-by-step increases in the levels of uncertainty; thus the overall objective function becomes increasingly comprehensive. These various levels of uncertainty are separated by dashed lines in Figure 11.2. The various model runs (labeled A to F) of Section 11.4 corresponding to these levels are also shown in Figure 11.2. For any level (and model run), the objective function includes all those components of the particular level plus all previous ones, as summarized in Figure 11.2.

At the most fundamental level, three different approaches and levels of comprehensiveness of the objective function can be differentiated. In the simplest case (level A in Figure 11.2), the objective is to minimize total discounted global energy systems costs under perfect foresight, or, more strictly speaking, to minimize the total expected value for these systems costs. This is the approach of traditional optimization models assuming a social planner and is considered here for comparison purposes only. In a second case (level B in Figure 11.2), the global optimum is replaced by a Pareto-optimal formulation, assuming no uncertainties (with "best guess" central values used as expected mean values). Each region acts as a cost-minimizing agent on its own regional energy systems costs; in other words, each has its own objective, all of which are summed into a global Pareto optimum based on regional weights. Finally, in a third category of cases (level C and onward in Figure 11.2), uncertainties are explicitly considered. The regional agents (typified, for instance, by departments of energy or similar national/international planning agencies) are assumed to have foresight, subject to stochastic uncertainties (demand, resource quantity–cost relationships, technology availability, costs, etc.). These uncertainties are quantified costwise in a risk function and are integrated into the objective function.

Representing Uncertainty

Following the approach described in Golodnikov *et al.* (1995), Messner *et al.* (1996), and Grübler and Gritsevskyi (1998), uncertainties are represented as follows. Unless otherwise indicated, the uncertainty space around all salient

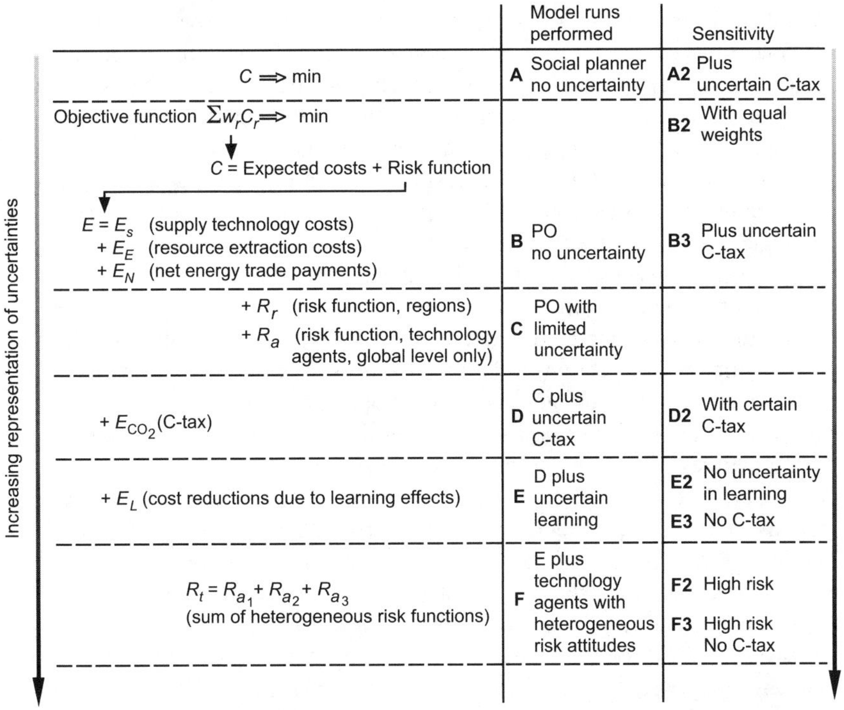

Increasing representation of uncertainties →

	Model runs performed	Sensitivity
$C \Rightarrow$ min	**A** Social planner no uncertainty	**A2** Plus uncertain C-tax
Objective function $\Sigma w_r C_r \Rightarrow$ min ↓ C = Expected costs + Risk function		**B2** With equal weights
$E = E_s$ (supply technology costs) $+ E_E$ (resource extraction costs) $+ E_N$ (net energy trade payments)	**B** PO no uncertainty	**B3** Plus uncertain C-tax
$+ R_r$ (risk function, regions) $+ R_a$ (risk function, technology agents, global level only)	**C** PO with limited uncertainty	
$+ E_{CO_2}$(C-tax)	**D** C plus uncertain C-tax	**D2** With certain C-tax
$+ E_L$ (cost reductions due to learning effects)	**E** D plus uncertain learning	**E2** No uncertainty in learning **E3** No C-tax
$R_t = R_{a_1} + R_{a_2} + R_{a_3}$ (sum of heterogeneous risk functions)	**F** E plus technology agents with heterogeneous risk attitudes	**F2** High risk **F3** High risk No C-tax

Figure 11.2. Overview of Representation of Increasing Sources of Uncertainty in the Model and Their Consideration in the Objective Function.

Corresponding model runs (from Section 11.4) are labeled A to F. E = expected costs, R = risk function term, PO = Pareto optimum; index r refers to components of regional objective functions, index a refers to components of the objective function that represent the technology agents.

variables is represented by a normal distribution function around a mean expected value. The model solution with all variables set at mean expected values yields the (minimized) expected costs (denoted as E in Figure 11.2; these are exactly the costs considered in traditional optimization models with perfect foresight). To these expected costs we add a risk function (denoted as R in Figure 11.2) to the objective function. (Note that for clarity of exposition we break the expected costs E into various components in Figure 11.2. Deviations around these expected cost items are not listed separately in Figure 11.2; they are all summarized in the regional risk functions denoted as R_r in the figure.)

Simultaneous random draws "poll" the uncertainty space, and a model solution is calculated for each draw. Deviations from the mean expected costs are summed into the risk function that is added to or subtracted from the objective function. Sampling continues until the uncertainty space is statistically sufficiently explored and an overall "robust" cost-minimal solution including the risk function can be calculated. Representative sampling of the uncertainty space around the mean expected value means that the risk term integrated into the objective function is the product of cost differences (compared with the expected costs) and their respective probability of occurrence (as reflected in the uncertainty distribution around the mean expected values). It is important to emphasize that the costs are discounted (at 5 percent in all calculations reported here)[2] and refer to the entire simulation horizon. The resulting changing energy systems structure and profiles of technology development over time are frequently referred to as "strategies" to emphasize their dynamic nature and the interdependence between important variables (such as cumulative demand and costs for technologies, in the case of increasing returns).

In other words, our modeling approach assumes that, during the planning process, the regional actors probe all of the most important uncertainties and contingencies of their region's energy development that deviate from their mean expected values to finally arrive at an optimal (cost-minimal) contingency or hedging strategy vis-à-vis multiple uncertainties. We assume that the planning process proceeds conservatively; that is, future realizations that lead to energy systems costs that are higher than mean expected values are considered to be more important for decision making than realizations that lead to lower-than-expected costs. In other words, risk aversion takes precedence over opportunity seeking.[3] For computational simplicity, we assume a linear risk function [see Grübler and Gritsevskyi (1998) for alternative, nonlinear formulations], with the risk-aversion gradient being twice as large as the opportunity-seeking one. We assume that a 10 percent cost overrun (costs turn out to be 10 percent higher than mean expected values) is weighted twice as high as a negative deviation of 10 percent (costs turn out to be 10 percent lower than expected). The risk function includes an additional weighting factor, representing different degrees of risk aversion. The risk factor is conceived as being different across regions and actors.

The agents interact through energy trade and technology flows. Regions interact via endogenously determined energy trade (for simplicity, only some energy products are traded in the model—coal, crude oil, natural gas, and methanol/ethanol). Regions "negotiate" an intertemporal energy trade flow pattern in which each region's total energy costs (including the net balance between exports and imports) can no longer be improved without a deterioration of the total energy costs of all regions (i.e., a Pareto optimum). Export prices are assumed to be close to marginal resource extraction costs; resource extraction for domestic use is determined based on average extraction costs. Financial transfers are accounted for: export revenues are added as negative energy systems costs to the producing regions, and import costs are added to the energy systems costs of importing regions.

The technology agents interact with one another and, if regionally based (i.e., all agents except "multinationals"), also with their respective regional agent. Because the technology agents are subject to laws and regulations, they cannot act as regional social planners. However, their relative success (or lack of it) in technology development and diffusion is assumed to be considered at least at the global level. The interaction between the technology agents and the regions proceeds via the interactions between technology demand (by regions) and technology supply (by agents). As the export market for energy technologies is essentially global, different technology demands (resulting from expected technology costs plus all other salient model drivers) across regions are not independent of one another. Realization of potential learning effects (cost reductions) depends on global technology demand rather than on regional demands alone. This again leads to a Pareto-optimal solution of the global demand for energy technologies. In other words, the technology actors and the energy-consuming/producing regions "negotiate" an intertemporal technology demand pattern, whereby each region's costs can no longer be improved without a deterioration of the costs of all regions.

Technology agents then supply that demand at *expected* cost values to the regions—subject, however, to uncertainty—with different technology agents exhibiting different degrees of risk aversion. Uncertainty is again represented via stochastic sampling, the results of which are incorporated into the overall objective function via a risk term, as explained above. These risk/benefit terms (i.e., technology actors supply technologies to the region at lower- or higher-than-expected costs) are then added to the sum of regional energy systems costs (i.e., at the global level), ensuring that the risks of the technology agents are considered in the Pareto-optimal global solution.

11.3.3 Energy demand

Two distinguishing features characterize the energy demand of our stylized model: first, future demand (growth) is *uncertain*, and, second, *quality matters*. The five regional energy economies demand three homogeneous energy goods, characterized by their energy form (and their exergetic form value): *solids, liquids*, and *grids*. Within each energy form, different combinations of primary resources and conversion technologies can satisfy the demand. In other words, we assume perfect substitutability within each energy form. For instance, the demand for high-quality energy carriers ("grids") can be satisfied via electricity generated from coal or from PVs or wind turbines, or alternatively by natural gas. We assume asymmetric (and partial) substitutability between energy forms; that is, substitutability is only possible in the direction from high to low exergetic energy forms. Electricity ("grids"), for instance, can replace coal ("solids") as a final energy carrier, but the reverse is not true. We consider this model feature to be an important improvement over conventional energy models which have focused only on quantities and prices, largely ignoring energy quality.

Quantitatively, the model is parameterized based on the long-term energy scenario study reported in IPCC (2000). These scenarios span a wide range of future uncertainties in energy demand, with the added benefit of having been carefully peer reviewed. Uncertainty in demand is represented by normal distributions around a mean expected value of energy demand per category, assumed to correspond to the numerical projections of the "middle ground" Scenario B2 in IPCC (2000). It must be emphasized that the demand uncertainty thus represented in our model is significantly smaller than the entire uncertainty range spanned by the full set of IPCC scenarios.

The treatment of demand uncertainty is analogous to the procedure outlined above. Supply expansion reflects successive random draws from the demand uncertainty space around a mean expected value. Lower-than-expected demand translates into excess supply (i.e., lower capacity utilization); higher demand translates into a supply shortfall. For the latter case, the model does not employ "hard" constraints, but instead uses a penalty function approach, analogous to the treatment of uncertainty in our model. The final result is again an optimal supply expansion path vis-à-vis an uncertain future evolution of energy demand.

11.3.4   Technologies

Systems Structure

The technologies represented in the model provide the necessary links (as a rule, conversion processes) between primary resources, on the one hand, and final energy demand, on the other hand (Figure 11.3). Three demand categories (solids, liquids, and grids) and five different primary resource categories (coal, oil, gas, biomass, and other zero-carbon energy sources) are modeled with 56 technologies in between. The reference energy system of our model includes 4 extraction technologies; 3 backstops; 23 transport/distribution, trade, and other energy systems balance technologies; and 26 energy-conversion technologies proper, 12 of which can exhibit increasing returns (indicated in Figure 11.3). The reference energy system of our model represents an intermediate level of complexity compared with the stylized three-technology model presented in Grübler and Gritsevskyi (1998) and the full-scale complexity of the model presented by Gritsevskyi and Nakicenovic (see Chapter 10 in this volume), which requires massive parallel computing.

Simplifications

To reduce model complexity and allow for analytical solutions within reasonable computing times, a number of simplifications were introduced. First, only technologies past their innovation stage are considered. Technologies therefore need to have demonstrated their technical feasibility and established a small commercial niche market, even if that market is infinitesimally small by the standards of the energy sector. Thus, cold or hot fusion is not considered, but wind turbines, fuel cells, etc., are. Second, technologies represented in the model are kept as generic as possible. Hydrogen-powered fuel cells are not distinguished from related technologies using natural gas, but the hydrogen generation and distribution technologies upstream in the energy chain are *included* in the model. Third, for simplicity the model considers only levelized costs: no distinctions are made between investment and operating costs. Expenditure "bulkiness" is, however, reflected by considering the different unit output sizes of technologies. In other words, a unit of output expansion leading to possible technological learning is

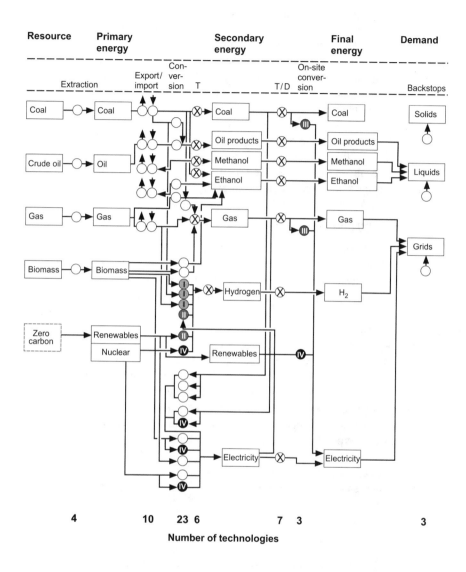

Figure 11.3. Reference Energy System of the Model Showing Energy Flows (Arrows and Boxes) and Conversion Technologies (Circles).

For each technology level, the number of technologies represented in the model is indicated at the bottom of the figure. All technology costs outside transport and distribution technologies are treated as uncertain. Technologies that can exhibit uncertain increasing returns are shown in full circles; roman numerals indicate the technology cluster within which spillover effects are assumed to occur. T = long-distance transport; T/D = transport and distribution.

more expensive for nuclear reactors than for solar PVs. Finally, a distinction is made between existing, mature technologies and new technologies. "Existing" technologies in the model mainly serve as accounting identities; costs are not treated as uncertain and technologies do not exhibit potential increasing returns. Typical examples are oil pipelines, international coal shipping, and conventional coal-fired power plants. For new technologies, both initial costs and future costs as a function of intervening efforts (investments in R&D and niche market installations) are treated as uncertain. Two generic categories of new technologies are considered: incremental and revolutionary. These two categories differ in terms of their initial costs, the potential for cost reductions with accumulated experience, the associated uncertainties, and the relative importance of disembodied versus embodied technological change (R&D intensity).

Disembodied versus Embodied Technological Change

The sources of potential increasing returns on new technologies are treated as interdependent and complementary. Thus, disembodied technological change ("learning" via R&D; i.e., new design blueprints) and embodied technological change (learning by doing and learning by using) are considered the necessary two sides of the same coin. Therefore, the model does not formally distinguish between R&D and plant expenditures: the necessary R&D component for new energy technologies is simply added as a fixed percentage to the technology costs. To avoid the criticism of underestimating R&D costs, we make the conservative assumptions of 20 percent for "revolutionary" technologies [based on Watanabe (1995)] and 10 percent for all other new technologies. Initial mean costs of new energy technologies are derived from a statistical analysis of the technology inventory CO2DB available at IIASA (see Strubegger and Reitgruber 1995). Formally, R&D expenditures are represented simply by increasing the intercept of initial technology costs by an R&D component in the functional formulation, where future costs can decline as a result of cumulative output. R&D expenditures are thus considered to be necessary as long as technologies exhibit increasing returns.[4]

Uncertain Increasing Returns

Learning rates (progress ratios) are treated as uncertain and are varied only parametrically and in an extremely simplified way. We assume mean expected learning rates of 10 percent (for a doubling of cumulative output) for all technologies that are classified as incremental innovations (e.g., a biomass-fired gas combined-cycle turbine compared with natural-gas-fired ones), or whose "lumpiness" precludes mass production (e.g., advanced reactor designs such as the high-temperature gas-cooled reactor, or HTGR). For "radical" innovations suitable for mass production (e.g., fuel cells, PVs), a mean expected learning rate of 20 percent is assumed. This figure is the empirical rule of thumb advanced in the literature; for example, it is the median of the sample of 108 learning rates reported in

Argote and Epple (1990). The above distinction is based on a taxonomy of empirically analyzed technological learning rates of energy technologies reported in Christiansson (1995) and Neij (1997). Uncertainty ranges around mean expected learning rates are higher for "revolutionary" technologies than for "incremental" ones.

Formally uncertain increasing returns are introduced into the model as follows: we assume that the progress ratio (the negative exponent in a classical learning curve formulation) is uncertain, reflected by a normal distribution around a mean expected value. The expected future costs for technologies with potential for learning effects (as supplied by the technology agents) are thus log-normally distributed, contingent of course upon the realization of the cumulative output growth demanded by the energy-consuming/producing regions. The uncertainty space of the learning curve's progress ratio is again sampled by random draws and deviations from the solution, with mean expected values added to the objective function via the risk function term, as explained above. The main difference, of course, is that, owing to nonconvexity, exploration of this uncertainty space is computationally much more demanding.

Spillovers

Technology spillover effects are also represented in the model, albeit in a rudimentary and simplified way. First, potential spillovers between energy sector technologies and other sectors are not considered (e.g., between aircraft and gas turbines, or spillovers from the semiconductor industry to PVs). For energy sector technologies, we adopt the concept of technology "family" learning: technologies that belong to a generic class of technologies (various gas turbines, different fuel cells, etc.)—irrespective of the primary energy resource used as input fuel—are assumed to be entirely substitutable as targets for technology improvements. This assumption postulates the rather optimistic viewpoint that whatever improvement is achieved for one technology within a technology "family" spills over to the other technologies within that "family." Thus, technological learning phenomena are treated as generic: improvements in gas-fired combined-cycle technology are assumed also to benefit gas combined cycles that use coal or syngas derived from biomass as fuel. Formally, this is done by adding all installations of technologies within one "family" to the cumulative installations of a combined technological learning curve. Altogether, four such technology "families" exhibiting joint technological learning via spillover effects are considered in the model (see Figure 11.3).

Causality

As mentioned, in addition to demand and resource quantity–cost uncertainties, the existence and magnitude of possible increasing returns from technological innovation are a main source of technological uncertainty in our model. Future technology costs can be influenced through different strategies; the outcome of such strategies is, however, uncertain. From the multiple dimensions of technology improvements, we consider only cost reductions that result from alternative research, development, and deployment strategies.

The causality mechanism of technological improvements (cost reductions) in our model is characterized by a twofold complementary relationship: between improvements in disembodied technological knowledge (R&D) and embodied technological knowledge (deployment),[5] and between supply (by technology agents) and demand (by regions). This twofold complementarity reflects our understanding of the literature (e.g., Freeman 1994): there are different sources of technological knowledge that are not perfectly substitutable. Increased R&D expenditures might yield improved technology blueprints, but they do not lead to lowered technology costs in the market place in the absence of demand; and no matter how high demand growth is, improved design blueprints will not emerge without R&D. Conversely, technology demand and supply are complementary sources of improved technological knowledge. Learning by doing in manufacturing cannot take place in the absence of technology demand (e.g., realization of corresponding "learning by using" by technology users), and supply and demand interact via prices (traditionally captured via demand elasticities). These distinctions are not important in social planner models,[6] but they are important in our multi-actor model (as well as in reality). Moreover, knowledge generation does not proceed in isolation from the economy or society at large. Changes in relative prices, environmental constraints, and even strategic considerations all play an important role and add to the complexity and uncertainty of the environment in which technology choices are made by a multitude of actors.

11.3.5   Resource and environmental constraints

Resource availability and environmental constraints are treated as uncertain. To some degree they represent two sides of the same coin: the uncertainty concerning the level of energy resources that can be extracted from nature and the level of resulting emissions that can be "disposed of" in the environment.

Resource availability and cost assumptions were derived from the "middle ground" Case B scenario described in Nakicenovic *et al.* (1998a), which we adopt for the mean expected values in our model. Corresponding resource extraction profiles and costs were obtained for the five regions of the model from that scenario [based on Rogner (1997)]. These assumptions ensure convex relationships between the quantity of resources available and their costs,[7] which are preferred for ease of numerical solution. As for other salient uncertain variables, here too normal distribution functions are constructed around mean expected values. The resulting uncertainty domain explored by the model is narrower than those explored within the entire range of scenarios of Nakicenovic *et al.* (1998a) and the IPCC (2000). The reason is that including enormous amounts of unconventional fossil resources at potentially very low costs (e.g., methane clathrates) in the uncertainty space would result in nonconvex resource extraction cost profiles around the mean values of the Case B scenario. Relaxing this simplifying assumption remains an important task for future model improvements.

Future environmental constraints might emerge, influencing technology choices. Such constraints could take the form of either "hard" quantitative limits or emissions taxes. The existence, magnitude, and timing of such emissions taxes can be treated with the approach used to treat technological uncertainty. For the model calculations reported here, we follow assumptions similar to those described in Grübler and Gritsevskyi (1998) and Grübler (1998) for a possible carbon tax. First, we assume a cumulative probability distribution of the occurrence of the emissions tax over the entire time horizon. Starting near zero in 2000, the starting year of our simulations, the cumulative probability distribution rises over time. The illustrative (conservative) distribution function assumed reflects only a 50 percent chance that a carbon tax is ever implemented. The probability of the tax's being introduced rises to 25 percent by 2030, reaching 50 percent toward the end of the model's time horizon. For the magnitude of the tax, we assume a distribution with a very small probability of a high carbon tax level, as represented by a Weibull distribution around a mean value of US$75 per ton carbon (C), with a 99 percent probability that the tax will not exceed US$150 per ton C *if it is implemented at all*. According to our understanding of the literature, such assumptions do not represent any particularly daunting outlook on possible future environmental constraints.

11.3.6   Computational implementation

Computational model solutions are obtained by applying a sequential optimization procedure to sub-problems that approximate the original problem in two respects. In the most general case, a highly nonlinear stochastic global optimization problem should be solved. Stochastic operators in the objective function are approximated by statistically calculated values using simultaneous random draws, where the dynamically adjusted "polling" size is set according to the required accuracy. Nonconvexities that arise from increasing returns and corresponding cost reductions are handled using an adaptive stochastic search technique. This approach is based on a modified "direct" search algorithm using the Monte Carlo technique, but in this case with a dynamical adjustment of the probability space depending on the results obtained in the previous steps. For sub-problems calculated during the adaptive stochastic search, a linear optimization technique is used. Because the linear optimization problem is rather large (10,000–15,000 variables and 5,000–7,000 constraints), the interior point method with simultaneously solved primary and dual problems was employed. This method has additional benefits, including accuracy and a computation time that is adjusted depending on the ratio of "the best solution obtained so far" to the "low-to-high" approximation for the current sub-problem.

The model is implemented in the Matlab 6.x computational environment using its Optimization and Statistical Toolbox and the advanced optimization toolbox

TomLab 3.0. The latter implements a large set of state-of-the-art optimization methods, including the Stanford Optimization Library.

11.4 Model Simulations

11.4.1 Introduction

Below, we discuss a number of quantitative model simulations, the results of which should be considered illustrative rather than definitive. The prime objectives of the model development and simulations reported here were to demonstrate the feasibility of an analytical solution of the proposed model and to gain qualitative insights into its behavior under an increasingly comprehensive treatment of uncertainty. To make the model's behavior transparent and to minimize computing time, for instance, no additional constraints beyond physical balance constraints were employed. Additionally, the use of levelized costs in this model version means that the dynamics of capital stock replacements or capacity underutilization cannot be represented. These factors need to be borne in mind when judging the "realism" of the model simulations reported here, especially in comparison with other modeling approaches that employ constraints for reproducing base-year energy flows or additional market penetration constraints to avoid "flip-flop" model behavior. The tentative model simulations presented here should therefore be viewed as a "work in progress" aimed at elucidating generic dynamic patterns of technological change under uncertainty and uncertain increasing returns in a multiregion, multi-agent model rather than results that can be taken directly "off the shelf" for energy or climate-change policy analysis.

To maintain comparability between model simulations, there are a number of important common assumptions for all model runs. First, energy demand, resource availability, and technology portfolio assumptions are shared by all simulations. The main differences are whether these assumptions are treated as certain or uncertain, whether technologies can exhibit (certain or uncertain) increasing returns to scale, and whether a (certain or uncertain) carbon tax is assumed in the model simulations. The uncertainty space around uncertain variables (demands, convex resource extraction cost curves, technology costs) is represented the same way in all model simulations. The uncertainty domain around each variable is defined as +/− three standard deviations around the mean expected value assuming a normal distribution. For the carbon tax case we assume a non-symmetric Weibull distribution. This reflects our interpretation of the literature that small-probability, high-risk events cannot be excluded from the climate-change problem. Hence, we consider the possibility of a low-probability but extremely high carbon tax level, as reflected in a Weibull distribution that otherwise shares the same variance as the other (normal) uncertainty distribution functions. Technically, all simulations also share the same discount rate (5 percent) and the same time horizon (eight 10-year time steps). Because of end-period truncation effects that diminish the significance of the model results for the last time period, in the subsequent discussion we report results only for the periods up to 2070.

11.4.2   A taxonomy of model simulations

Altogether six main types of model simulations were performed. These are denoted by the letters A to F in Figure 11.2 in the previous section and in Figures 11.4 to 11.7 in this section. In addition, a number of sensitivity runs for particular assumption permutations within scenario clusters were performed. These are denoted by consecutive numbers (2, 3) in combination with the letters of the main scenario clusters. Altogether 14 different scenarios (model runs) are reported in the subsequent sections.

Scenarios in cluster groups A and B embrace the traditional deterministic model framework. Key variables are not assumed to be uncertain; in other words, we assume perfect *ex ante* foresight. Scenario A represents the traditional social planner perspective; it postulates the existence of a single global social planner that optimizes regional resource allocation under a global (discounted) cost-minimum criterion. Scenario A2 is a sensitivity run assuming an uncertain carbon tax for the social planner, no-uncertainty scenario A. Group B scenarios contrast this scenario with a Pareto-optimal formulation. Scenario B2, where regions have equal weights, is for all practical purposes identical to the social planner perspective of scenario A and serves only as a control scenario calculation. Scenario B3 is equivalent to the main scenario B (Pareto optimum), but in addition assumes an uncertain carbon tax (comparable to scenario A2).

Scenarios in groups C and D systematically introduce additional uncertainties. Scenario C, a kind of initial base case for the subsequent simulations, uses a Pareto-optimal formulation (like the scenarios in group B), but treats energy demand, resource availability, and technology costs as uncertain. No carbon taxes or increasing returns are assumed. Scenario D adds an uncertain carbon tax (see Section 11.3 and discussion below) to Scenario C; the sensitivity run of Scenario D2 assumes a carbon tax with an *ex ante* known probability of occurrence and a perfectly known value. The difference between scenarios D and D2 illustrates the effect of treating future carbon tax levels as uncertain (assuming an asymmetrical uncertainty distribution, in this case, a Weibull distribution).

Scenarios in group E model the impacts of assuming increasing returns to scale in new energy technologies as a combined result of investments in research, development, and deployment; learning by doing; and learning by using (see Section 11.3). Scenario E (scenario with full uncertainty) assumes uncertainty in both increasing returns and a carbon tax. Scenarios E2 and E3 provide sensitivity runs for this scenario. In scenario E2, increasing returns are treated as certain (i.e., potential future cost reductions as a function of intervening investments are assumed to be perfectly known *ex ante*). Scenario E3 models uncertain increasing returns in the absence of an uncertain carbon tax.

Finally, scenarios in group F model heterogeneous agents on top of the scenarios in group E. Whereas in the previous scenarios, regions and technology agents all share the same risk function, in scenario group F the risk functions differ. Scenario F assumes (conservatively) only a small dispersion of risk factors. The small-scale, local technology agents (constrained to a maximum of 5 percent market share in each region) are assumed to be high risk takers, whereas the global

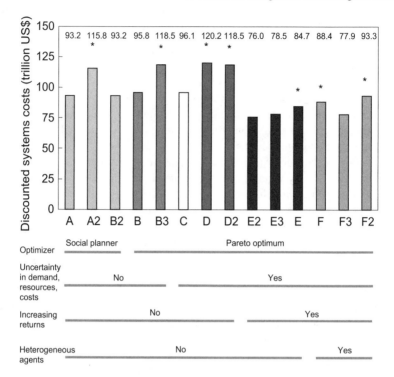

Figure 11.4. Overview of Scenarios and Their Global Total Expected Energy Systems Costs (Discounted 5 Percent), in Trillion (10^{12}) US$ for the Model Run Scenarios.

For explanation of individual scenarios, see Figure 11.2 and text. Scenarios assuming a carbon tax are indicated with an asterisk.

technology agents (maximum 30 percent market share in any region) are assumed to be risk averse. Regional technology agents are assumed to share the same risk function as the energy-producing/consuming regions, as in the previous scenarios. Scenario F2 assumes that all technology agents are highly risk averse; that is, cost overruns are weighted three times as much as the energy regions. Scenario F3 is identical to scenario F2, except for the absence of the uncertain carbon tax.

Figure 11.4 summarizes the scenarios that form the basis of the subsequent discussion in this section in terms of their global total, discounted expected energy systems costs. Costwise, the greatest differences between scenarios concern assumptions regarding the introduction of a future carbon tax that raises total expected costs (e.g., scenario B2 versus scenario B, or scenarios D and D2 versus scenario C), and assumptions about possible increasing returns to technological innovation that lower total expected costs (e.g., scenario E3 versus scenario C, or scenarios E2 and E versus scenario D). Conversely, cost differences between social planner (A, B2) and Pareto-optimal scenarios (e.g., B), between scenarios of deterministic versus uncertain carbon taxes (D2 versus D), between certain

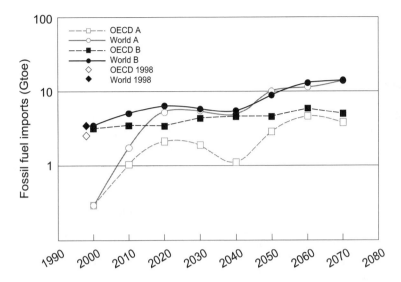

Figure 11.5. Energy Trade Flows (Imports of Fossil Fuels, in Gtoe) at the Global Level (Solid Lines) and for the OECD Regions (Dashed Lines) for Two Optimization Scenarios: Social Planner Perspective (Scenario A) and Pareto-Optimal Formulation (Scenario B).

All other salient scenario variables (demand, resource availability, technology costs, absence of environmental constraints) are identical. For comparison, 1998 trade flows for the world (black diamonds) and the OECD (white diamonds) are also shown [based on IEA (2001)].

versus uncertain increasing returns (E3 versus E2), and between homogeneous versus heterogeneous risk aversion (scenarios F2 or F versus E) are small. Scenario differences in terms of trade flows, technology portfolios, or emissions will be much larger than this simple comparison of total expected costs might suggest. Nonetheless, the cost differences of the scenarios provide a useful basis for structuring the following discussion into scenario differences under the presence (or absence) of increasing returns or the existence (or absence) of carbon taxes, or a combination of these factors. Before doing so, however, we will address the scenario differences between a social planner and a Pareto-optimal formulation.

11.4.3   Social planner versus Pareto optimum

Scenario A embraces the traditional social planner perspective of long-term energy-optimization models.[8] Conversely, Scenario B replaces the traditional global optimization with a "negotiated" Pareto optimum among the five energy-consuming/producing regions of the model. As expected, scenario differences are greatest in terms of energy trade flows, particularly in the short to medium term (Figure 11.5).

Short- to medium-term trade flows are substantially lower in the social planner scenario. If there were a single global social planner deciding on energy supply options, supply security concerns and import diversification would not be factored into the global optimization. Instead, regions focus on domestic energy production where they have the highest comparative cost advantages (e.g., coal-based electricity as opposed to imported liquefied natural gas in OECD countries outside Europe and Japan), with trade flows restricted to non-substitutable commodities (e.g., oil in the transport sector) and between regions with high resource extraction cost differences (e.g., Middle East versus North Sea oil). The generally low energy trade flows in scenario A also reflect our use of marginal rather than average extraction costs for the initial, flat part of the upward-sloping regional resource extraction cost curves [increasing resource extraction costs, particularly for oil in the "rest of the world" region and for gas in the "reforming economies" (EEFSU) region]. They also reflect the assumption of partial substitutability between the energy demand categories "liquids" (e.g., oil products) and "grids" (electricity, gas). Recall also that base-year energy flows are not constrained in the scenario calculations reported here.

The Pareto-optimal solution of scenario B is radically different. Short- to medium-term trade flows are much larger, both for the OECD and for the world as a whole. Simulated trade flows are also in much better agreement with actual trade statistics both quantitatively and qualitatively (Figure 11.5). Whereas after 2020, global trade patterns of fossil fuels are quite similar in both scenarios on aggregate, regional differences persist. In particular, fossil fuel imports to the OECD region are much larger in the Pareto-optimal scenario than in the social planner scenario.

The results suggest that, over the long term and at the global level, the differences between a social planner and a Pareto-optimal perspective could be less pronounced than one might expect *a priori*. Given a limited number of negotiating parties[9] and decisive regional differences in the long-term availability of fossil resources, *global* trade patterns are almost identical between the two approaches. On the one hand, this finding is welcome news for global, long-term energy and emission scenarios, which have almost invariably embraced a global social planner perspective. On the other hand, the results also indicate that short-term and regional trade patterns are very different in the two approaches. This finding cautions against the use of global social planner models in policy analysis of "winners" and "losers" of climate policies, for instance, when identifying the impacts of carbon taxes on future oil export revenues.

11.4.4   Technology choice under uncertainty

To compare the differences in technology choice under various conditions of uncertainty, we first reduce the comparatively high technological complexity of the model results to a manageable number of variables. To that end, Figure 11.6 compares the diffusion of various energy technologies aggregated into three clusters with two technology groups each (i.e., six groups altogether). The technology clusters are differentiated with respect to technological maturity. Within each cluster, we then differentiate between two groups of technologies: fossil-fuel based and non-fossil based. The first cluster aggregates currently available "off-the-shelf" technologies such as conventional coal-fired power plants, gas combined cycles, or conventional nuclear power plants. The second technology cluster comprises fossil and non-fossil technologies that represent incremental improvements over existing technologies or that can be expected to become commercially available in the medium term, such as advanced coal-fired power plants, gas-based fuel cells, wind turbines, or hydrogen from steam reforming of natural gas, etc. Finally, the third cluster represents long-term technology innovations: coal-based fuel cells, non-fossil hydrogen production and end-use technologies, advanced decentralized renewable electricity options (e.g., PVs), or inherently safe advanced reactor designs.

The top left-hand panel in Figure 11.6 shows technology choice without considering any uncertainty at all (Pareto-optimal scenario B). The patterns of technology choice are characterized by total reliance on currently known "off-the-shelf" technologies, which are almost exclusively based on fossil fuels. As such, the scenario mimics perfectly the static technology picture of traditional "business-as-usual" scenarios available in the literature.

More interesting are the patterns that emerge when salient energy systems uncertainties—such as demand, resource availability, or technology costs (scenario C, middle panel, left), or an uncertain carbon tax (scenario D, middle panel, right)—are introduced in the absence of possible increasing returns. Although existing fossil-fuel-based technologies remain dominant over the entire simulation horizon, there is nonetheless a substantial diversification of technology choice, including long-term "third-generation" technology options. Technology diversification and up-front investments in technology options that may have a significant market impact only after 2050 are thus an economically rational hedging strategy in view of uncertainty. The possibility of a carbon tax biases the technology portfolio more in the direction of non-fossil technologies, albeit not exclusively. The general pattern that emerges is that, with additional uncertainties (such as an uncertain carbon tax on top of all other uncertainties), the technology portfolio relies even less on conventional fossil-based energy technologies and becomes even more diversified.

Introducing possible increasing returns to technological innovation yields yet another distinguishing feature of technology choice. The dominance of current, fossil-based technologies disappears entirely. The direction and breadth of diversification in turn are a direct function of the extent to which uncertainty is considered. Making the rather unrealistic assumption that the magnitude of possible increasing returns is known with perfect foresight (scenario E3, bottom panel, left) and assuming no carbon tax reduces technological diversity in favor of rapid diffusion of "third-generation" fossil technologies (e.g., coal-based fuel cells). This focus on a few technologies is a direct result of the hypothetical assumption of no uncertainty and perfect foresight. Under these conditions, there is no economic logic for incurring additional costs[10] for diversifying technology portfolios

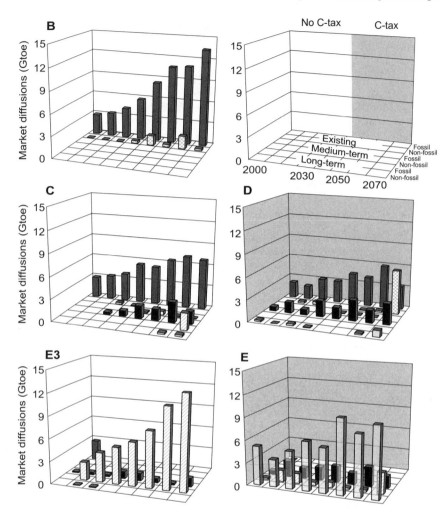

Figure 11.6. Global Diffusion of Six Groups of Energy Technologies (in Gtoe) under Different Types of Uncertainty.

Scenarios grouped to the right (gray background) include an uncertain carbon tax; scenarios grouped to the left (white background) exclude a carbon tax. For explanation of scenarios, see Figure 11.2 and text.

and experimenting in realizing potential increasing returns across a broad front of technology options. The fact that long-term fossil-based technologies are favored in this particular simulation of scenario E3 is not particularly significant, as it is entirely dependent on input assumptions (in this case, initial start-up costs of fossil versus non-fossil "third-generation" technologies). More interesting is the case with full uncertainty (scenario E, bottom panel, right). Here, a much broader diversification takes place, and the presence of an uncertain carbon tax shifts the direction of technology choice in favor of both medium- and long-term non-fossil technologies. As in scenario E3, this swift shift occurs to the detriment of existing technologies (technology group one in Figure 11.6).

Considering that our simplified model does not employ base-year constraints and that it uses levelized costs and optimizes over an entire time horizon spanning many decades, the dynamics of technology "switch-over," particularly over the short term, are neither plausible nor realistic. As we have shown elsewhere (Grübler and Gritsevskyi 1998), moving toward an explicit representation of capital stock (i.e., also of sunk investments) alleviates this problem of an overly dynamic technology response, especially in the case of potential increasing returns.

Nonetheless, a number of general conclusions can be drawn from the model simulations shown in Figure 11.6. Foremost, while a radical transformation of the energy technology landscape is possible, it requires consideration of all salient uncertainties. Thus, postulating increasing returns, general scenario uncertainty, or carbon taxes alone is not sufficient to move away from a "lock-in" situation of the dominance of current fossil technologies. A combination of all these factors, however, can result in a shift from the current lock-in on fossil technologies and a radical transformation of the technological landscape. Second, radical technological change (i.e., toward "third-generation," long-term technology options) seems possible only by considering some mechanism of increasing returns on technological innovation, as uncertain as it may be. Third, there is an inherent trade-off between phenomena of increasing returns and uncertainty: assuming perfect foresight and no uncertainty leads to a pronounced "crowding-out" effect in the technology portfolio. Innovation efforts (investments in R&D and niche market deployments) concentrate on a few technologies in order to maximize learning effects at the lowest possible costs. Conversely, treating the potential benefits from innovation as uncertain requires hedging and technology portfolio strategies that increase diversity but come at additional costs. Finally, independent of the scenario considered, in the model simulations one technology group—fossil-based "incremental" innovations (cluster 2)—systematically has the lowest option value and diffusion potential. Advanced coal-fired power plants, but also gas-based fuel cells (even though they look promising from a medium-term perspective), do not succeed in the model simulations. Whereas this result should not be interpreted as a definitive judgement of the potential merits of these technologies, it does point to yet another trade-off. Ultimately, technology policy will have to move from the easy decision to "let one hundred flowers bloom" to tough choices of whether to further near- to medium-term incremental innovation or concentrate on radical, long-term options. Because in our model we have embraced a time frame that extends many decades into the future, the answer we can report here is predetermined. A definitive answer on the option value of near- versus long-term technology portfolios awaits further methodological improvements in our model—for example, moving from devising an optimal hedging strategy for

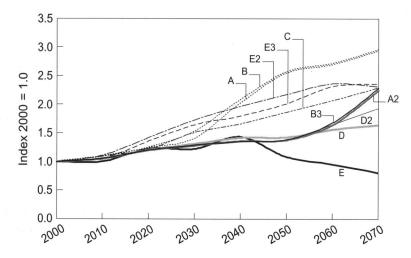

Figure 11.7. Carbon Emissions for Different Scenarios with Varying Degrees of Representation of Uncertainties (Index, 2000=1.0).

Scenarios including a carbon tax are denoted by solid lines; scenarios without a carbon tax, by dashed lines. For an explanation of the scenarios, see Figure 11.2 and text.

a single long time horizon to some other decision-making paradigm, such as sequential decision making under a rolling time horizon.

11.4.5   Environmental constraints

Figure 11.7 reports the resulting carbon emissions of the various scenarios simulated by the model. Because of unconstrained model simulations, slight scenario differences emerge in the base year.[11] To clarify the dynamics of carbon emissions under increasing treatment of uncertainty, Figure 11.7 shows emissions as an index starting at base-year values of one.

The highest, continuously rising carbon emissions result from the scenarios embracing the traditional deterministic modeling framework without uncertainty and static technology (scenarios A and B). In these scenarios, emissions rise threefold by 2070, in line with typical "business-as-usual" scenarios available in the literature that embrace identical modeling perspectives and assumptions. It is interesting to note that moving from a social planner (scenario A) to a Pareto-optimal (scenario B) model formulation has little impact on future carbon emissions. Similarly, the responses of the social planner and Pareto-optimal scenarios to a given carbon tax are almost identical (see scenario A2 versus scenario B3).

Including scenario uncertainties (demands, resource availability, costs), as in scenario C, lowers emissions somewhat, but the continuous upward trends persist. Conversely, adding a carbon tax on top of a deterministic model formulation (scenario B3) yields much lower (but still rising) medium-term emissions, as in the scenario considering uncertainty in the absence of a carbon tax (scenario C).

By 2070, however, emissions in both scenarios are of comparable magnitude. Evidently, combining these two uncertainties—even in the absence of possible increasing returns—yields an even more dramatic effect on future carbon emissions. Combining uncertainty in demand, resource availability, and technology cost, as well as a possible carbon tax (scenario D), results in near stabilization of global carbon emissions at levels that are about 50 percent higher than in 2000.

An interesting finding concerns the emissions impacts of treating *ceteris paribus* the carbon tax as certain (scenario D2) as opposed to treating it as uncertain (scenario D). Were the timing of the introduction and the level of a carbon tax known with certainty (scenario D2), emissions would be *higher* than in the uncertain tax case (scenario D). The reason for lower emissions in the uncertainty case lies in the model formulation of uncertainty along a Weibull distribution. In other words, the even very low probability of a high carbon tax underlying scenario D (recall that the mean and variance of the uncertain carbon tax are identical, unlike the case with a symmetrical distribution of uncertainty, as in a normal distribution) leads to hedging strategies away from fossil-based technologies that result in higher emission reductions than with a deterministically known carbon tax.

Thus, the type and extent of uncertainty of future environmental limits are as important as *ex ante* known absolute levels of these limits, as represented, for instance, by carbon tax levels. This observation has important implications for the climate debate. Any discussion of "optimal" levels of future carbon taxes—however defined—cannot be separated from a discussion of the uncertainties influencing such tax levels, and especially from a discussion of the extent and shape of such uncertainties. In other words, the impact on optimal technological hedging strategies in the face of uncertain climate damages will be influenced as much by the absolute magnitude of such damages [be it 1 or 3 percent of future gross domestic product (GDP)], as by the possibility of low-probability, extreme events. Admitting a less than 1 percent chance that climate-change damages could be substantially higher (say, 10 percent of GDP) leads to much higher anticipatory hedging strategies than assuming a "well-behaved" uncertainty distribution (e.g., when a symmetrical distribution of climate damages around a mean value of a GDP loss of 1–2 percent is assumed).

The emissions picture changes dramatically if the possibility of even uncertain increasing returns is added to all other uncertainties. In such a case (scenario E in Figure 11.7), emissions could actually decline below current levels in the long term (post 2040), yielding the lowest cumulative emissions of all scenarios considered. The scenarios of possible increasing returns are also the three cheapest in terms of total discounted energy systems costs (see Figure 11.4), with a significantly lower cost penalty of a carbon tax identical to those in the other scenarios. This result suggests that any discussion of quantifying climate-change externalities—for example, in the form of carbon taxes—cannot be separated from a discussion of the possible increasing returns an induced technological change perspective entails. In the presence of increasing returns, not only is much more technological diversification optimal for a given level of quantification of environmental externalities (a carbon tax), but the emissions reduction impact of any given level of carbon taxes is also substantially higher. However, the conclusions on technological diversification given above also hold for emissions. Any individual measure, be it stimulating increasing returns to technological innovation or quantifying environmental externalities (via taxes), is unlikely to yield the kind of drastic emission reductions that could yield "stabilization of atmospheric concentration of CO_2 below dangerous levels" (UNFCC 1992). A combination of all factors might.

11.4.6   Heterogeneous agents

Figure 11.8 illustrates the different patterns of technological choice when heterogeneous agents are represented in the model. To this end, in addition to the assumptions of scenario group E (the scenario with full uncertainty including increasing returns and a possible carbon tax), we assume that the technology agents that supply new energy technologies to the five energy-producing/consuming regions in our model differ from the regional actors in their risk attitudes (i.e., in their risk function). In scenario F, only about one-third of the actors on the market are assumed to differ from the regional actors in their risk function. In scenarios F2 and F3, all of them differ markedly. For illustrative purposes, we assume that the technology agents are three times as risk averse as the regional agents in the presence (scenario F2) or absence (scenario F3) of an uncertain carbon tax.

To simplify the scenario comparison, we focus on scenario differences in energy-conversion technologies (hydrogen production, various electricity-generation technologies, etc.), which we aggregate into nine different clusters (designated by roman numerals in Figure 11.8).[12] Scenarios are then ranked on a linear scale based on their cumulative (2000–2070) deployment of these nine energy-conversion technology clusters. Interconnecting the nine technology clusters gives a simple "spider graph" overview of the different technology portfolios of the scenarios.

Interestingly, in the absence of a carbon tax, even introducing the heterogeneous and highly risk-averse behavior of the technology agents (scenario F3 versus scenario D, shown as dashed lines in Figure 11.8) does not lead to a narrower technology portfolio. Scenario F3 even has a slightly more diversified portfolio (cf. the gains in technology clusters VIII and IX) compared with scenario D, which assumes the same risk attitude for all agents in the model. Apparently, the higher risks of the technology agents considered in the global Pareto optimum lead to a higher degree of regional specialization, and hence to a more differentiated technology demand from the regions. At the global level, these regional differences compensate for the otherwise-expected loss in technological variety under heterogeneous and risk-averse behavior of the technology agents.

Conversely, with a carbon tax, even an uncertain one, technological variety seems to be more difficult to maintain with rising risk aversion. Scenario F assumes (only slightly) heterogeneous risk behavior of the technology agents combining risk-averse, risk-taking, and risk-"neutral" (in comparison to

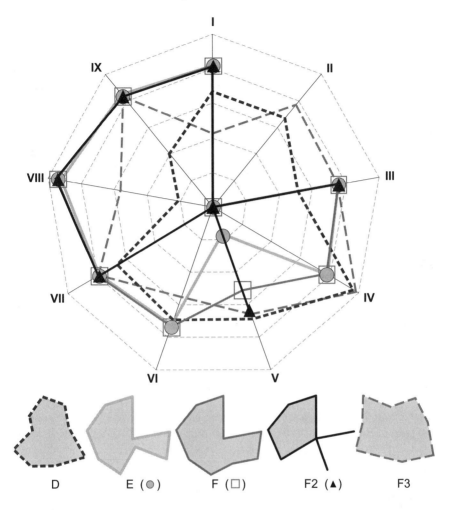

Figure 11.8. "Spider Graph" of Technology Portfolio of Nine Clusters of Energy-Conversion Technologies (Denoted Clockwise by Roman Numerals).

The graph plots relative proportions of cumulative market deployment between 2000 and 2070 for scenarios of homogeneous technology agents versus scenarios of heterogeneous technology agents. Scenarios assuming an uncertain carbon tax are denoted by solid lines; scenarios without a tax, by dashed lines. For an explanation of scenarios, see Figure 11.2 and text.

the energy-producing/consuming regions) behavior. The scenario shows a portfolio almost identical to that of the comparable scenario E (with homogeneous risk). However, scenario F2, in which the technology agents are three times as risk averse as the energy-producing/consuming regions, experiences a drastic collapse of the richness of the technology portfolio. The "spiderweb" of technology diversification in Figure 11.8 almost collapses to a star-like structure. Scenario

F2, for instance, entirely dispenses with the deployment of three technology clusters (II, IV, and VI) that were a prominent part of the technology portfolio in the homogeneous, risk-"neutral" scenario (scenario E). In a high-cost (carbon tax) scenario, increasing the risk aversion of the technology agents thus leads them to focus on fewer technologies as targets of their forward-looking innovation investments. The result is a "crowding out" of technological variety.

Despite the considerable work that remains to be done on further model testing and sensitivity analysis (not to mention on developing plausible model parameterizations), two conclusions can be drawn at this stage. First, methodologically, a feasible approach has been developed that allows us to represent heterogeneity between different agents acting under deep uncertainty while maintaining the tradition of an optimization framework. Second, policywise, the simulations illustrate the inherent trade-off between maintaining a broad technological portfolio that is optimal in the face of the numerous uncertainties the future inevitably holds and the desire to provide market signals factoring in environmental externalities. Single-purpose policy signals such as a carbon tax ultimately lead to a "crowding-out" effect in narrowing technology portfolios that may be more optimal for facing uncertainties beyond climate change. Technological diversity and a "healthy" heterogeneity between actors appear to go hand in hand. How these two intertwined factors can be nurtured by creative policies remains a key issue for further research.

11.5   Conclusion and Outlook

Owing to the new nature of the model, we do not wish to overemphasize conclusions from the first, illustrative simulation runs reported in this chapter. Rather, we focus here on open research questions and the future agenda for model development.

Nonetheless, two conclusions appear possible at this early stage of the model's life cycle. First, the proposed approach and model formulation of uncertain increasing returns in a multiregion, multi-actor, multi-technology model are feasible and solvable. Even if simplified, the model's representation of the energy system is sufficient for analyzing the most important technological linkages and interdependencies while maintaining a minimum degree of technological (innovation) variety. Second, the model simulations add to earlier results suggesting that radical technological change and long-term reversals of increasing emission trends are only feasible endogenously via a mechanism of increasing returns—even if these returns are uncertain. Incremental or "routine" innovations, or the consideration of various types of uncertainties (including environmental uncertainties), yield individual changes in technology adoption decisions, but no radical technological change or "big hits." Radical change is possible over the long term, provided the necessary intervening efforts (R&D efforts and niche market deployment of new technologies) are taking place, initiating a stream of continuous improvements and innovations that can translate into increasing returns, albeit with no guarantee of success. As innovation is highly uncertain, such strategies need to be spread over a broad portfolio of technological options rather than directed at winners "picked" prematurely. But as our model simulations indicate, such innovation strategies, even if uncertain, make eminent economic sense.

A number of items remain on the agenda for future research. First and foremost, the model awaits further detailed sensitivity analysis and further scrutiny of the plausibility of both behavior and results. Further work is also required on the model's Pareto-optimal trade formulation, even if it is not at the core of our investigations on technological innovation and dynamics under uncertain increasing returns. The concept of different levels of risk aversion for different agents also has not yet been explored in detail in model simulations and sensitivity tests. Finally, representation of past decisions (i.e., modeling of *existing* energy systems structures) that limit near- to medium-term decision flexibility also remains an important task for improving both model plausibility and policy relevance.

Concerning methodological extensions, three important unresolved issues remain on the research agenda. First, improvements are needed in the representation of uncertainties. The current model version and calibration unduly "compresses" the uncertainty space to small random variations around mean expected values. Representation of the possibility of large "surprises" with potentially large consequences (e.g., availability of enormous amounts of methane clathrates at competitive costs) awaits definitive methodological and algorithmic solutions. Simply changing the functional form of uncertainty representation—for example, using distribution functions with long tails—might not be sufficient when dealing with possible long-term "surprises," as low probability combined with discounting might not significantly change the model's behavior.

Second, improvements in the model's oversimplified treatment of the innovation process are required. Inclusion of the extremely high uncertainty of technological invention (i.e., option generation) and better representation of the interplay between embodied and disembodied aspects of technological change and knowledge generation (as well as knowledge depreciation) remain important tasks. We do not adhere to the opinion that disembodied and embodied aspects of technological change are substitutable (no matter how large, R&D efforts can never replace the ultimate test of market applications and substitute for the costly and time-consuming efforts of learning by doing and learning by using). Important lags in knowledge transfer and spillover effects also need to be recognized in more detail. Ultimately, the model will also have to keep track of investments and capital flows that are treated extremely simplistically in the current model version.

Third, the biggest remaining challenge is to allow for some endogenous mechanism of uncertainty reduction and information updating, even if this implies a departure from "once-through" analytical resolutions within an optimization framework. Rolling time windows instead of century-long optimization periods and/or their combination with simulation techniques with informational updating (e.g., on realized technological progress, environmental limits, etc.) need to be explored.

Thus, the tasks ahead are huge. But the challenge is worth the effort. After all, it is only through improved technology that humans can prepare for the numerous contingencies that an uncertain future holds. "Unlike resources found in nature, technology is a man-made resource whose abundance can be continuously increased, and whose importance in determining the world's future is also increasing" (Starr and Rudman 1973:364).

Acknowledgments

The model presented here builds on earlier work performed at IIASA dealing with stochastic uncertainties in future technology costs (Golodnikov *et al.* 1995; Grübler and Messner 1996; Messner *et al.* 1996) the possibility of increasing returns via learning curve effects (Messner 1995, 1997), and treating technological learning curves as stochastic (Grübler and Gritsevskyi 1998; Gritsevskyi and Nakicenovic 2000; and Chapter 10 in this volume). We wish to thank in particular Yuri Ermoliev for his continuous help in clarifying conceptual and mathematical issues as well as for proposing solutions to them. Our IIASA colleagues are to be absolved of blame for the shortcomings of this chapter.

Notes

1.　While it is impossible to estimate the relative contributions of embodied and disembodied knowledge in the technological knowledge stock as a whole, it is possible to look at least at the relative knowledge acquisition costs, or expenditures. Even in the most R&D-intensive industries, such as the aerospace industry, R&D generally does not exceed 20 percent of sales; the average for the manufacturing industry does not exceed 5 percent (including R&D funded by industry as well as from other sources; see NSF 1998). Development expenditures account for the majority of R&D expenditures (typically about two-thirds in the United States; NSF 1998). Thus, measured by expenditures, disembodied technological change in its purest form (i.e., research) accounts for less than 2 percent on average. This average, however, masks huge differences between sectors and technologies.

2.　The sensitivity of technology choice and diffusion profiles to variations in the discount rate is explored in more detail in Grübler and Messner (1998).

3.　Alternatively, one could consider risk profiles derived from standard utility functions. However, this would require a more comprehensive treatment of heterogeneous agents in our model (e.g., also representing consumers). Choosing empirically based parameter assumptions also represents a formidable challenge.

4.　In a future version of the model, we plan to relax the rather unrealistic assumption that technological learning can go on forever by explicitly including knowledge depreciation. In such an improved model version, learning is assumed to be contingent on a continuously increasing knowledge stock. That is, once installation expansion rates cease to grow, the technology-related knowledge stock begins to depreciate. This assumption reflects the empirical observations reported in Michina (1992, 1999) and Irwin and Klenow (1994).

5.　This technology knowledge stock is represented simply by cumulative expenditures. We recognize the importance of "forgetting by not doing," but in this first model version we have not included a knowledge-depreciation parameter.

6.　Consequently, learning-curve types of models have been justifiably criticized on the basis that they ignore the important difference between who pays for new technological knowledge (learning) and who benefits (via lowered prices) from it.

7. This is the reason why we adopted this particular scenario and not the structurally and quantitatively very similar IPCC-B2 scenario. In terms of cumulative "call on resources," the two scenarios are very similar to each other.

8. A confirmation sensitivity run of scenario B2 (Pareto optimum with equal weights) yields identical results and thus need not be discussed separately here.

9. Open research questions concern the stability of the Pareto optimum under a large number of regions/countries and the issue of coalition formation in such a model formulation. Considering the long time horizon of our model and the possibility of increasing returns, it remains doubtful if stable Nash equilibria could be identified under such conditions.

10. The additional costs of a scenario of uncertain increasing returns compared with those of the certain scenario are US$2.5 trillion (see Figure 11.4), or some 3 percent of the total discounted energy systems costs of the uncertainty scenario.

11. Differences are greatest for scenarios considering (uncertain) increasing returns.

12. Given the illustrative nature of these model simulations, we refrain from reporting the hardware equivalents of these nine energy-conversion technology clusters, not least to avoid giving the impression of drawing premature policy conclusions on relative technological merit. Hence, our discussion focuses simply on differences in technology structure or portfolios between the scenarios.

References

Abramovitz, M., 1993, The search for the sources of growth: Areas of ignorance old and new, *Journal of Economic History*, **52**(2):217–243.

Alchian, A.A., 1963, Reliability of progress curves in airframe manufacturing, *Econometrica*, **31**:679–693.

Argote, L., and Epple, D., 1990, Learning curves in manufacturing, *Science*, **247**:920–924.

Arrow, K.J., 1962, The economic implications of learning by doing, *Review of Economic Studies*, **29**:155–173.

Arthur, W.B., 1983, On Competing Technologies and Historical Small Events: The Dynamics of Choice under Increasing Returns, Working Paper WP-83-90, International Institute for Applied Systems Analysis, Laxenburg, Austria.

Arthur, W.B., 1989, Competing technologies, increasing returns, and lock-in by historical events, *The Economic Journal*, **99**:116–131.

Ayres, R.U., 1987, *Industry Technology Life Cycles: An Integrating Meta Model?*, Research Report RR-87-3, International Institute for Applied Systems Analysis, Laxenburg, Austria.

Cantley, M.F., and Sahal, D., 1980, *Who Learns What? A Conceptual Description of Capability and Learning in Technological Systems*, Research Report RR-80-42, International Institute for Applied Systems Analysis, Laxenburg, Austria.

Christiansson, L., 1995, Diffusion and Learning Curves of Renewable Energy Technologies, Working Paper WP-95-126, International Institute for Applied Systems Analysis, Laxenburg, Austria.

Cohen, W.M., and Levinthal, D.A., 1989, Innovation and learning: The two faces of R&D, *The Economic Journal*, **99**:569–596.

Cowan, R., and Foray, D., 1995, The Changing Economics of Technological Learning, Working Paper WP-95-39, International Institute for Applied Systems Analysis, Laxenburg, Austria.

Denison, E.F., 1962, The Sources of Economic Growth in the United States and the Alternatives Before Us, Supplementary Paper 13, Committee for Economic Development, New York, NY, USA.

Denison, E.F., 1985, Trends in American Economic Growth 1929–1982, The Brookings Institution, Washington DC, USA.

Dosi, G., 1982, Technological paradigms and technological trajectories: A suggested interpretation of the determinants and directions of technical change, *Research Policy*, **11**:147–162.

Epple, D., Argote, L., and Murphy, K., 1996, An empirical investigation of the microstructure of knowledge acquisition and transfer through learning by doing, *Operations Research*, **44**(1):77–86.

Fishlow, A., 1965, *American Railways and the Transformation of the Antebellum Economy*, Harvard University Press, Cambridge, MA, USA.

Foray, D., and Grübler, A., 1990, Morphological analysis, diffusion and lock-out of technologies: Ferrous casting in France and the FRG, *Research Policy*, **19**(6):535–550.

Foray, D., and Grübler, A., 1996, Technology and the environment: An overview, *Technological Forecasting & Social Change*, **53**(1):3–13.

Frankel, M., 1955, Obsolescence and technological change in a maturing economy, *American Economic Review*, **45**:296–319.

Freeman, C., 1989, The Third Kondratieff Wave: Age of Steel, Electrification and Imperialism, Research Memorandum 89-032, MERIT, Maastricht, Netherlands.

Freeman, C., 1994, The economics of technical change, *Cambridge Journal of Economics*, **18**:463–514.

Godet, M., 1997, *Manuel de Prospective Stratégique*, Vol. I and II., Dunod, Paris, France.

Golodnikov, A., Gritsevskyi, A., and Messner, S., 1995, A Stochastic Version of the Dynamic Linear Programming Model MESSAGE III, Working Paper WP-95-94, International Institute for Applied Systems Analysis, Laxenburg, Austria.

Griliches, Z., 1996, The discovery of the residual: A historical note, *Journal of Economic Literature*, **34**:1324–1330.

Gritsevskyi, A., and Nakicenovic, N., 2000, Modeling uncertainty of induced technological change, *Energy Policy*, **28**: 907–921.

Grübler, A., 1998, *Technology and Global Change*, Cambridge University Press, Cambridge, UK.

Grübler, A., and Gritsevskyi, A., 1998, A model of endogenous technological change through uncertain returns on learning, http://www.iiasa.ac.at/Research/TNT/WEB/Publications/

Grübler, A., and Messner, S., 1996.,Technological uncertainty, in N. Nakicenovic, W.D. Nordhaus, R. Richels, and F. Toth, eds, *Climate Change: Integrating Science, Economics, and Policy*, Conference Proceedings CP-96-1, International Institute for Applied Systems Analysis, Laxenburg, Austria.

Grübler, A., and Messner, S., 1998, Technological change and the timing of mitigation measures, *Energy Economics*, **20**:495–512.

Ha-Duong, M., Grubb, M., and Hourcade, J.-C., 1997, Influence of socio-economic inertia and uncertainty on optimal CO_2 abatement, *Nature*, **390**:270–273.

IEA (International Energy Agency), 2001, Key World Energy Statistics 2000, Paris, France, http://www.iea.org/statist/keyworld/keystats.htm

IPCC (Intergovernmental Panel on Climate Change), 2000, *Emissions Scenarios*, Cambridge University Press, Cambridge, UK.

Irwin, D.A., and Klenow, P.J., 1994, Learning-by-doing spillovers in the semiconductor industry, *Journal of Political Economy*, **102**(6):1200–1227.

Katz, M., and Shapiro, C., 1985, Network externalities, competition and compatibility, *American Economic Review*, **75**:424–440.

Kline, S.J., 1985, What is technology? Bulletin of Science, *Technology and Society*, **5**(3):215–219.

Maddison, A., 1991, *Dynamic Forces of Capitalist Development: A Long-Run Comparative View*, Oxford University Press, Oxford, UK.

Mahajan, V., Muller, E., and Mass, F.M., 1991, New product diffusion models in marketing: A review and directions for research, in N. Nakicenovic and A. Grübler, eds, *Diffusion of Technologies and Social Behavior*, Springer Verlag, Berlin and Heidelberg, Germany, and New York, NY, USA.

Mansfield, E., 1968, *The Economics of Technological Change*, W.W. Norton & Co, New York, NY, USA.

Mansfield, E., 1977, The diffusion of eight major industrial innovations in the United States, in N.E. Terleckyj, ed., *The State of Science and Research: Some New Indicators*, Westview Press, Boulder, CO, USA.

Mansfield, E., 1985, How rapidly does new industrial technology leak out?, *Journal of Industrial Economics*, **34**(2):217–223.

Messner, S., 1995, Endogenized Technological Learning in an Energy Systems Model, Working Paper WP-95-114, International Institute for Applied Systems Analysis, Laxenburg, Austria.

Messner, S., 1997, Endogenized technological learning in an energy systems model, *Journal of Evolutionary Economics*, **7**(3):291–313.

Messner, S., Golodnikov, A., and Gritsevskyi, A., 1996, A stochastic version of the dynamic linear programming model MESSAGE III, *Energy*, **21**(9):775–784.

Michina, K., 1992, Learning by New Experiences, Working Paper 93-084, Harvard Business School, Harvard, Cambridge, MA, USA.

Michina, K., 1999, Learning by new experiences: Revisiting the Flying Fortress learning curve, in N.R. Lamoreaux, D.M. Raff, and P. Temin, eds, *Learning by Doing in Markets, Firms and Countries*, National Bureau of Economic Research Conference Report, Chicago University Press, Chicago, IL, USA.

Mokyr, J., 1990, The Lever of Riches: *Technological Creativity and Economic Progress*, Oxford University Press, Oxford, UK.

Nakicenovic, N., Grübler, A., and McDonald, A., eds, 1998a, *Global Energy Perspectives*, Cambridge University Press, Cambridge, UK.

Nakicenovic, N., Victor, N., and Morita, T., 1998b, Emissions scenario data base and review of scenarios, *Mitigation and Adaptation Strategies for Global Change*, **3**(2–4):95–120.

Neij, L., 1997, Use of experience curves to analyze the prospects for diffusion and adoption of renewable energy technology, *Energy Policy*, **23**(13):1099–1107.

Nordhaus, W.D., 1973, The allocation of energy resources, *Brookings Papers on Economic Activity*, **3**:529–576.

Nordhaus, W.D., 1997, Modeling Induced Innovation in Climate Change Policy, Paper presented at the Workshop on Induced Technological Change and the Environment, 26–27 June, International Institute for Applied Systems Analysis, Laxenburg, Austria.

NSF (National Science Foundation), 1998, Science and Engineering Indicators 1998, NSF, Washington DC, USA, http://www.nsf.gov/ sbe/srs/seind98/start.htm

Pavitt, K., 1993, What do firms learn from basic research?, in D. Foray and C. Freeman, eds, *Technology and the Wealth of Nations: The Dynamics of Constructed Advantage*, Pinter Publishers, London, UK.

Rogers, E., 1983, *Diffusion of Innovations*, 3rd edition, The Free Press, New York, NY, USA.

Rogers, E., and Shoemaker, F., 1971, *Communication of Innovations*, The Free Press, New York, NY, USA.

Rogner, H.H., 1997, An assessment of world hydrocarbon resources, *Annual Review of Energy and the Environment*, **22**:217–262.

Rosegger, G., 1996, *The Economics of Production and Innovation: An Industrial Perspective*, Butterworths, Oxford, UK.

Rosenberg, N., 1990, Why do firms do basic research (with their own money)?, *Research Policy*, **19**(2):165–174.

Rosenberg, N., 1996, Uncertainty and technological change, in R. Landau, T. Taylor, and G. Wright, eds, *The Mosaic of Economic Growth*, Stanford University Press, Stanford, CA, USA.

Ruttan, V.W., 1959, Usher and Schumpeter on invention, innovation, and technological change, *Quarterly Journal of Economics*, **73**:596–606.

Scherer, F.M., 1980, *Industrial Market Structure and Economic Performance*, 2nd edition, Rand McNally College Publ. Co., Chicago, IL, USA.

Schumpeter, J.A., 1911, *Theorie der wirtschaftlichen Entwicklung*, Duncker & Humbolt, Leipzig, Germany.

Schumpeter, J.A., 1934, *The Theory of Economic Development: An Inquiry into Profits, Capital, Credit, Interest, and the Business Cycle*, Harvard University Press, Cambridge, MA, USA.

Schumpeter, J.A., 1942, *Capitalism, Socialism and Democracy*, Harper & Brothers, New York, NY, USA.

Silverberg, G., 1991, Adoption and diffusion of technology as a collective evolutionary process, *Technological Forecasting & Social Change*, **39**(1–2):67–80.

de Solla Price, D.J., 1965, Is technology historically independent of science? A study in statistical historiography, *Technology and Culture*, **6**(4):553–568.

Solow, R.M., 1957, Technical change and the aggregate production function, *Review of Economics and Statistics*, **39**:312–320.

Starr, C., and Rudman, R., 1973, Parameters of technological growth, *Science*, **182**:358–364.

Strubegger, M., and Reitgruber, I., 1995, Statistical Analysis of Investment Costs for Power Generation Technologies, Working Paper WP-95-109, International Institute for Applied Systems Analysis, Laxenburg, Austria.

Tinbergen, J., 1942, Zur Theorie der langfristigen Wirtschaftsentwicklung, *Weltwirtschaftliches Archiv*, **1**:511–549.

Toynbee, A.J., 1962, *A Study of History*, collected works of A.J. Toynbee, Vols. I–XII, New York, (abridged) paperback reprint: Oxford University Press (2 Vols.), 1987.

von Tunzelmann, N., 1982, Structural change and leading sectors in British manufacturing 1907–1968, in C.P. Kindleberger and G. di Tella, eds, *Economics in the Long View, Vol. 3 Part 2: Applications and Cases*, New York University Press, New York, NY, USA.

UNFCC (United Nations Framework Convention on Climate Change), 1992, Convention Text, Climate Change Secretariat, Geneva, Switzerland.

Watanabe, C., 1995, Identification of the role of renewable energy, *Renewable Energy*, 6(3):237–274.

Watanabe, C., 1999, Inducing Technology Substitution for Energy, Mimeo, Department of Industrial Engineering and Management, Tokyo Institute of Technology, Tokyo, Japan.

Wright, T.P., 1936, Factors affecting the costs of airplanes, *Journal of the Aeronautical Sciences*, 3:122–128.

Chapter 12

Modeling Induced Technological Change: An Overview

Leon E. Clarke and John P. Weyant

12.1 Introduction

This chapter explores induced technological change (ITC) in the context of research and policy models of energy, the environment, and climate change. What elements of ITC can and should be included in these models? What should be left out and considered qualitatively in interpreting the models? How have modelers thus far included ITC? Are certain modeling approaches more amenable to ITC than others?

Almost everyone—researchers and policy makers alike—agrees that the response of technology to economic incentives and to policy over the coming decades may be crucially important in designing energy and environmental policies. Thus, questions about the optimal timing and stringency of greenhouse gas abatement policies have become increasingly concerned with assumptions about technological change in economic models. In addressing questions about the optimal timing of carbon abatement, Grubb *et al.* (1995) helped focus attention on the need to fully endogenize the rate and direction of technological change. Since then, Goulder and Schneider (1997, 1999), Goulder and Mathai (1998; also Chapter 9 in this volume), Grübler and Gritsevskyi (1998), Gritsevskyi and Nakicenovic (2000; also Chapter 10 in this volume), Nordhaus (Chapter 8 in this volume), and others have constructed policy models that include ITC.

The purpose of this chapter is not to critically evaluate particular model findings or to draw policy conclusions from them. It is to examine the methodology by which ITC has been modeled and to find inherent modeling limitations and opportunities for improvement, drawing on insights from the existing economic literature on technological change. We do not attempt a comprehensive literature review, only a selective one to highlight relevant ideas and present important distinctions. [Other reviews and related discussions can be found in Carraro (1997), Weyant (1997); Grübler *et al.* (1999); and Jaffe *et al.* (2000).] A critical question concerns how much confidence we can expect to place in models of long-term technological change, and how much of the analysis should rest on the qualitative

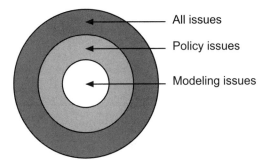

Figure 12.1. Set of Issues Appropriate for ITC Models, as a Subset of a Larger Set of Issues.

insights of policy makers. Policy makers should ideally consider all important elements of ITC, but only some of these elements are amenable to inclusion in climate-change models.

Figure 12.1 is a simple visual representation of how one might consider the elements of ITC, just as modelers have already had to consider the complex elements of global social, economic, and climate systems. First, there is a set of possible phenomena and dynamics that are relevant to ITC, represented by the largest circle. But policy makers cannot consider everything: they must condense the system into something more manageable, even at a qualitative and intuitive level. Hence, the set of phenomena that are policy relevant is a subset of the larger set of issues, as illustrated by the mid-sized circle. Further, only a subset of the policy-relevant phenomena and issues is amenable to mathematical modeling: modelers cannot include everything that is important, just as economic and climate-change modelers more generally do not include everything that is potentially important. The smallest circle captures the set of policy-relevant phenomena amenable to mathematical modeling.

Hence, two issues present themselves to ITC modelers: first, which phenomena should modelers consider and which should they leave out as qualitative adjustments to models or as irrelevant, and second, how should modelers incorporate the elements they decide are important and amenable to modeling? An important implicit conclusion of this review is that communication between modelers and policy makers is essential. It is incumbent upon policy makers to understand that models capture only a portion of reality, to ascertain what is in and what is out, and to interpret model results accordingly. It is incumbent upon modelers to clearly convey this information to policy makers.

The remainder of this chapter is divided roughly into two parts. The first part reviews current thinking about technological advance in energy and environmental models. We begin with a definition of ITC in Section 12.2. We then present a selective review of the economic literature on technological advance in Section 12.3; discuss spillovers, including the empirical evidence on their magnitude, in Section 12.4; elaborate on the notion of innovation market failures more generally and the implications for climate-change modeling in Section 12.5; discuss prominent examples of state-of-the-art climate-change models with ITC in Section 12.6; and conclude with a summary discussion of ITC in energy and environmental models, based on the previous sections, in Section 12.7. The second part of the chapter, Section 12.8, looks at the limitations of current ITC modeling approaches and potential extensions. We explore notions such as complementary sources of technological change, heterogeneity, uncertainty, path dependence, and diffusion. Concluding remarks are presented in Section 12.9.

12.2 Definitions

What is ITC? This is actually a two-part question: what is "technological change"—what is it that is changing—and what does it mean for technological change to be "induced"? Broadly defined, a society's technology refers to the set of goods and services the society can produce and the methods, or the input combinations, by which it produces these goods and services. The introduction of the internal combustion engine was technological change: a new good and associated services became available. Subsequent improvements in the internal combustion engine are also technological change: cars are faster, more powerful, more comfortable, and more reliable than 100 years ago, and the methods of production are more advanced and efficient. And technological change is not just a matter of machinery. Fast food and the "Wal-Mart method" of retailing also constitute technological change.

Climate-change models capture technology through mathematical functions that specify the possible input–output combinations available to a society, a country, an industry, or a representative firm. Our definition of technology here should be consistent with this mathematical approach. In climate-change models, and economic models more generally, mathematical representations of possible input–output combinations are often referred to as "production functions." This is the definition we will use here: technology is the production function. The essential point to remember is that for technology to change, the production function must change. Changing the set of feasible products and processes is technological change, choosing among products and processes is not.

What does it mean for technological change to be induced, or endogenous? One requirement is that technological change must respond to more than just the passage of time, for that would be *exogenous* technological change. But this requirement does not sufficiently pin down the notion of *induced* technological change. In one common paradigm, ITC refers to the fact that the research and innovation decisions of firms and individuals are influenced considerably by the private costs and rewards of innovation. For example, dreams of enormous wealth, some of them realized, are responsible for many of the technological advances coming out of Silicon Valley. Hence, technological change does not fall like manna from heaven, but is endogenous to the social and economic system. In this paradigm, technological advance is strictly a private sector phenomenon associated with private sector incentives. Governments can alter the rate and direction of technological advance by altering these incentives. Emissions taxes, for example, change the relative prices of polluting and nonpolluting technologies, and therefore the incentives to improve these technologies.

In this chapter we use a broader notion of ITC: *technological advance responds to policy*. The goal in taking this broader perspective is to emphasize that our ultimate concern is policy and, further, that policies affect technological advance through a multitude of mechanisms, not just through changes in the relative prices of polluting and nonpolluting goods brought on by emissions taxes. Governments have an extensive history of acting directly on technological advance through research and development (R&D) at national laboratories and universities, through research grants to private firms, and, more recently, through public–private research partnerships. Policies can also directly target the diffusion of technology—a salient concern in the context of global warming because technology in less developed countries lags significantly behind that in developed ones.

12.3 A Selective Review of the Economic Literature on Technological Change

The economic literature on technological change is vast and diverse. It includes traditional neoclassical work, historical studies, "appreciative" theory, econometric efforts, and less mainstream mathematical work such as that in evolutionary economics. [See Ruttan (2001) for a thorough discussion of the ITC literature.] In this section, we will conduct a brief review of two particular strands of economic theorizing about technological change. First, we refer to "innovation theory" as the microeconomic study of innovation—roughly defined as the development of new products and processes—using neoclassical economic concepts. Innovation theory focuses on firms and industries. It explores firms' *incentives* to improve technology ("We'll make millions!") and the *inefficiencies* that result from firms' failure to share the gains of their innovative endeavors ("We've got to patent this thing!"). The second category of conventional economic theorizing concerns macroeconomic growth. "Endogenous growth theory," or "new growth theory," borrows insights and methods from innovation theory and applies them within the context of neoclassical growth models. It looks at how innovation investments by private actors might prove to be a source of long-term economic growth. The new growth theory models of technological change are further abstracted from firm and industry behavior than is innovation theory, but as macroeconomic models, they are more consistent with climate-change models.

To be clear, innovation theory and endogenous growth theory are not the only important lines of economic research into technological change, but they are a good starting point for looking at efforts to date to include ITC in climate-change models. One reason is that they use the traditional neoclassical analytical structures that underlie much of the state-of-the-art climate-change work—they are working from the same manual as many climate-change modelers. Further, they are bound together by a common focus on one particular aspect of technological advance, namely, that technology is associated with knowledge, knowledge is a public good, and, hence, we should expect firms to spend less on improving technology than we as a society would like. In Section 12.8 we make up somewhat for our selectivity here. There, we give a broader view of technological change and the associated literature and discuss possible extensions to, and limitations of, current ITC models raised by the wider body of relevant research.

12.3.1   Innovation theory

Innovation theory asks why and how firms invest in innovation, how effectively they appropriate the fruits of their innovation investments through profits, and how efficient the resulting system is at the industry level. Technology and knowledge are often viewed as synonymous in this paradigm; since technological information is a public good, the private sector probably underinvests in innovation. The main contributions of innovation theory have been to emphasize both the importance of knowledge in markets for innovation and the private profit incentive as a key source of innovation activity, and to clarify the complex and imperfect competition in which innovations arise and are put to use.

In his seminal work, Arrow (1962a) demonstrated that neoclassical mathematical economic models could be used to explore firm behavior and incentives to innovate. He showed that an innovating monopolist and an innovating firm licensing technology into a competitive market have different incentives to innovate, and both have weaker incentives than would be socially optimal, even if none of their knowledge leaks out. In other words, market structure is important in technological advance, and private actors probably underinvest in innovation. Arrow further noted that a firm's ability to appropriate the gains from innovative activity may spur innovation, but it does so at the price of inefficiencies following innovation: "in a free enterprise economy, the profitability of invention requires a nonoptimal allocation of resources" (p. 617). If everybody shared everything they know, it would dilute incentives to innovate. These dual results of appropriability, its countervailing incentive and efficiency effects, are often referred to as a tension between dynamic and static efficiency. From Arrow's simple model has grown an extensive and increasingly complex line of mathematical theoretical research, relying heavily on game theory, into the interactions between firms before and after innovation. [Good reviews of seminal work in this literature can by found in Kamien and Schwartz (1982) and Reinganum (1989).]

Arrow also characterized technological information as being freely available and generally applicable, and he asserted that it is the primary *resource* for innovation. From this and related work has emerged a notion that basic science provides the technological paradigms, or raw material, for major cycles of innovation. Applied research draws from a pool of readily available technological information derived from basic science or basic research [see Evenson and Kislev (1975) for early work in this paradigm]. For example, basic research in physics helped set the stage for the development of transistors and, consequently, the computer on which this chapter was composed. The more basic the research, the less appropriable the resulting information and the lower the private incentive for investment. Consistent with this notion, the innovation literature frequently distinguishes between technological breakthroughs and the resulting, more incremental, product and process improvements. Innovation theory implicitly focuses on the second stage of the innovation process because it is the most predictable and it most clearly responds to market forces.

Innovation theorists have developed various notions of the interaction between market forces and technological advance. On the one hand, market-pull theory, generally attributed to Schmookler (1966), supposes that technological advance is largely the result of conscious responses to market forces: firms identify opportunities and then create the products or processes to take advantage of the opportunities. Schmookler hypothesized that major innovations create a new product frontier that is common knowledge. Identifying and bringing innovative products to market under the new paradigm is straightforward; the challenge to entrepreneurs is gauging and responding to market needs. Alternatively, technology-push theory, stated formally by Rosenberg (1976), asserts an opposite causal direction: technology evolves over time as the result of unpredictable product and process innovations. Technology-push theory emphasizes the uncertainty involved in innovation. Firms cannot predict the results of their innovative endeavors and therefore respond less directly to market forces than market-pull theory asserts.

In the 1980s, economic theorists were confronted by analytical barriers in trying to extend neoclassical models, and a number of researchers shifted their focus to empirical work. A primary new result of this generation of empirical work was the identification of a surprisingly large range of methods by which firms appropriate the returns from innovation. Instead of patents, the most commonly cited method of protecting innovations was reported to be a combination of learning curves, lead-time effects, and trade secrets.

In summary, innovation theory has demonstrated that profit incentives account for a major source of innovative activity, but that appropriating these profits leads to inefficient monopolistic behavior. Further, because knowledge is not fully appropriable, private markets probably underinvest in innovation. The simple lesson for climate modelers is that technological change is endogenous to the economic system—that is, it can be induced by policy—but the market response to technological opportunities is probably less than socially optimal.

12.3.2   Endogenous growth theory

Researchers in endogenous growth theory were interested in understanding how the neoclassical growth model might be extended or modified to better capture two important empirical observations: that economies have been able to maintain extended periods of economic growth and that a wide gap remains between rich and poor nations. Researchers in endogenous growth theory absorbed innovation theory's lessons about knowledge spillovers and incorporated them into the traditional growth models by making these models more rigorous about the manner in which human capital and technology change over time (Romer 1986, 1990; Grossman and Helpman 1991b; Aghion and Howitt 1992).

The new growth models, reviewed by Romer (1994), Grossman and Helpman (1994), and Jorgenson (1996), differ from traditional growth models in that they no longer assume exogenous technological advance. Instead, purposive, profit-seeking investments in knowledge are critical to the long-run growth process. In particular, the nonappropriable aspects of new technology created by profit-seeking firms create spillovers, or positive externalities. Endogenous growth theory has shown how positive externalities of this kind can lead to steady, long-run economic growth. A major contribution of Romer's important work (Romer 1990) was the integration of a neoclassical innovation theory model in the spirit of Arrow into a neoclassical growth model in the spirit of Solow. He demonstrated theoretically one combination of technological opportunities and market structure that might result in sustainable, long-run growth.

Because endogenous growth theory is still a relatively young field, it remains largely a theoretical structure, and computable models based on it are not yet available. Still, it provides important lessons for climate-change modeling. Endogenous growth theory emphasizes the importance of spillovers in modeling technological change and points to how such spillovers might be incorporated into aggregate economic models. From the premise that spillovers are a fundamental source of economic growth, it follows that any model of long-term technological change needs to incorporate spillovers.

12.4   Spillovers and Appropriability

Economists have long recognized that markets invest inefficiently in innovation. In Section 12.3 we looked at two particular strands of economic research held together by a common focus on one cause of this inefficiency, namely, that technology is associated with knowledge, and knowledge is a public good. There are other reasons for inefficiency, and we will touch on these in Section 12.5 and discuss them in more detail in Section 12.8. Nonetheless, because the notion of knowledge as a public good has taken on such a prominent role in economic theorizing about technological change, it is useful to explore the notion in more detail.

The "technology-is-knowledge-is-a-public-good" notion is captured by the dual phenomena of appropriability and spillovers. While appropriability and spillovers are widely recognized as fundamental aspects of technological change and economic growth, they are talked about and measured in many different ways. Here, we draw some important distinctions in the discussion of appropriability and spillovers, and then summarize the lessons from associated empirical work.

12.4.1 Spillovers versus appropriability

Spillovers and appropriability are two sides of the same coin: what innovators do not appropriate spills over. When we discuss one, we are implicitly discussing the other. But there are philosophical and practical reasons to distinguish between the two, and these reasons help to explain why innovation theory concentrates on appropriability and endogenous growth theory focuses on spillovers. The distinction has much to do with the underlying research agenda.

Endogenous growth theory observes economies at the macroeconomic level, developing models to account for aggregate-level growth. Growth is viewed as a *consequence* of firm innovation. Hence, spillovers are viewed as positive externalities: society gets more from innovation than do firms because of spillovers. As endogenous growth models develop more microeconomic rigor and detail, they are more explicit in modeling the individual investment incentives of firms and how innovation comes about (e.g., Romer 1990). The primary research focus, however, remains on innovation and spillovers as a combined source of macroeconomic growth, and, hence, on spillovers as positive externalities.

Innovation theory, on the other hand, examines firms at the microeconomic level, trying to understand their innovation incentives and therefore to describe their behavior. Innovation theory therefore places its emphasis on appropriability's implications for investment incentives *prior* to innovation and on its market structure effects immediately following innovation—the tension between static efficiency and dynamic efficiency. In this sense, spillovers are no longer a strictly positive externality because they hold back innovation.

12.4.2 Forms of spillovers

A second source of confusion about spillovers involves the form of the spillover. Just what is spilling over? (For simplicity, hereafter we refer simply to spillovers rather than to spillovers and appropriability.) At least as far back as Griliches (1979), many economists have—sometimes explicitly and sometimes implicitly—organized spillovers roughly into two categories: "rent spillovers" and "knowledge spillovers."

The distinction is clearest if we imagine that research has two products: "blueprints" and "knowledge." Blueprints are designs of specific products or processes that can be used to generate profits. For example, research at Goodyear might result in a new tire material that nets Goodyear a solid profit, justifying their research expenditure. Rent spillovers refer to firms' inability to appropriate all the benefits associated with the use of the specific product or process described in the blueprint. Use of the knowledge embodied in the blueprint to create new blueprints or new knowledge falls under the heading of knowledge spillovers, which we will discuss shortly.

There are a number of ways rents might spill over. For one, innovators cannot appropriate the full social benefits of their blueprints unless they can perfectly price discriminate, even if they are able to hide the blueprint from their competitors and appropriate all the knowledge for themselves (Arrow 1962a). Further, other firms may imitate the product in the blueprint. For example, Pirelli may copy the Goodyear tire, thereby cutting into Goodyear's profits. If the two firms compete aggressively, the price of the new tire may be driven down, and not only will Goodyear fail to appropriate the full social returns, but the returns may accrue largely outside the tire industry. A portion of the returns will be embodied in quality improvements to tires, which might, in turn, increase the productivity of the trucking industry. This latter concern, where the innovative rents end up, is an important issue in statistical work on spillovers (again, see Griliches 1979). Empirical case studies (e.g., Mansfield *et al.* 1977; Trajtenberg 1990) focus largely on rent spillovers, attempting to determine the private and social returns from the use of particular innovations.

Blueprints are not the only product of research efforts. Research creates new knowledge, some embodied in blueprints and some not, that can be used by others as an input to new research. For example, inspection of the Goodyear tires may give Pirelli researchers ideas of how to improve their own tires without actually copying the Goodyear tires. Or researchers in a less closely related field might read in a professional journal about a coincidental discovery made while developing the new tires and use the ideas to improve their products or processes. In the new growth literature, the idea that researchers can build on the knowledge of others takes a number of forms. For example, it often takes the form of "quality ladders," in which each innovation is an improvement to an existing product, as in Grossman and Helpman (1991a) and Aghion and Howitt (1992). In Romer (1990), it takes the form of an aggregate knowledge parameter—actually a count of the number of blueprints in the economy—that is added to with each new blueprint. Jones and Williams (1996) use a combination of these two approaches.

The distinction between rent spillovers and knowledge spillovers is an issue of some concern in statistical work. Griliches (1979) refers to the short-term productivity effects of rent spillovers as "pecuniary" externalities from declining real prices. He argues that their social product should be computable in principle from declining real factor prices, and, hence, that such spillovers are not knowledge spillovers and thus are not really spillovers. Regardless, rent spillovers are important to understand because they help to drive a wedge between social and private returns to R&D. They are also complex: the availability of new technologies may carry implications that go beyond mere cost reductions and quality improvements. New or different goods not only reduce other firms' production costs, they alter other firms' costs and opportunities for innovation. For example, the development of increasingly complex and higher-speed microprocessors by the semiconductor industry has had a profound impact on the types of products offered by the telecommunications and computer industries. In this way, the knowledge "embodied" in the new or cheaper good, the blueprint, does serve as a positive externality on innovation productivity, even if firms only use the good as an input and do not directly exploit the new knowledge embedded in it.

Regardless of the type of spillover, there are many ways that knowledge may be transmitted through an economy. Levin *et al.* (1987) conducted a particularly ambitious and important study of spillover channels (see also Levin 1988). They surveyed 650 R&D executives in 130 industries regarding the methods by which they learn about competitors' technology. They included seven transmission mechanisms in the survey: licensing technology, patent disclosures, publications or technical meetings, conversations with employees of innovating firms, hiring of employees of innovating firms, reverse engineering, and independent R&D. Broadly speaking, they found that all methods of transmission were important under certain circumstances, although some were more important on average than others. Of interest, many of the most important modes of transmission required imitating firms to allocate significant resources to imitation, contradicting the notion that knowledge is "free." [Cohen and Levinthal (1989) discuss this issue in more detail.]

12.4.3   Level of spillover

A third set of distinctions concerns the level at which spillovers occur. A first distinction is between *intrasectoral* and *intersectoral* spillovers. Intrasectoral spillovers occur within a particular industry as firms benefit from the innovation and development activities of competitors. If Intel discovers a way to pack more into their chips, they will care most about hiding the information from other chip manufacturers, AMD, for example. They will care little what Ford or General Motors might do with the information. This is the sort of spillover considered in innovation theory models because it is the most clearly associated with appropriability and it is the most predictable to innovators. Intersectoral spillovers, on the other hand, occur between industries, which may borrow products or ideas, or be stimulated by developments in related fields. Intersectoral spillovers are important for assessing the economy-wide impacts of innovation, but they are a more difficult modeling challenge, and one that may not be immediately appropriate for climate-change modelers.

Spillovers may also be distinguished by their geographic spread. *Local* spillovers occur within regions or countries. *International* spillovers work across national or regional boundaries. International spillovers are starting to be seen as a potentially positive feedback for R&D on environmental control technologies. For example, renewable energy technologies that are competitive in the markets of the Organisation for Economic Co-operation and Development (OECD) economies may also have large global benefits, allowing low-cost emissions reductions in developing countries. However, empirical studies provide evidence that many spillover externalities tend to be limited locally or nationally. A number of authors have suggested that institutional, political, and even cultural factors may prevent innovating firms or industries in one country from collecting innovative rents in another, especially in industries, such as energy, with a "national" character (see, e.g., Grossman and Helpman 1991b; Fagerberg 1994; Abramovitz and David 1995).

12.4.4   Empirical evidence: Spillovers and returns to R&D

In new growth theory, spillovers are the essential phenomenon that allows economies to maintain long-run growth. In innovation theory, spillovers are the main cause of private underinvestment in innovation, and appropriability is an important cause of monopoly behavior. Theory aside, spillovers are only a concern for energy and environmental models if there is empirical evidence that they are important. A large body of work has developed around this issue [see Griliches (1992) and Nadiri (1993) for surveys of the literature]. As one might expect, measurement of such a complex phenomenon has been difficult, and results have varied by method and application.

Nonetheless, there is a consensus among researchers that spillovers are significant and play an important role in technological advance. In the words of one expert,

> In spite of these difficulties, there has been a significant number of reasonably well done studies all pointing in the same direction: R&D spillovers are present, their magnitude may be quite large, and social rates of return remain significantly above private rates. (Griliches 1992:S43)

Nadiri, in his survey of the literature, finds that estimates of firm-level returns average around 20 to 30 percent, whereas estimates of social returns average around 50 percent. Mansfield *et al.*, in their seminal 1977 case study work, find average social returns to R&D of over 50 percent and private returns of about half that.

Two important lessons for climate modelers emerge from the empirical spillover literature. First, private returns to R&D are probably appreciably smaller than social returns, confirming the intuition from theoretical work. This means that the private sector R&D response to environmental policy—for example, an emissions tax—will be less aggressive than we as a society would like. It also opens the door for consideration of technology approaches to environmental problems, an option that has received only limited attention in climate-change models to date.

The second lesson is that a full accounting of spillovers in climate-change models is asking too much. The empirical work makes clear that spillovers are not confined within industries or countries, and, therefore, that models including ITC only within particular sectors or countries may miss intersectoral or international effects. Although these effects are significant, it may be inappropriate for particular climate-change models to include them. For example, if a climate-change modeler includes ITC in the energy industry and only the energy industry—a good first step to be sure—the modeler will neglect the effects of such R&D outside the industry. The point works in reverse as well: some level of technological advance within the energy sector will be exogenous, because it will derive from R&D elsewhere (we will have more to say on these complementary sources of R&D in Section 12.8).

12.5   Innovation Market Failures in ITC Models

There is a consensus among historians, empiricists, and theorists that markets do not invest efficiently in innovation and that underinvestment is significant enough to deserve the attention of policy makers (hence, the patent system). In other words, there are innovation market failures (IMFs), and these failures are important. By IMFs, we mean any characteristic of the market that causes private innovation investment or the results of innovation investments to be different from what would be most beneficial from a society-wide perspective. In the previous two sections we highlighted one cause of IMFs: because knowledge is a public good, markets are prone to underinvest in knowledge creation. But there are other reasons why markets invest inefficiently in innovation. One reason is that managers in firms may be more risk averse or prefer payoffs sooner than would be optimal from a social perspective. Few CEOs are excited about high-cost, high-risk projects that may not pay off for 20 years. Another reason is the limitations to rational behavior, as emphasized by evolutionary economic theory (see Nelson and Winter 1982). More will be said on these topics in Section 12.8, but readers should keep in mind that, henceforth, when we refer to IMFs, we are referring to the totality of factors that render market innovation decisions inefficient. We assume that IMFs most often reduce innovation investment below what would be best for society.

What are the implications of IMFs for climate-change modeling and policy? One implication is that optimal carbon taxes from models without IMFs—or without ITC at all, for that matter—may be biased. Another implication is that technology-based policy measures may be an important or useful component of climate policy. The strength of these implications rests largely on two issues: (1) how large the opportunities for technological change are and (2) how large the IMFs are. Neither of these is well understood at this juncture, but they are potentially important enough to justify the inclusion of ITC and IMFs in climate-change models. Below, we discuss some of the important challenges to incorporating IMFs in climate-change models. In Sections 12.6 and 12.7, we will discuss how ITC modelers have gone about tackling these and other challenges.

First, IMFs are complex. Even simple, two-period innovation theory models that focus exclusively on knowledge appropriability have enough complexity to limit strong general conclusions. In climate-change modeling the situation is many times more complex. For one, the time frames are often on the order of 100 to 200 years. Further, the scope of analysis is large enough to include intersectoral and international effects. Moreover, appropriability is not the only factor in markets for innovation. There are issues of private discounting and risk aversion that may be crucial for high-risk, high-cost, long-lead-time technologies—those "backstop" technologies that may ultimately play a large role in pollution reduction. Hence, a first challenge for climate-change modelers is to incorporate the enormous complexity of innovation systems without overwhelming the models or the intuition of those interpreting the models. It is our belief that climate-change modelers will be most effective if they capture this complexity in simple,

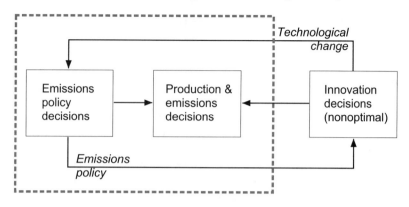

Figure 12.2. Stylized Representation of Decision Making in Energy and Environmental Models.

heuristic, approximate models of nonoptimal innovation behavior rather than by attempting to capture all its nuances. More will be said in this regard in Section 12.7.

A second modeling challenge entails an asymmetry between control variables. Modelers often wish to ascertain *optimal* profiles for the carbon tax; this has been a primary focus of climate-change modeling to date. On the other hand, in the presence of IMFs, the time profile of innovation expenditures should be a socially *nonoptimal*, and likely a socially suboptimal, response to market prices including any environmental taxes. This asymmetry creates modeling problems. How can modelers determine optimal emissions policy while at the same time incorporating nonoptimal behavior with respect to innovation decisions? For example, how can a social planner act simultaneously optimally and nonoptimally?

Figure 12.2 is a stylized representation of the situation. Climate-change models have traditionally taken technological change as exogenous: they have worked within the dashed box of Figure 12.2, looking only at production and emissions decisions. How should researchers incorporate ITC into the system? One choice is to take innovation decisions as optimal—that is, under the control of a social planner. But this approach neglects IMFs. Another option is to take emissions policy (e.g., a carbon tax) as exogenous and then to model the nonoptimal market response. But this option does not provide optimal policy information. A third option is to draw a box around the whole system—that is, to determine optimal policy given nonoptimal innovation behavior. Modelers might use an iterative procedure: set a policy profile, model the system, change the carbon tax, model the system, and so forth until they reach a solution. But this approach is computationally challenging. Hence, modelers wishing to capture the full system face either this computational challenge or the challenge of developing an integrated framework in which a single decision maker acts both optimally and inefficiently.

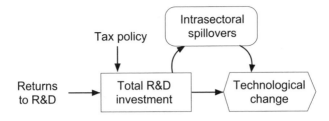

Figure 12.3. Linear, Deterministic Model of Technological Change.

12.6   State-of-the-Art ITC Modeling

How have modelers incorporated ITC into climate-change models? It is constructive to classify ITC models into four types: cost-function models, neoclassical growth models, intertemporal general equilibrium models, and bottom-up energy systems models. This list runs in order of increasing technological detail, with cost-function models being the most abstract and bottom-up energy systems models being the most specific and rigorous about technology. In this section we discuss the mechanisms, purposes, and strengths and weaknesses of each type, along with prominent examples. We focus here on modeling methodologies and the importance of ITC, not on modeling extensions or implications for particular policy arguments. Table 12.1, at the end of this section, summarizes and compares the characteristics of five prominent models.

The innovation and growth literature emphasizes two forces behind technological change: technology evolves largely as a result of private investment incentives and appropriability of innovations, and it is enhanced by spillovers, or positive externalities, from these investments. Figure 12.3 is a stylized representation of this linear model of technological change. The models in this section do not all precisely follow the dynamic in Figure 12.3. For example, some models use experience-based technological advance instead of R&D, and some models do not include IMFs. Nonetheless, the dynamic in Figure 12.3 is a good starting place for thinking about state-of-the-art ITC modeling.

As we proceed through the types and examples of climate-change models in this section, we will focus on three key dimensions of ITC modeling. First, who are the decision makers? Most important in this regard, who makes the emissions decisions and who makes the innovation decisions? For example, social planning models assume an omniscient social planner, whereas private sector actors make decisions in general equilibrium models. Second, how are IMFs included in models? Do models explicitly capture interactions between firms, or are approaches more stylized? Third, how does technology improve? Do firms invest in R&D, or do they simply learn how to do things better the more often they do them? In drawing conclusions about the importance of ITC, we must keep in mind basic assumptions about these three dimensions.

12.6.1   Cost-function models

Cost-function models are the most abstract of the ITC models. They contract all emissions decisions and their economic consequences into a single abatement decision: how far should emissions be reduced from an exogenously given, baseline emissions path? Technological advance entails changes in the abatement cost function, which derives from the production function, making it cheaper to reduce emissions from the baseline path. With ITC, changes in the abatement cost function can be induced, or controlled, along with period-to-period abatement levels. The simplicity of cost-function models allows them to focus on general and theoretical conclusions.

Goulder and Mathai's (Chapter 9 in this volume) set of models is a prime and important example of cost-function modeling. [Tol (1996) uses a similar modeling framework.] As is typical in theoretical work, Goulder and Mathai capture ITC through changes in a "knowledge" parameter in the abatement cost function. Abatement costs at each point in time are a function of abatement levels *and* the knowledge level. They use two formulations to capture how knowledge increases over time: R&D and experience. In the R&D formulation, explicit expenditures on research increase the knowledge stock and thereby reduce future abatement costs. In the experience formulation, technological advance derives simply from the act of abating: the more we abate, the better we are at it. (Both of these mechanisms are discussed in more detail in Section 12.7.) An important characteristic of Goulder and Mathai's model is that an omniscient social planner makes all decisions: there are no IMFs. (Below, we discuss Nordhaus' efforts to include IMFs in a social planning framework.)

Goulder and Mathai explore optimal abatement and emissions tax trajectories under both a cost-effectiveness criterion (how to achieve a target most cheaply) and a benefit–cost criterion (what is the optimal level of abatement?). Their work is a response to the "timing issue"—assertions that ITC calls for greater near-term abatement activity or tax levels relative to models in which technological change is exogenous. From first and second derivatives, Goulder and Mathai draw strong theoretical conclusions about ITC's effect on optimal abatement decisions, challenging assumptions that ITC calls for more aggressive up-front actions.

12.6.2   Neoclassical growth models

Neoclassical growth models are the most consistent with work in endogenous growth theory, as they typically employ the same aggregate formulation of technology based on inputs of capital and labor. Neoclassical growth models of climate change add an input to capture the impacts of emissions or emissions control. The social planner makes innovation decisions in neoclassical growth models.

The DICE model (Nordhaus 1994) is the most widely known of the neoclassical models. Nordhaus builds on DICE to create the R&DICE model, which incorporates ITC (see Chapter 8 in this volume). Both DICE and R&DICE represent production as a function of labor and capital, and both include a compact climate-change model that translates carbon outputs into a damage function reducing total output, and thus per capita consumption. The objective is to maximize discounted per capita consumption by controlling capital investment, carbon emissions, and, in the R&DICE model, R&D.

Nordhaus splits technological change into two components. The first component, which Nordhaus takes as exogenous, affects the productivity of labor and capital (it is the standard technology term in neoclassical growth models). The second component, technological advance in the carbon intensity of production, is endogenous. Nordhaus captures this endogeneity through a research productivity equation.

Because a social planner makes all decisions in neoclassical growth models, it is natural to assume that such models cannot include IMFs. Nordhaus demonstrates that this need not be the case. He captures IMFs in an approximate, heuristic fashion that maintains the computational advantages of the social planning framework: he simply raises the cost of research resources by a factor of four. This serves two purposes simultaneously. First, it puts a brake on carbon energy research; the market invests at a lower level than would be called for by the price of R&D services. Second, by including the inflated R&D cost in the objective function, Nordhaus accounts for the opportunity costs of carbon energy innovation—such innovation may pull resources from other sectors where their impacts are also supernormal. Although Nordhaus combines these two effects, they are potentially separable. The model could be solved using the inflated R&D costs and the final summary statistics calculated on uninflated values. This approach would capture intrasectoral IMFs, but not the opportunity costs of R&D.

Of importance, Nordhaus endogenizes only one particular sort of technological advance: departures from the assumed path of energy-efficiency improvements. This allows Nordhaus to isolate the effects of ITC in the energy sector by comparing his DICE and R&DICE results. He compares the results from a calibrated version of the R&DICE model (endogenous technological change but no substitution) with those from the DICE model (substitution and exogenous technological change).

12.6.3 Intertemporal general equilibrium models

Intertemporal general equilibrium models divide an economy into distinct sectors and then model economic activity, including interactions between the sectors, over time. The approach is computationally demanding and requires detailed data on the chosen sectors. A strength of general equilibrium models is that they allow researchers to explicitly consider interactions *between* sectors, something that is missed in the more abstract cost-function and neoclassical growth approaches. In addition, general equilibrium models are a more explicit representation of real market behavior than cost-function or neoclassical growth models, because the different sectors do not work together as an integrated entity. Two notable intertemporal general equilibrium models of the US economy related to pollution policy and ITC are those of Jorgenson and Wilcoxen (1993) and Goulder and Schneider (1997, 1999).

Jorgenson and Wilcoxen's model is distinguished by the econometric methods used to calibrate the model coefficients. Coefficients are based on extensive time-series data (1947–1985) of interindustry transaction tables. Jorgenson and Wilcoxen use the econometrically calibrated model to develop quantitative estimates of the impacts of environmental policy up to 2020. But the Jorgenson and Wilcoxen model does not include ITC as we have defined it in this chapter; in their model, changes to the production function(s) are *exogenous*. Prices and policy affect the economy's use of the changing production function, but not the production function itself.

Goulder and Schneider, on the other hand, do include ITC as we have defined it in this chapter. Representative firms can enhance their production capabilities by investing resources in knowledge accumulation. A strength of Goulder and Schneider's approach is their explicit inclusion of IMFs. They capture IMFs by modeling intrasectoral spillovers in a manner consistent with endogenous growth models. The accumulation of knowledge for representative firms is only partly appropriable, so markets as a whole underinvest relative to the social optimum. Goulder and Schneider do not attempt to ascertain optimal carbon taxes. Instead, they consider the effects of different carbon tax and R&D subsidy levels.

Another strength of the Goulder and Schneider model is the distinction they make between an alternative, or carbon-free, energy sector and a conventional energy sector. This distinction allows them to begin to address issues of technological heterogeneity, a topic we will return to in Section 12.8. In doing so, Goulder and Schneider must make assumptions about the direction of technological change in each industry, assumptions that influence their results in an interesting fashion. They assume that technological advance in the conventional energy industry cannot be energy/carbon reducing, only productivity improving. Hence, such advance *increases* carbon emission levels through cheaper, dirty fuels.

A final interesting component of the Goulder and Schneider formulation is the inclusion of an R&D services sector that supplies all R&D resources to innovating industries. Increases in alternative fuel research may have indirect effects on research in other sectors by reallocating a limited supply of R&D services. This raises the opportunity costs of environmental R&D. Nordhaus pursued the same notion, but less rigorously, in the R&DICE model by raising the costs of carbon energy R&D above the market price.

12.6.4   Bottom-up energy systems models

Whereas all the models discussed thus far capture technology through parameterized production functions of one sort or another, bottom-up energy systems models are specific about the individual technologies that serve the energy sector. They explicitly represent individual technologies or clusters of technologies, sometimes hundreds of them. Bottom-up models typically seek to minimize the costs of serving an exogenous energy demand by choosing which technologies to install, where, and when. A number of researchers have extended bottom-up models to include experience-based ITC. Three of these models are reviewed in

Seebregts *et al.* (1999). All fall within a social planning framework and do not include IMFs.

The defining characteristic of bottom-up energy systems models is their technological detail and the resulting ability to explore the impacts of technological heterogeneity. By distinguishing between individual technologies, they demonstrate discontinuities in aggregate energy systems characteristics as previously uncompetitive technologies—emerging technologies such as photovoltaic cells or fuel cells—reach market thresholds and begin to contribute to the energy system. They show that investment in a subset of technologies, rather than investment across the whole spectrum of technologies, drives long-term technological change. In other words, the *allocation* of innovative activity may be as important as the *level* of innovative activity. Investment in emerging environmental technologies, however it comes about, can also be viewed as an important hedging strategy against uncertainty.

Several authors have attempted to include uncertainty in bottom-up energy systems models, although the computational challenges are daunting. Grübler and Gritsevskyi (1998) minimize the computational requirements by simplifying down to three technologies: existing, incremental, and revolutionary. They go on to jettison the social planning assumption, and instead assume a set of regional actors and energy suppliers that must plan their technology deployment decisions under uncertainty (see Chapter 11 in this volume). They allow for ITC in 12 technologies. Gritsevskyi and Nakicenovic (Chapter 10 in this volume) include uncertainty in the MESSAGE model framework by reducing the full possible uncertain space to a set of 250 scenarios, and the possible technology installation scenarios to 520 "technology dynamics."

12.6.5 Importance of ITC: Conclusions from state-of-the-art models

How important is ITC in climate-change models? Does the inclusion of ITC significantly alter the results from models with exogenous technological change? Most important, are the policy prescriptions different? Does ITC imply greater up-front abatement than in the exogenous models? Does ITC imply greater emphasis on R&D support? The results to date are less than definitive.

At a minimum, we should expect emissions to be more responsive to policy with ITC than without it. But how much more responsive? Nordhaus' R&DICE results indicate that ITC is less important than technological substitution: reductions in carbon dioxide concentrations and in the global mean temperature with ITC alone are about half those from substitution alone. In Goulder and Schneider's general equilibrium model, a US$25/ton carbon tax noticeably increases alternative energy R&D, especially in the early years, but also dramatically decreases conventional energy R&D. In fact, conventional energy R&D drops to zero in the first three years of their model run! Goulder and Schneider find that ITC noticeably increases the emissions reductions from emissions taxes, thereby increasing gross costs, but also increasing net benefits because abatement is cheaper. The effect is significant: the gross domestic product (GDP) sacrificed to achieve the same cumulative abatement is approximately 25 percent lower with the presence of ITC than without it. However, the market is not as responsive as it might be because of limitations on the level of R&D resources.

How dramatically does the presence of ITC alter the optimal emissions and carbon tax profiles? In models with exogenous technological change (e.g., Wigley *et al.* 1996) it is best to push abatement toward the future because abatement costs will be lower the longer we wait. Some authors have suggested that ITC might alter this result. Goulder and Mathai, using their cost-function model, arrive at two strong theoretical conclusions in this regard. First, if innovation derives from R&D—and assuming no IMFs—the larger the potential for innovation, the further abatement should be pushed into the future. This is an intuitive result; the greater the potential to improve technology is, the better technology will be in the future—assuming we spend on the necessary R&D, and this is an important assumption—and the cheaper future abatement will be. If technological advance results solely from experience, on the other hand, the timing result is ambiguous. The lower future abatement costs are, the further abatement should be pushed into the future; but future abatement costs are a function of near-term abatement levels, which calls for more near-term abatement. Nordhaus's R&DICE model also addresses the timing issue. He finds that ITC has a negligible impact on the optimal carbon tax profile because ITC has a negligible impact on the path of climate change.

Another relevant policy issue is the importance of near-term government technology support. The Goulder and Schneider model directly compares R&D subsidies and emissions taxes. The authors find that the GDP costs of a 15 percent emissions reduction over the next century are approximately nine times higher with a targeted R&D subsidy alone than with a carbon tax alone. Hence, Goulder and Schneider indicate that R&D subsidies should be used as a *complement* to emissions taxes, not as a *substitute* for them.

The energy systems models also address issues of policy, albeit less directly. The models in Seebregts *et al.* (1999) make clear that the *allocation* of inventive effort is an important concern, something that is missed in models based on parameterized production functions. What matters in these models is the emergence of technologies that heretofore were uncompetitive. Whether these emerging technologies come to market through experience effects, as in the energy systems models, or through R&D investment, the conclusion remains the same: large technological changes will come from the emergence of new technologies.

An important, related result of Gritsevskyi and Nakicenovic's analysis is that there are many potential "local equilibrium" technology paths that might result from market forces, and hence significant uncertainty as to where the market may go. They found that 53 of their 520 technology dynamics result in risk-adjusted expected outcomes that are essentially identical in terms of energy systems costs. However, the dynamics are very different from a climate-change perspective, with some resulting in significantly greater carbon concentrations than others. This

Table 12.1. Characteristics of Five State-of-the-Art ITC Models.

	Goulder and Mathai (1998)	Goulder and Schneider (1999)	Nordhaus (Chapter 8, this volume)	Seebregts et al. (1999)	Grübler and Gritsevskyi (Chapter 11, this volume)
Time frame, years	200	60	100	60	70
ITC model type	Cost function	General equilibrium	Neo-classical growth	Bottom-up energy systems	Bottom-up energy systems
Innovation decision maker(s)	Social planner	Market actors	Social planner	Social planner	Market actors: Special
Geographic scope	Ambiguous	USA	World	World and Western Europe	World and five regions
Control parameters	Abatement levels, R&D levels	Carbon tax, R&D subsidy	Emission levels, R&D levels	Technology installations	Technology installations
Innovation market failures	No	Yes	Yes	No	No
Experience learning	Yes	No	No	Yes	Yes
R&D learning	Yes	Yes	Yes	No	Limited
Complementary sources of advance	No	No	No	No	No
Technological heterogeneity	No	Limited	No	Yes	Yes
Technologies with ITC	1	2	1	5–10	12
Diffusion	No	No	No	No	No
Major innovations	No	No	No	No	No
Path dependence and inertia	No	No	No	Limited	Limited

Note: ITC = induced technological change; R&D = research and development.

implies that a small initial push one way or another may have significant long-term effects on technology paths and climate change.

All in all, it is difficult to interpret the early ITC models because so much rests on modeling assumptions—most notably, the level of IMFs and the potential for technological advance—and on the mathematical structures of the models. Further, the sample size is still relatively small. What we can say is that the early models do not provide strong evidence for significantly lower abatement costs, different optimal carbon taxes, or changes in the optimal timing of abatement from models with exogenous technological change. However, there is more work to be done.

12.7   Fundamentals of ITC Modeling

Whether implicitly or explicitly, all models of ITC must address three central modeling issues: the innovation decision maker(s), IMFs, and the characteristics of technological advance. This section reviews the manner in which the state-of-the-art models deal with each of these elements.

12.7.1   Decision makers

While it may seem obvious, every ITC model needs at least one innovation decision maker. Models fall into one of two camps in this regard: the decision maker may be the social planner or it may be market actors. The nature of the incentives will vary depending on the decision maker. In social planning models such as those of Goulder and Mathai and Nordhaus, and the energy systems models, the incentive relates to social welfare. In models that attempt to capture market behavior, such as that of Goulder and Schneider, the market actors choose innovation levels, so innovative rents—profits—provide the incentive for technological advance.

12.7.2   Innovation market failures

IMFs are another essential matter in ITC modeling. Somehow, models must capture the fact that markets invest inefficiently in innovation. To fail to do so is to miss a phenomenon that historians and theorists alike deem critical to technological advance. How IMFs find their way into climate-change models depends on who is making the innovation decisions and what the models are trying to determine.

Social planning models aim to ascertain *optimal* policy; they put innovation and production decisions in the hands of the social planner. As we touched on in Section 12.5, a major challenge for these models is how to maintain the social planning framework's simplicity while including inefficient innovation markets. Goulder and Mathai and most energy systems models sidestep this issue by neglecting IMFs. Nordhaus, on the other hand, includes IMFs through a simple model adjustment. He maintains a central planner throughout, but approximates the vast and complex array of IMFs by increasing the cost of carbon-reducing R&D.

Market-based models, such as Goulder and Schneider's general equilibrium model, focus on market responses to policy rather than optimal policy itself. Because they try to model real market behavior, including inefficiencies, market-based models are conceptually more amenable to IMFs: there is no conflict between optimality in policy and nonoptimality in innovation decisions because optimal policy is not the goal.

So there is a challenge. Modelers wishing to determine optimal policy in the presence of IMFs must somehow combine two actors, the optimal policy maker and the suboptimal innovation decision maker, into a single integrated model. Coming from the social planning perspective, Nordhaus has attempted such a combination. Models coming from the market-based perspective could attempt such a combination through nested optimization, but no modeler has yet attempted this computational challenge.

A Proposal. IMFs are complex. We believe that it is acceptable to be approximate about IMFs: *modelers should accept and use approximate, heuristic models of IMFs that result in socially nonoptimal, and typically suboptimal, innovation behavior.* In other words, models should concentrate on the observation that markets underinvest in innovation and worry less about the economic rigor of the modeling approach used to achieve the underinvestment.

Nordhaus's R&DICE model is a good example of this approach. By increasing innovation costs in his social-planning framework, he forces the market to underinvest in innovation relative to the social optimum (if the price really were the opportunity cost, that is). The tie to the complexity of real-world innovation dynamics is tenuous at best, but the result ties well to the empirical observation that markets underinvest in innovation. Although the model of Goulder and Schneider is more rigorous than that of Nordhaus, it can also be viewed as using an approximate, heuristic tool. All IMFs in Goulder and Schneider's model boil down to "spillover" parameters, one for each industry. The higher the value of these parameters, the lower the investment in innovation, and vice versa. In the end, the result is identical to that of Nordhaus: individual industries underinvest in innovation and a single parameter controls the level of underinvestment. Whether researchers prefer one approach to the other depends largely on which better fits their intuition and their model structure.

It is tempting to try to bring the rigor of innovation theory into climate models, perhaps including more complex market and appropriability structures. We suggest, however, that attempts to model the complexity of IMFs in climate-change models may add confusion rather than clarity. In the end, we care most that markets *do* underinvest in innovation, not *why* they do so. We are not suggesting that the models of innovation theory and other explorations into the microstructure of innovation are not important concerns—they most certainly are, in a wide range of policy contexts, including climate change. We are saying that in long-time-frame, high-level analysis such as integrated assessment modeling, the goal is to capture the observations of these explorations.

12.7.3   Characteristics of technological advance

Climate-change modelers wishing to include ITC must be clear about two characteristics of technological advance: the mechanism by which technology advances and the manner in which advance alters technology.

R&D versus Experience

Technology advances through innumerable interactions between producers, users, designers, researchers, and so forth. For the purposes of modeling, however, a simple representation must be chosen. Two approaches have been used to date: R&D and experience.

In the R&D approach, technological advance occurs through the costly allocation of resources specifically to the task of innovation—that is, through R&D. If society allocates all its resources to production and nothing to R&D, then technology will stand still. The Goulder and Schneider and the Nordhaus models, and one of the models in Goulder and Mathai use the R&D approach. A difficulty with this approach is that the real-world effects of R&D are very difficult to discern with any precision, so modelers find themselves in a tight spot when they estimate parameters. [Recent econometric work by Popp (2001a, 2001b) attempts to make parameters less speculative using patent data as a link between R&D expenditures and industrial energy consumption.]

In the experience approach, technological advance is a happy consequence of the production and use of technologies (see Arrow 1962b). Technological advance is "free" in the sense that nobody needs to do R&D. For example, the more photovoltaic cells we produce, the less costly photovoltaic cells will become. An advantage of the experience approach is simplicity. Whereas decision makers in the R&D approach must decide on both production/abatement levels and on R&D expenditures, decision makers in the experience approach need only consider production/abatement levels. The experience approach is used in the energy systems models in Seebregts *et al.* (1999), Grübler and Gritsevskyi (Chapter 11 in this volume), and Gritsevskyi and Nakicenovic (2000; see also Chapter 10 in this volume), and in one of the models in Goulder and Mathai (Chapter 9 in this volume). (Chapter 7 in this volume provides a more thorough discussion of experience curves in energy models.)

Both the R&D and experience approaches tie well to real-world phenomena. On the one hand, governments and private firms spend billions of dollars on R&D, and it seems likely that this money is not entirely wasted. On the other hand, reductions in production costs do occur through "accidental" learning on the shop floor from routine, day-to-day operations: workers do simply learn how to do things better and faster. In fact, Kline and Rosenberg (1986) discuss industry studies that indicate that, in some cases, learning-by-doing improvements to processes contribute more to technological progress than the initial process development itself.

But neither approach is a complete picture of reality, so models based exclusively on one or the other are bound to miss something important. The experience approach, in particular, exudes a false veneer of precision because experience curves are so easy to estimate; all that is needed is a production history and a cost history. But experience curves miss an extensive history of public and private research expenditures and therefore tend to underestimate the costs of technological advance. Further, there are questions about the direction of causality. The experience literature has viewed the correlation between declining costs and increased production as evidence that the latter causes the former, but causality goes the other way, too. Decreasing costs from R&D spur technologies into new markets, thereby increasing cumulative production. Further, cost reductions may exhibit a strong correlation with the passage of time, and this applies to R&D as well.

There are similar problems on the R&D side. For one, the R&D approach misses experience effects. Moreover, investment in research should not be interpreted strictly as emanating from an R&D lab. Kline and Rosenberg (1986) argue that the notion that innovation is initiated by research of the scientific sort is wrong most of the time. Innovations evolve through cycles of design, testing, production, and marketing, all of which may draw on state-of-the-art knowledge and interact with research initiatives.

The point here is not that R&D is better than experience or vice versa: there is no "right" way to capture the determinants of advance. The point is that both approaches, R&D and experience, are approximations to aid in analysis, both are useful, and both miss important phenomena. Modelers can use the R&D approach to great effect knowing full well that experience is also important, and the same can be said about the experience approach. It is essential that modelers be clear on the implications of what is included in their models and what is not.

The Change in Technology

Kline and Rosenberg (1986) emphasize that there is no simple, single measure or dimensionality to innovation. We might think of innovations as new products or processes, substitution of inputs, or reorganization of production and distribution arrangements. How should technology be represented in ITC models and how should it change over time? This is as fundamental a question as any in ITC modeling.

At one extreme, the bottom-up models contain information about the manner in which individual technologies advance. A benefit of this approach is that the models provide information about which technologies might be important over the long haul, and therefore might be good candidates for government support. Further, modelers often receive information in terms of individual technologies; for example, the prospects for advance in clean coal. Bottom-up models are well placed to use this information. In addition, bottom-up models are explicit about heterogeneity (see Section 12.8). On the negative side, the detail of bottom-up models often comes at a severe computational price and may cloud intuition with complexity. Further, the technological detail may give the illusion that we know more about individual technologies than we really do, and therefore, that we can effectively "pick winners."

At the other extreme, neoclassical growth models, such as Nordhaus' R&DICE model, and stylized cost-function models, such as that of Goulder and Mathai, capture environmental technological advance through changes in a single parameter of the production function or the cost function. This approach greatly simplifies analysis and macroeconomic intuition at the expense of a keener understanding of which technologies emerge to cause the reductions.

Goulder and Schneider cut a middle ground between the two approaches by separately modeling distinct industries representing distinct sorts of technologies: an alternative, or emissions-free, energy industry and a conventional energy industry. Innovation in either industry reduces the costs of producing the particular sort of energy. The notion of heterogeneous technological advance, and particularly the notion of singling out innovation in backstop technologies, seems eminently worthwhile.

Regardless of the model type and innovation mechanism, it seems wise to start with ITC in the energy industry, leaving other technological change as exogenous. While it is true that intersectoral spillovers are real and important, models trying to include the complex interrelations between energy technologies and other technologies would probably be too cumbersome or abstract to be useful. It may also be worthwhile to consider two sources for energy- or carbon-saving improvements: decarbonization of energy services and reductions in the energy intensity of economic activities. The second source of technological advance is more troublesome to incorporate into ITC models, since it involves R&D efforts in sectors outside the energy industry—for example, energy-reducing improvements in transportation technology. Because endogenizing technological change across all industries might be too ambitious in many models, ITC modelers may consider improvements in the energy intensity of technologies outside the energy sector as exogenous.

12.8   Extensions and Limitations of ITC Modeling

Climate-change modelers have not attempted to incorporate the entire complex system of technological change into their models. Instead, they have narrowed their focus to a particular set of observations and phenomena, as every modeler must do to keep analysis manageable. ITC models in climate change have thus far focused on (1) deterministic innovation (2) in the energy or carbon sector (3) that results from private sector actions and (4) that is too slow as a result of IMFs. Many of the state-of-the-art models discussed in Section 12.6 deviate from this simple model in one way or another, but it still describes the main thrust of ITC modeling efforts to date and is a useful starting point for discussing extensions to and limitations of ITC modeling.

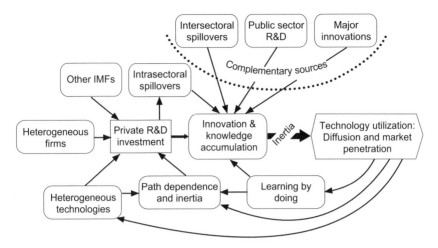

Figure 12.4. A More Complete Framework for Interpreting and Analyzing Technological Change.

One of the main conclusions to come out of the discussion here is that this simple conception is only one piece of a dynamic, uncertain, heterogeneous, context-sensitive, and path-dependent system, not all of which is amenable to modeling. This section, drawing on "appreciative theory" beyond conventional equilibrium economics, seeks to understand which extensions of the simple model are most helpful in improving upon exogenous models of technological change. Here, we discuss six sorts of extensions: complementary sources of technological change, heterogeneity, uncertainty, other IMFs, path dependence, inertia, lock-in, and diffusion. Figure 12.4 proposes a more complete framework for thinking about ITC. Though all these factors are critical in the formulation, analysis, and interpretation process, not all can be included explicitly in models.

12.8.1   Complementary sources of technological change

ITC in climate-change models is not an all-or-nothing proposition. Part of the modeler's task is to decide what should be endogenous and what should remain exogenous. The simple model described above focuses on private, energy sector innovation. Sources of change that are complementary to, or outside of, the simple ITC model may have complex complementary or feedback relationships with energy sector innovation. Some may be induced—that is, they may respond to policy—and are therefore sources and opportunities for technological change in their own right. Here we will highlight three complementary sources of technological advance.

The first complementary source is *public sector R&D*. ITC models spring from a set of innovation literature, innovation theory and endogenous growth theory, that focuses on private sector innovation. It is therefore natural for ITC models to neglect public sector sources of change. But publicly financed basic research as well as subsidies to private R&D have been discussed as central pieces of near-term climate-change policy. [If we think broadly enough, public sector R&D may also be induced. For example, Hayami and Ruttan (1985) show that public sector agriculture R&D was induced by differences and changes in relative factor endowments and prices.] The second complementary source is *intersectoral spillovers*. For example, metallurgical improvements in the past century made possible gradual improvements in electric power generation by allowing a steady rise in operating temperatures and pressures. To models that focus exclusively on the energy sector or restrict spillovers to within industries, such changes will be exogenous. The third complementary source of technological change is groundbreaking *major innovations*. At first glance, these major innovations may seem to fit well within the simple model: are they not simply bigger versions of the average innovation? To some extent they are, but there are also legitimate questions about the degree to which they respond to market forces. To the extent that they do not, they, too, are exogenous to the simple model, and they are highly uncertain.

ITC Modeling Implications. A first note is that modelers should not exclude technology policies—for example, government R&D and R&D subsidies—from the set of options they consider in their models. A more fundamental point in this section, and one that is fundamental to this chapter as a whole, is that modelers should not attempt to make everything endogenous, but rather should pick and choose what is most appropriate given their objectives and their model structures. Complementary sources of technological change tell us that some technological advance must ultimately remain exogenous in ITC models. What is exogenous and what is induced is a matter of model construction: it is a matter of how far we cast our net.

   The empirical work of Newell (1997) and Newell *et al.* (1998; see also Chapter 5 in this volume) supports this notion. The authors used data from 1958 through 1993 to estimate the cost reductions and changes in energy-efficiency characteristics of domestic air conditioning and water heating equipment. They found evidence that the direction of technological advance responded to policy, and that about one-quarter to one-half of the improvement in mean energy efficiency since 1973 was associated with rising energy prices. At the same time, though, the authors found evidence that a large component of the cumulative energy-efficiency improvements occurring over the three decades consisted of "proportional," or neutral, improvements in technology that were largely autonomous, or unresponsive to the set of policies they considered.

12.8.2   Heterogeneity

To ease analysis, economic modelers often simplify the diverse set of firms and technologies into representative firms and aggregate production functions. Here we discuss the implications of these simplifications for ITC modeling.

Heterogeneity in Firm Behavior

Evolutionary economics, drawing inspiration from the process of "creative de-struction" outlined by Joseph Schumpeter over 50 years ago, highlights the im-portance of differences in firm behavior (see Nelson and Winter 1982; Nelson 1995). The issue is not how capitalism administers existing structures, but how it creates and destroys them in the dynamic process of growth and change. The neo-classical focus on the general equilibrium allocation of wealth takes a distinctly secondary role in the Schumpeterian world.

In evolutionary economic models, firms are heterogeneous, most notably in size and in organizational or behavioral structure. Firms are carriers of technolo-gies (or routines or customs) that determine firm performance as a function of their environment. Routines represent at any time the best that firms know and can do in terms of standard operating procedures, investment behavior, innova-tion behavior, and so forth. An important implication of this focus on routines is that optimization is myopic and local. Routines are analogous to genes; the role of the market is to select from among the various routines. Consequently, the rise and fall of firms and nations, and changes in technology, are explained by the survival and proliferation of routines. Conditioned by a changing selection environment, firms evolve in a process that is partly stochastic, but not wholly random. As a consequence, the process of technological advance is strongly path dependent, with no unique, optimized equilibrium.

The point here is not that ITC modeling should be based on evolutionary eco-nomic models, but rather that a whole class of economic models, with a tradition dating back to before Schumpeter, has as its fundamental premise the *differences* between firms. Firm heterogeneity can also be gathered from a more traditional business strategy perspective. The notion of routines is analogous to the conscious differentiation of firms seeking competitive advantage over rivals.

The potential importance of firm heterogeneity implies that ITC models based on homogeneous firms may misrepresent industry innovative behavior. Imagine, for example, that carbon taxes change the selection environment. Those firms that are well positioned to capitalize on price changes will invest heavily in alternative energy or emissions-reducing technologies, while other firms will invest very lit-tle. The sum total of industry investment may easily be greater than if we assumed that numerous identical, "average" firms only invested a small amount: one firm in 10 that is well placed to respond may spend more than would 10 "average" firms in total.

ITC Modeling Implications. The implications of firm heterogeneity for ITC models are unclear. It is tempting to assert that the use of representative firms will underestimate industry responses to policy changes, as in the example just discussed. But the validity of this assertion depends crucially on what the repre-sentative firm represents. If it represents something resembling an "average," then the assertion may be true. If, on the other hand, it is a representation of aggregate industry response, then there is less reason to expect bias one way or the other.

Technological Heterogeneity: Discontinuity in Technological Advance

Just as the representative firm hides heterogeneity in firm behavior, aggregate production functions hide technological heterogeneity. One potential implication of technological heterogeneity is discontinuous technological advance. Technological advance in parameterized production functions takes on only as many dimensions as there are parameters. For example, if a single knowledge parameter captures technological advance, then there is a single dimension for advance. In reality, though, advance has at least as many dimensions as there are individual technologies. As individual technologies advance, the aggregate production function changes.

Even if innovation is continuous and incremental in *individual* technologies, the aggregate production function's response to innovation investment may be nonlinear and may exhibit significant discontinuities. Discontinuities arise when previously uncompetitive technologies reach important market thresholds through cumulative incremental improvements and rapidly diffuse into the market. Prior to reaching a threshold, innovation investment in emerging technologies has a limited impact on aggregate production; after reaching a threshold, innovation expenditures have a more direct and significant impact. In economic terms, it is as if technologies sit on the interior of the production function and R&D efforts to bring them out toward the production function have little discernible impact at an aggregate level. To give a hypothetical example, R&D investment in photovoltaic cells over the past half century has had a relatively small impact on aggregate energy efficiency, but photovoltaic cells may continue to improve and someday compete with other electricity resources. If they do, R&D occurring at a point of rapid market acceptance will appear dramatically more efficient in reducing aggregate efficiency characteristics than early photovoltaic cell R&D. The relationship between aggregate efficiency and R&D, and the energy-efficiency characteristics in the energy sector as a whole, will undergo a discontinuous shift.

Chakravorty *et al.* (1997) emphasize the importance of heterogeneous technologies. Using a framework of optimal natural resource extraction, they study endogenous substitution between the energy resources of coal, oil, gas, and solar power, and the implications for climate policy. They find that carbon emissions will experience a sharp drop around 2050 as the costs of solar generation become more competitive. Although the results depend crucially on assumptions about solar development, they make clear that cost reductions in the backstop technology relative to those of fossil fuels may be at least as important as technological change in the energy sector as a whole.

The notion of heterogeneous technologies raises another notion: that of asymmetric IMFs. Improvements in sulfur dioxide scrubber technology may be less risky and may pay off sooner than investments in photovoltaic cells. To the extent that IMFs are based on technology characteristics such as risk and development time, and that technologies differ in these characteristics, the impacts of IMFs will be asymmetric: some technologies will be more susceptible than others. In particular, one would expect that emerging technologies would be disproportionately susceptible to these IMFs and, therefore, that advancement in emerging technologies might respond less vigorously to emissions taxes than would advancement in more mature and widespread technologies. We will have more to say on these topics later in this section.

ITC Modeling Implications. What do ITC models miss when they aggregate technology? First, these models will ascribe too much predictability and consistency to the innovative process by glossing over the continual process of emergence and obsolescence of technologies. Second, they will ascribe too great a private sector response to development of emerging environmental technologies, because these technologies are typically more susceptible to IMFs such as high private risk aversion and high private discounting. Third, and this seems the crucial point, aggregate models miss the central importance of emerging technologies and the associated notion that the *allocation* of innovative effort is important, not just the absolute level.

It seems feasible for ITC models to consider heterogeneous technologies, and some already do. Bottom-up models can explicitly consider hundreds of technologies and therefore give detailed representations of changes in aggregate production characteristics, including discontinuities and nonlinearities. Top-down models, such as the general equilibrium model of Goulder and Schneider, can distinguish between carbon-intensive and non-carbon-intensive industries. The level of aggregation is higher than in the bottom-up approach, but this is a great leap forward from models with a single, parameterized production function. As Chakravorty *et al.* demonstrate, technological advance in backstop (alternative energy) technologies may be more important than changes in conventional fossil fuel technologies. Hence, we believe that a good first step for modelers is to make a distinction between fossil-fuel-based and alternative energy industries or technologies.

12.8.3 Technological uncertainty

Technological change is an uncertain process. Uncertainty arises not only because the consequences of individual technological changes are so difficult to predict (see Rosenberg 1986), but also because we do not know what changes the future holds. One way uncertainty enters ITC models is by holding back private sector innovation: if private actors are excessively risk averse, they may bias their activities toward less risky projects. We will leave this issue for later, when we discuss "other IMFs." Here we concentrate on uncertainty in the rate and direction of technological advance: we are unsure how technology will evolve in the future, and the further into the future we go, the less we know.

The fundamental concern for modeling the induced portion of technological change is, *How will the production function respond to innovation investment (or cumulative experience)?* There is enormous uncertainty about this response and it is not just a matter of "the right number"; the response will change over time as technologies come and go in the market. We suggest here three aspects of

350 Modeling Induced Technological Change: An Overview

uncertainty in technological change that ITC models need to consider: uncertainty in the potential for individual technologies, heterogeneity and discontinuity in technology development, and major innovations.

The first uncertainty refers to our inability to predict how *individual* technologies will respond to R&D and to experience. The second uncertainty is at a higher level. When we use aggregate production functions, we miss the dynamic competition between technologies and the discontinuities when one technology displaces another. Both of these are "parameter" uncertainties, where the parameters refer to the response of technology to innovative effort or R&D. Both are important in ITC modeling.

The third uncertainty is different in nature. It concerns major, groundbreaking, Schumpeterian-style innovations. Mokyr (1990) draws a useful distinction between *microinventions* and *macroinventions*. Microinventions are "small, incremental steps that improve, adapt, and streamline existing techniques already in use, reducing costs, improving form and function, increasing durability, and reducing energy and raw material requirements." Microinventions are consistent with the models of innovation theory and new growth theory because predictability is relatively high. On the other hand, macroinventions are those inventions in which radical new ideas emerge without precedent. These macroinventions do not seem to obey obvious laws, are not necessarily preceded by profit incentives, and "defy most attempts to relate them to exogenous economic variables." In this sense, macroinventions may be exogenous in ITC models. Also, according to Mokyr, "the essential feature of technological progress is that the macroinventions and microinventions are not substitutes but complements."

In the past, conventional energy-policy-oriented models have focused on time frames of up to 50 years, depending on the scope of analysis. In these time frames, it was perhaps justified to consider only continuous, incremental improvements in technology, such as those implicitly addressed in innovation theory and endogenous growth theory. In climate-change models, though, the scope is often extended to 2100 or beyond. Extrapolating the focus on microinventions into long-term models of technological change may introduce significant error. Witness, for example, the unprecedented emergence of nuclear power during the mid-twentieth century after, as late as the 1930s, leading scientists claimed power could never be harnessed from the atom. While microinventions and the "D" of R&D account for the majority of technological activity, long-term models of technological change are incomplete without consideration of macroinventions.

The distinction between microinventions and macroinventions concerns not only the magnitude of the technological response to innovative activity, but also the degree of causality. ITC models typically assume that technological advance will respond to the economic climate. For example, technological advance should respond to emissions taxes. This assumption is perhaps valid for microinventions, and we might expect an induced change in macroinventions as well, but historical observation indicates that major innovations are not necessarily preceded by a vast commitment of resources directed exclusively toward their development. They often arise as an unexpected byproduct of other innovative endeavors or from basic research. In a sense, they occur more or less randomly, though their occurrence gives rise to a subsequent large-scale commitment of scientific and technological resources to complementary microinventions. A classic example is the invention of the transistor (see Rosenberg 1994). Before the advent of the transistor in 1948, solid-state physics was an obscure subdiscipline. After this macroinvention, R&D communities in both universities and the private sector made large-scale commitments to exploit this new path of innovation.

ITC Modeling Implications. What uncertainties should ITC models include, and how should models include them? A first answer is that, again, some technological uncertainty pertains to technological advance that is largely exogenous in the context of climate models. Macroinventions may fall mainly into this category. On the other hand, some uncertainty pertains to the induced portion of technological advance. Most notable in this regard is uncertainty about innovation production function parameters and experience curve parameters that arises from uncertainty about individual technologies and from the use of aggregate production functions.

One approach to including this parameter uncertainty would be to base R&D production functions on expected values of uncertainty distributions. But this could be a mistake; point estimates can lead to erroneous conclusions in nonlinear systems. In the climate-change system, damage is a nonlinear function of climate change. Nordhaus (1994) showed the importance of accounting for uncertainty in climate-change models. Rather than using the expected values of the uncertain parameters, Nordhaus considered probability distributions on the major uncertain parameters in his DICE model. Using Monte Carlo simulation, he found that the optimal carbon tax more than doubles when uncertainty is taken into account, and the optimal control rate increases by slightly less than half. Similar phenomena may exist with respect to technological advance. Nordhaus' approach could serve as a model for dealing with technological uncertainty in ITC models. Technological uncertainty is especially difficult, though, because it may only be resolved if we attempt to advance technologies: resolution may not be simply a function of time, as is often assumed in analytical models of uncertainty.

Regardless, the importance and prevalence of technological uncertainty indicate that energy policy decisions should be made in an incremental manner, making use of the gradual reduction of uncertainty and preserving options. Projecting technological characteristics far into the future is a daunting task, but this does not mean that we cannot make decisions in the face of uncertainty, for example, by using the principles of decision analysis.

12.8.4 Other IMFs

Innovation theory and new growth theory are based on the dual notions of spillovers and appropriability. Empirical and historical evidence tells us that these are, indeed, important aspects of technological change. There are, however, other

IMFs that should be and are considered by policy makers and, therefore, should be addressed by policy modelers.

Risk Aversion and Discounting

Innovation takes time and is risky. To the extent that markets behave with different preferences for risk and time than we as a society would like, markets will invest in innovation differently than we as a society would like. These differences are not new concepts, and they are not limited to technological change. However, because of the time frames and risk associated with technological change, they may play a disproportionately important role. At the simplest level, in models with only a single dimension for advance—that is, where technology is represented by a single parameter, such as an aggregate knowledge parameter—these IMFs should simply hold back the rate of innovation.

Risk aversion and discounting begin to play a more important role when we consider technological heterogeneity, and emerging environmental technologies in particular. Some technologies will take longer to become competitive than others and some have greater risk. Differences between private and social preference for time and risk therefore affect not only the rate of technological change, but also its direction: markets choose inefficient levels of innovation effort *and* an inefficient allocation among potential technologies. Government energy research funding is often targeted along these lines, attempting to identify high-risk, long-time-frame technologies.

Because price-based policies such as emissions taxes cannot differentiate between technologies, they are unable to fully alter the allocation of private investment. This is one justification for technology instruments as part of the climate-change policy portfolio. At the same time, though, it is reasonable to ask whether these asymmetries should really be under the purview of climate policy. Why should governments target only environmentally related technologies that are high risk and far from competitive and not emerging nonenvironmental technologies?

ITC Modeling Implications.   How can models address these deviations of private risk aversion and time preference from the socially preferred values? One way is simply to include them implicitly in approximate, heuristic models of IMFs. For example, we might increase the price of R&D resources in Nordhaus' approach or we might adjust the spillover parameter upward in the Goulder and Schneider approach.

These two IMFs are challenging, though, because they may distinguish between heterogeneous technologies. If models differentiate between emerging and mature technologies, there may be ways to incorporate the effects. For example, modelers might ramp up the IMFs for emerging technologies relative to those for mature technologies, perhaps correlating IMFs to market share or some other measure of market position.

The Limits of "Rational" Modeling of Innovation

Not all investment activity can be captured by models of rational behavior. Some assert that a spirit that defies rational behavior often guides the entrepreneur, accepting a high probability of failure for a low-probability shot at success. Schumpeter went so far as to argue that the innovation process cannot be characterized by rational behavior:

> [T]he assumption that business behavior is rational and prompt, and also that in principle it is the same with all firms, works tolerably well only within the precincts of tried experience and familiar motive. It breaks down as soon as we leave those precincts and allow the business community under study to be faced by—not simply new situations, which occur as soon as external factors unexpectedly intrude, but by—new possibilities of business action which are as yet untried and about which the most complete command of routine teaches nothing. (Schumpeter 1939:98–99)

Rosenberg, drawing on his experience as an economic historian, makes an equally strong statement:

> The nature of the innovation process, the drastic departure from existing routines, is inherently one that cannot be reduced to mere calculation, although subsequent imitation of the innovation, once accomplished, can so be reduced. Innovation is the creation of knowledge that cannot, and therefore should not, be "anticipated" by the theorist in a purely formal manner, as is done in the theory of decision-making under uncertainty. (Rosenberg 1994:53–54)

While an important realization for the limits of ITC modeling, these observations are also discouraging because they call into question the predictive power of traditional economic models.

ITC Modeling Implications. While important, it seems a bit ambitious to incorporate notions like "entrepreneurial spirit" into ITC models. A more appropriate approach is for modelers to be clear that their models do not include such behavior and therefore greater uncertainty surrounds their results than otherwise would. It also seems, at this point at least, too ambitious to include models of routine-based behavior, such as those in evolutionary economics, in climate models. Routine-based behavior does, however, have one clear implication for ITC that might be amenable to models. Because firms generally search out routines that are similar to those that they already use, private markets will tend to innovate on technologies already in use. Hence, the market may have a sticky innovation response with respect to technological direction. The effect is similar to that of private risk aversion and time preference: it biases private sector innovative behavior toward dominant technologies.

12.8.5   Path dependence, "lock-in," and inertia

Most economic historians and devotees of evolutionary economics will argue that there is much more involved in the evolution and diffusion of technology than merely prices, production functions, and knowledge stocks. In reality, technological change is highly conditioned on the past paths of major and discontinuous innovations, the development activities of firms, and existing capital stocks.

Technically, by path dependent we mean that a process is *nonergodic*—the sequence of historical events conditions future possibilities. Rosenberg explains the notion of path dependence applied to technology:

> [T]he main features of the stock of technological knowledge available at any given time can only be understood by a systematic examination of the earlier history out of which it emerged. There is ... a strong degree of path dependence, in the sense that one cannot demonstrate the direction or path in the growth of technological knowledge merely by reference to initial conditions. (Rosenberg 1994:10)

The notion of path dependence is important for making policy, on the one hand, because it suggests that technological change evolves with a great deal of inertia. On the other hand, path-dependent activity may propagate small changes in the system, so that small policy-induced changes today can result in substantial changes in the future.

There are a number of reasons for path dependence and inertia. At a first level, technological capital and R&D organizations are costly to redirect or replace. Capital stock turnover is another source of inertia. Even if less costly or more efficient technologies are available, old technologies may still be competitive because of sunk costs. Grubb (1996) estimates that for various components of the power generation and energy use infrastructure, capital stock turnover cycles range from 20 to 100 years. Grubb *et al.* (1995) and Ha-Duong *et al.* (1996) have emphasized the importance of this aspect of inertia in energy models. They argue that inertia tends to increase the optimal near-term abatement. Moreover, development activities tend to focus on existing capital, biasing development toward older technologies.

At a deeper level, inertia is generated by the costly nature and limited mobility of development activities. Rosenberg argues that there is an often underappreciated distinction between the availability and implementation of publicly known knowledge or information:

> Development activities accounted for approximately 67 percent of total R&D spending (in the US, according to 1991 *Science and Engineering Indicators*). These figures, at the very least, suggest great skepticism about the view that the state of *scientific* knowledge at any time illuminates a wide range of alternative techniques from which the firm may make cost-less, off-the-shelf selections. It thereby also encourages skepticism toward the notion that is so deeply embedded in the neoclassical theory of the firm, that one can draw a sharp and well-delineated distinction between technological change and factor substitution. (Rosenberg 1994:13)

Indeed, technological change depends greatly on how firms have already geared up, or in the language of economics, the point on the production function at which they currently operate. Firms cannot instantaneously shift to alternative technologies, even if the shift involves only exploiting available but unfamiliar knowledge. This is similar to the notion of routine-based firms making localized searches for new knowledge in evolutionary economic models.

Arthur (1989) has demonstrated how the stochastic nature of the innovation process may lead to technology "lock-in," even by inferior technologies. Development activities reinforce the lock-in, as they focus on existing technologies. The existing technologies suggest certain directions where research efforts can be usefully exercised, resulting in a series of minor improvements that may amount to significant change over the long term.

A prominent example of a path-breaking major innovation that shaped future development is the internal combustion engine. Its rapid development in the early twentieth century made possible numerous other innovations in the automotive and aircraft industries. Moreover, soon after its introduction, the internal combustion engine dominated research and engineering efforts in propulsion devices, even though the engine may not have been inherently technologically superior to the competing technologies of electric and steam power for its initial use in cars. Another example is the steam turbine's domination of electricity generation and the resulting focus of R&D on incremental improvements in that technology.

ITC Modeling Implications. An important implication of path dependence and associated phenomena is that the rate, and especially the direction, of innovation may respond more sluggishly to the economic climate than the neoclassical model of the firm would predict. But there is a deeper and more problematic concern associated with path dependence: actions and technological choices today are more important than conventional economic models would indicate because today's actions not only advance current technological characteristics, they also redirect the future path of technological change. What we do today affects how the economy will respond in the future.

Incorporating path sensitivity into ITC models is a challenge that is perhaps not worth the added complication. Realizing the distinction between changes in technology characteristics and technology paths, however, is important in *interpreting* model results and analyzing policy. Still, ITC models can be extended to better complement qualitative insights about path dependence. Models with experience effects already include inertia, because experience effects reinforce the lead associated with dominant technologies, making it harder for new technologies to emerge. Models with heterogeneous technologies might include asymmetric IMFs to produce a similar effect, and models may be more specific about capital stocks. Models might also include time lags to account for technological expectations, diffusion of innovations (discussed below), and the costliness of development activities. Lags in technology development would be particularly relevant to sudden changes in policy or carbon taxes, to developments following upon major innovations, or in the exploitation of spillovers.

12.8.6 Diffusion

In the simple model of ITC, firms invest in R&D to improve the production function. But what does the production function represent? If it represents the capital stock, and not the technologies society *could* use, then it includes the choice of technologies: it includes diffusion. The choice of ostensibly available technology may be as important as the frontier. The essential point here is that innovation does not in and of itself lead to the *use* of technology; the use of technology will lag behind innovation. [For a review of diffusion literature and models, see Karshenas and Stoneman (1995).]

But diffusion is not simply a matter of time lags. Rates of diffusion and the factors underlying diffusion vary by country and region, and policies should be tailored geographically to reflect this variation. The importance of diffusion is particularly dramatic in developing countries, since their technology lags well behind that of the developed countries. The gap might be closed, but the convergence process depends on numerous social factors (Fagerberg 1994).

A number of studies have looked at the spread of energy-efficiency technologies under the rubric, "the energy-efficiency paradox"—the slow diffusion of apparently cost-effective energy-efficiency technologies or, put another way, the apparently high discount rates used in evaluating energy-efficiency investments (see, e.g., Jaffe and Stavins 1994a, 1994b, 1995; and *Energy Policy*, issue No. 10, 1994). Studies show that market failures are not entirely to blame for the paradox—much of it is due to rational behavior in the face of uncertainty and advancing technology—but almost all studies point to some market failures. Notable among these are information market failures—equipment users may be unaware of the efficiency characteristics, the importance of the efficiency characteristics, or the durability of more efficient equipment. Appliance efficiency labeling and automobile mileage labeling are good examples of government policies aimed directly at information market failures.

Further, one consequence of "quasi-rational" behavior is the impediment of the diffusion of new technologies. Rosenberg (1982) has suggested several reasons why expectations about technology act to slow diffusion. Such expectations generally fall outside the scope of most economic models of ITC, in which firms decide to invest in knowledge without consideration of technology diffusion. First, expectations about the continued improvement and refinement of a technology, particularly the arrival or development of a major innovation, may lead to postponement of innovation activities or adoption. Firms are reluctant to invest in a fledgling technology when they expect substantial improvements to be forthcoming. As anyone who has bought a PC can attest, no one wants to feel burned by investing in a technology that is immediately rendered obsolete by subsequent improvements. Second, competition from new technology sometimes spurs development in old technology, making it more competitive and thus slowing diffusion of the new entrant. Similarly, because single breakthroughs seldom constitute complete innovations, decisions to adopt an innovation are often postponed "in situations that might otherwise appear to constitute irrationality, excessive caution, or over-attachment to traditional practices in the eyes of uninformed observers" (Rosenberg 1982).

ITC Modeling Implications. A first lesson for modelers is to be clear about just what the production function represents. If it represents the capital stock, then it implicitly includes diffusion. A second lesson is that technology diffusion can be induced in addition to innovation, but that diffusion of energy-efficiency improvements may respond more slowly to carbon taxes or other market-based climate policies than might be expected. Modelers may capture this effect through time lags. Further, other policies, most notably information diffusion policies, may play an important and effective role in climate policy.

12.9 Conclusions from a Broader Perspective

We can be confident about two aspects of ITC. First, it has the potential to be exceedingly important. A world with low-cost photovoltaic cells, solar hydrogen, fuel cell cars, and so forth will be very different from a world based largely on fossil fuels, as is today's world. ITC is important to the extent that today's policies influence which future world will emerge. Second, the real-world mechanisms underlying ITC are enormously complex. Together, these two aspects of ITC call for its consideration in climate policy, and environmental policy more generally, and present a modeling challenge.

In Section 12.6, we saw that current state-of-the-art climate-change models do not immediately suggest dramatic impacts from including ITC. In considering possible extensions to and limitations of the current models, this result needs to be qualified. A number of considerations, summarized in Table 12.2, suggest that current ITC models may miss important, policy-relevant nuances of technological advance. Current models may overestimate the speed of technological responses to policy; they may overestimate the flexibility of technological systems to change direction; they may miss important biases for and against particular technology paths (e.g., biases against emerging technologies); and they may miss the value of policies that directly stimulate innovation and diffusion. How might modelers respond to the challenges presented by the considerations in Table 12.2? What is important, what is best left to model interpretation, and what should be included in ITC models?

The most crucial element of ITC is that technological change responds to policy. All of the ITC models we reviewed in Section 12.6 have this fundamental element. It is in movements beyond that point that modelers must make more difficult choices. And no model can include everything. In considering what elements to include, we suggest that four extensions would be the most productive in the near term.

The first extension is the notion that private markets invest inefficiently in innovation. We suggest that it is inappropriate for large-scale models, such as

Table 12.2. Impact of Model Extensions and Limitations on the Current Generation of ITC Models.

Model extensions and limitations	Implications for ITC models
Complementary sources of technological advance	Some technological advance should remain exogenous in energy and environmental models. Government R&D is a source of technological advance, and therefore a policy option.
Firm heterogeneity	Firm heterogeneity implies greater complexity and uncertainty in market innovative behavior than representative firm models indicate.
Technological heterogeneity	Aggregate models may miss important discontinuities and nonlinearities in technological advance. Aggregate models underestimate importance of emerging technologies.
Uncertainty in rate and direction of technological advance	Models should explicitly consider technological uncertainty because climate systems are nonlinear.
"Other" IMFs: risk aversion and discounting	Impacts of these "other" IMFs may be asymmetric. Models without asymmetric IMFs will overestimate technological advance in emerging technologies.
Less-than-rational innovative behavior	Quasi-rational, routinized behavior implies less flexibility and responsiveness than traditional economic models indicate. "Entrepreneurial spirit" decreases the link between incentives and technological advance, causing more technological change to be exogenous in models.
Path dependence, inertia, lock-in	Path-dependent phenomena imply less flexibility and responsiveness in technological systems than traditional economic models indicate. Path-dependent phenomena point to the value of policies to maintain technological diversity.
Diffusion	Models without diffusion may overestimate the speed of ITC. Models without diffusion overlook the importance of policies to spur diffusion.

Note: IMF = innovation market failure; ITC = induced technological change; R&D = research and development.

climate models, to delve deeply into the microeconomic foundations of this inefficiency. Rather, we suggest that modelers use approximate, heuristic representations of IMFs that capture the fact that markets do invest inefficiently rather than worrying about *why*. The microeconomic foundations are important, and continued research in innovation theory may provide additional insights into the aggregate behavior of markets, but that is not the charge of the climate modeler. At least two modeling efforts to date—Nordhaus, and Goulder and Schneider—have explicitly included IMFs.

The second extension is technological heterogeneity. If technological advance is to play an important role in the global society's response to climate-change concerns, it will most likely be through the development of emerging, low-emissions technologies that are currently unproven or uncompetitive with dominant, more-polluting technologies. We believe that it would be productive to explicitly capture this dynamic and its policy implications. To date, energy systems models [e.g., those models reviewed in Seebregts *et al.* (1999); Grübler and Gritsevskyi (1998; see also Chapter 11 in this volume); and Gritsevskyi and Nakicenovic (2000; see also Chapter 10 in this volume)] all explicitly consider technological heterogeneity. Top-down models generally do not. A notable exception is the Goulder and Schneider model, which makes the important distinction between technological advance in alternative energy and conventional energy. We think future top-down modelers would be wise to refine and expand upon this approach.

The third extension is uncertainty in how far and fast technology will advance, and how costly will it be. (We are not referring to the impact of uncertainty on markets for innovation; this is best captured through models of IMFs.) Uncertainty is a thorny issue. We know it is exceedingly important, whether it is in climate damages, costs, or technological advance, but it has proved difficult to model. The resolution of uncertainty over time—that is, learning—has proved especially difficult to handle using standard economic concepts. Nonetheless, it appears crucial enough that modelers would be well advised to consider it explicitly. One possibility might be to reduce the uncertain space into a manageable number of discrete possibilities and then to apply the tools of decision analysis or stochastic control.

The fourth, and last, extension is technological diffusion, both in time and between regions and countries. Empirical studies have shown that technology in developing countries often lags behind that in developed countries, and implementation in all countries lags behind R&D. Diffusion is important because it alters the responsiveness of technology to emissions or R&D policies, and it highlights the value of policies to speed diffusion.

In closing, we would like to emphasize again the crucial importance of communication and interpretation. It is essential for modelers to communicate to policy makers the scope of their models—what is in and what is out—and how reality might differ from model results. Similarly, policy makers must understand that models are models: they do not capture the full scope of reality. Models are an input to decision making, they are not the answer.

References

Abramovitz, M., and David, P., 1995, Convergence and deferred catch-up: Productivity leadership and the waning of American exceptionalism, in R. Landau, T. Taylor, and G. Wright, eds, *Growth and Development: The Economics of the 21st Century*, Stanford University Press, Stanford, CA, USA.

Aghion, P., and Howitt, P., 1992, A model of growth through creative destruction, *Econometrica*, **60**(2):323–351.

Arrow, K., 1962a, Economic welfare and the allocation of resources for invention, in National Bureau of Economic Research, *The Rate and Direction of Innovative Activity*, Princeton University Press, Princeton, NJ, USA.

Arrow, K., 1962b, The economic implications of learning by doing, *Review of Economic Studies*, **29**:155–173.

Arthur, B., 1989, Competing technologies, increasing returns, and lock-in by historical small events, *Economic Journal*, **99**(March):116–131.

Carraro, C., 1997, Induced Technological Change in Environmental Models: Theoretical Results and Implementations, Working Paper, Department of Economics, University of Venice, prepared for the IIASA meeting on Induced Technological Change and the Environment, held in Laxenburg, Austria.

Chakravorty, U., Roumasset, J., and Tse, K., 1997, Endogenous substitution among energy resources and global warming, *Journal of Political Economy*, **105**(6):1201–1234.

Cohen, W.M., and Levinthal, D.A, 1989, Innovation and learning: The two faces of R&D, *The Economic Journal*, **99**:569–596.

Evenson, R.E., and Kislev, Y., 1975, *Agricultural Research and Productivity*, Yale University Press, New Haven, CT, USA.

Fagerberg, J., 1994, Technology and international differences in growth rates, *Journal of Economic Literature*, **32**(3):1147–1175.

Goulder, L.H., and Mathai, K., 1998, Optimal CO_2 Abatement in the Presence of Induced Technological Change, NBER Working Paper 6494, National Bureau of Economic Research, Cambridge, MA, USA.

Goulder, L.H., and Schneider, S., 1997, Achieving low-cost emissions targets, *Nature*, **389**(4):13–14.

Goulder, L.H., and Schneider, S., 1999, Induced technological change and the attractiveness of CO_2 abatement policies, *Resource and Energy Economics*, **21**:211–253.

Griliches, Z., 1979, Issues in assessing the contribution of research and development to productivity growth, *Bell Journal of Economics*, **10**:92–116.

Griliches, Z., 1992, The search for R&D spillovers, *Scandinavian Journal of Economics*, **94**(Supplement):29–47.

Gritsevskyi, A., and Nakicenovic, N., 2000, Modeling uncertainty of induced technological change, *Energy Policy*, **28**:907–921.

Grossman, G., and Helpman, E., 1991a, Quality ladders in the theory of growth, *Review of Economic Studies*, **58**:43–61.

Grossman, G., and Helpman, E., 1991b, *Innovation and Growth in the Global Economy*, MIT Press, Cambridge, MA, USA.

Grossman, G., and Helpman, E., 1994, Endogenous innovation in the theory of economic growth, *Journal of Economic Perspectives*, **8**(1):23–44.

Grubb, M., 1996, Technologies, energy systems, and the timing of CO_2 emissions abatement: An overview of economic issues, in N. Nakicenovic, W. Nordhaus, R. Richels, and F. Toth, eds., *Climate Change: Integrating Science, Economics, and*

Policy, IIASA Workshop Proceedings, International Institute for Applied Systems Analysis, Laxenburg, Austria.

Grubb, M., Chapuis, T., and Ha-Duong, M., 1995, The economics of changing course: Implications of adaptability and inertia for optimal climate policy, *Energy Policy*, **23**(4/5):417–432.

Grübler, A., and Gritsevskyi, A., 1998, A model of endogenous technological change through uncertain returns on learning. http://www.iiasa.ac.at/ Research/TNT/WEB/ Publications/

Grübler, A., Nakicenovic, N., and Victor, D., 1999, Dynamics of energy technologies and global change, *Energy Policy*, **27**:247–280.

Ha-Duong, M., Grubb, M., and Hourcade, J.C., 1996, Optimal Emission Paths Towards CO_2 Stabilization and the Cost of Deferring Abatement: The Influence of Inertia and Uncertainty, Working Paper, CIRED, Montrouge, France.

Hayami, Y., and Ruttan, V., 1985, *Agricultural Development: An International Perspective*, The Johns Hopkins University Press, Baltimore, MD, USA.

Jaffe, A., and Stavins, R., 1994a, The energy paradox and the diffusion of conservation technology, *Resource and Energy Economics*, **16**:91–122.

Jaffe, A., and Stavins, R., 1994b, Energy-efficiency investments in public policy, *The Energy Journal*, **15**(2):43–65.

Jaffe, A., and Stavins, R., 1995, Dynamic incentives of environmental regulations: The effects of alternative policy instruments on technology diffusion, *Journal of Environmental Economics and Management*, **29**:43–63.

Jaffe, A., Newell, R.G., and Stavins, R., 2000, Technological Change and the Environment, Resources for the Future Discussion Paper 00-47, RFF, Washington, DC, USA.

Jones, C., and Williams, J., 1996, Too Much of a Good Thing: The Economics of Investment in R&D, Working Paper, Department of Economics, Stanford University, Stanford, CA, USA.

Jorgenson, D., 1996, Technology in growth theory, in J. Fuhrer and J.S. Little, eds, *Technology and Growth*, Conference Proceedings, Federal Reserve Bank of Boston, Boston, MA, USA.

Jorgenson, D., and Wilcoxen, P.J., 1993, Energy, the environment and economic growth, in A. Kneese and J. Sweeney, eds, *Handbook of Natural Resources and Energy Economics*, North Holland, Amsterdam, Netherlands.

Kamien, M., and Schwartz, N., 1982, *Market Structure and Innovation*, Cambridge University Press, Cambridge, UK.

Karshenas, M., and Stoneman, P., 1995, Technological diffusion, in P. Stoneman, ed., *Handbook of the Economics of Innovation and Technological Change*, Blackwell Publishers, Oxford, UK.

Kline, S.J., and Rosenberg, N., 1986, An overview of innovation, in R. Landua and N. Rosenberg, eds, *The Positive Sum Strategy: Harnessing Technology for Economic Growth*, National Academy Press, Washington, DC, USA.

Levin, A., 1988, Appropriability, R&D spending, and technological performance, *American Economic Review, Papers and Proceedings*, **78**:424–428.

Levin, A., Klevorick, R., Nelson, R., and Winter, S.G., 1987, Appropriating the returns from industrial research and development, *Brookings Papers on Economic Activity*, **3**:783–820.

Mansfield, E., Rappaport, J., Romeo, A., Wagner, S., and Beardsley, G., 1977, Social and private rates of return from industrial innovations, *Quarterly Journal of Economics*, **77**:221–240.

Mokyr, J., 1990, *The Lever of Riches: Technological Creativity and Economic Progress*, Oxford University Press, New York, NY, USA.

Nadiri, M., 1993, Innovations and Technological Spillovers, NBER Working Paper 4423, National Bureau of Economic Research, Cambridge, MA, USA.

Nelson, R.R., 1995, Recent evolutionary theorizing about economic change, *Journal of Economic Literature*, **33**:48–90.

Nelson, R.R., and Winter, S.G., 1982, *An Evolutionary Theory of Economic Change*, Harvard University Press, Cambridge, MA, USA.

Newell, R.G., 1997, Environmental Policy and Technological Change: the Effects of Economic Incentives and Direct Regulation on Energy-Saving Regulation, PhD dissertation, Harvard University, Cambridge, MA, USA.

Newell, R.G., Jaffe, A., and Stavins, R., 1998, The Induced Innovation Hypothesis and Energy-Saving Technological Change, Resources for the Future Discussion Paper 98-12, Resources for the Future, Washington, DC, USA.

Nordhaus, W., 1994, *Managing the Global Commons: The Economics of Climate Change*, MIT Press, Cambridge, MA, USA.

Popp, D., 2001a, Induced Innovation and Energy Prices, NBER Working Paper W8284, National Bureau of Economic Research, Cambridge, MA, USA.

Popp, D., 2001b, The effect of new technology on energy consumption, *Resource and Energy Economics*, **23**:215–239.

Reinganum, J., 1989, The timing of innovation: Research, development, and diffusion, in R. Schmalensee and R. Willig, eds, *Handbook of Industrial Organization,* Vol. 1, Elsevier Science Publications, Amsterdam, Netherlands.

Romer, P., 1986, Increasing returns and long-run growth, *Journal of Political Economy*, **94**(5):1002–1037.

Romer, P., 1990, Endogenous technological change, *Journal of Political Economy*, **98**:S71–S102.

Romer, P., 1994, The origins of endogenous growth, *Journal of Economic Perspectives*, **8**(1):3–22.

Rosenberg, N., 1976, *Perspectives on Technology*, Cambridge University Press, Cambridge, UK.

Rosenberg, N., 1982, *Inside the Black Box: Technology and Economics*, Cambridge University Press, Cambridge, UK.

Rosenberg, N., 1986, The impact of technological innovation: A historical view, in R. Landua and N. Rosenberg, eds, *The Positive Sum Strategy: Harnessing Technology for Economic Growth*, National Academy Press, Washington, DC, USA.

Rosenberg, N., 1994, *Exploring the Black Box: Technology, Economics, and History*, Cambridge University Press, Cambridge, UK.

Ruttan, V.W., 2001, *Technology, Growth, and Development: An Induced Innovation Perspective*, Oxford University Press, New York, NY, USA.

Schmookler, J., 1966, *Invention and Economic Growth*, Harvard University Press, Cambridge, MA, USA.

Schumpeter, J., 1939, *Business Cycles*, Vol. I, McGraw-Hill, New York, NY, USA.

Seebregts, A., Kram, T., Schaeffer, G., Stoffer, A., Kypreos, S., Barreto, L., Messner, S., and Schrattenholzer, L., 1999, Endogenous Technological Change in Energy Systems Models: Synthesis of Experience with ERIS, MARKAL, and MESSAGE, Netherlands Research Foundation ECN, Petten, Netherlands.

Tol, R., 1996, The Optimal Timing of Greenhouse Gas Emission Abatement, the Individual Rationality and Intergenerational Equity, Working Paper, Institute for Environmental Studies, Vrije Universiteit, Amsterdam, Netherlands.

Trajtenberg, M., 1990, *Economic Analysis of Product Innovations*, Harvard University Press, Cambridge, MA, USA.

Weyant, J., 1997, Technological Change and Climate Policy Modeling, Stanford University, Paper prepared for the IIASA meeting on Induced Technological Change and the Environment, International Institute for Applied Systems Analysis, Laxenburg, Austria.

Wigley, T., Richels, R., and Edmonds, J., 1996, Economic and environmental choices in the stabilization of atmospheric CO_2 concentrations, *Nature*, **379**(6582):240–243.

Chapter 13

Induced Institutional Innovation

Vernon W. Ruttan

13.1 Introduction

Technical change is increasingly generated by activities carried out in industrial research laboratories, agricultural experiment stations, and research universities—institutions that have become pervasive during the past century. Changes in resource endowments, cultural endowments, and technology have been an important source of institutional change. In attempting to understand institutional change, I employ a model that is similar to the model of induced technical change discussed in Chapter 2 of this volume. Institutional change, like technical change, is viewed as largely endogenous—as induced by changes in physical, social, and economic environments.[1]

The purpose of this chapter is not to understand how institutions evolve from a primitive institutional "state of nature." Rather, it is to better understand how agents, acting individually and collectively, redesign existing institutions (such as land tenure or labor relations) or design new institutions (such as constructed markets to manage atmospheric pollution) that have a reasonable chance of success. Successful institutional design or redesign cannot simply be the product of the designers' objective function or negotiations among interested groups and representative bodies. If institutional design or redesign is to be successful, it must respond to the changes occurring in the environment in which the institution will exist—such as increases in the price of labor relative to capital, or a rise in the relative value of open access or common property environmental resources or services relative to other factors.

In this chapter I elaborate a theory of institutional innovation in which shifts in the demand for institutional innovation are induced by changes in relative resource endowments and by technical change. The impact of advances in social science knowledge and of cultural endowments on the supply of institutional change are also considered. In a final section I discuss an important recent institutional innovation—the constructed market for sulfur dioxide (SO_2) emissions.

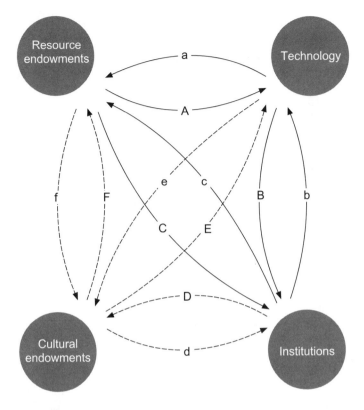

Figure 13.1. Interrelationships between Changes in Resource Endowments, Cultural Endowments, Technology, and Institutions.
Source: Fusfeld (1980).

13.2 A Pattern Model

The elements of a model that maps the relationships among resource endowments, cultural endowments, technology, and institutions are presented in Figure 13.1.[2] The model goes beyond the conventional general equilibrium model in which resource endowments, technologies, institutions, and culture (conventionally designated as tastes) are given.[3] In the study of long-term social and economic change, the relationships among the several variables must be treated as dialectical or recursive. In my own work (with Hayami, Binswanger, and Thirtle), I have given major attention to the relationships among changes in resource endowments, technology, and institutions.

The changes in Japanese and US industrial technology and organization since World War II illustrate the dialectical or recursive interaction between technical and institutional change. The early postwar productivity gap in automobile assembly induced innovations in inventory control, assembly methods, and vehicle quality in Japan. These in turn induced the development of "lean production" in the US automobile industry. US innovations in information and communication technology led to institutional changes in the form of network-based coordination beyond the boundaries of the traditional firm. Japanese firms responded by adapting similar institutional arrangements in an attempt to overcome the US lead in high-technology sectors such as biotechnology (Aoki 1996:17).

The formal economic models that are employed to analyze the supply and demand for technical and institutional change can be thought of as being "nested" within the framework of Figure 13.1. One advantage of the "pattern model" outlined in Figure 13.1 is that it helps to identify areas of ignorance. Our capacity to model and test the relationships between resource endowments and technical change is relatively strong. Our capacity to model and test the relationships between cultural endowments and either technical or institutional change is relatively weak. A second advantage of the model is its usefulness in identifying the components that enter into other attempts to account for secular economic and social change. Failure to analyze historical change in a general equilibrium context tends to result in a unidimensional perspective on the relationships bearing on technical and institutional change.

For example, historians working within the Marxist tradition often tend to view technical change as dominating both institutional and cultural change. In his book *Oriental Despotism* (1957), Wittfogel (mistakenly) views the irrigation technology used in wet rice cultivation in East Asia as determining political organization. As it applies to Figure 13.1, his primary emphasis is on the impact of resources and technology on institutions (lines B and C). A serious misunderstanding can be observed in neo-Marxian critiques of agricultural developments associated with the "green revolution." These criticisms have focused attention almost entirely on the impact of technical change on labor and land tenure relations. In terms of Figure 13.1, both the radical and populist critics have emphasized relation B, but they have tended to ignore relations A and C.[4] This bias has led to repeated failure to identify the separate effects of population growth and technical change on the growth and distribution of income. The analytical power of the more complete induced innovation model is illustrated in this chapter in the work on the impact of both technical change and population growth on changes in land tenure and labor market relationships in the Philippines (see Section 13.3).

Alchian and Demsetz (1973) identify a primary function of property rights as a source of incentives to achieve greater internalization of externalities. They consider that the clear specification of property rights reduces transaction costs in the face of growing competition for the use of scarce resources. North and Thomas, building on the Alchian–Demsetz paradigm, have attempted to interpret the economic growth of Western Europe between 900 and 1700 primarily in terms of changes in property institutions (1970, 1973).[5] During the eleventh and thirteenth centuries, the pressure of population against increasingly scarce land resources induced innovations in property rights that in turn created profitable opportunities for the generation and adoption of labor-intensive technical changes in agriculture. The population decline in the fourteenth and fifteenth centuries was viewed as a primary factor leading to the demise of feudalism and the rise of the nation-state (line C in Figure 13.1). These institutional changes in turn opened up new possibilities for economies of scale in nonagricultural production and in trade (line B).

Olson (1982) has emphasized the proliferation of institutions as a source of economic decline. He also regards broad-based encompassing organizations as having incentives to generate growth and redistribute incomes to their members with little excess burden. For example, a broadly based coalition that encompasses the majority of agricultural producers is likely to exert political pressure for growth-oriented policies that will enable its members to obtain a larger share of a larger national product. A smaller organization that represents the interests of the producers of a single commodity is more likely to pursue the interests of its members at the expense of the welfare of other producers and the general public. In contrast, an even more broadly based farmer–labor coalition would be more concerned with promoting economic growth than would an organization representing a single sector. But large groups, in Olson's view, are inherently unstable because rational individuals will not incur the costs of contributing to the realization of the large group program—they have strong incentives to act as free riders. As a result, organizational "space" in a stable society will be increasingly occupied by special interest "distributional coalitions." These distributional coalitions make political life divisive. They slow the adoption of new technologies (line B) and limit the capacity to reallocate resources (line C). The effect is to slow economic growth or in some cases to initiate a period of economic decline (Olson 1982).[6]

The relationships in the lower left-hand corner of Figure 13.1 (dashed lines) have received relatively little attention from economists. An important exception is Greif's analysis of how the differential impact of the collectivist cultural endowments of Maghrebi traders and the individualistic cultural endowments of Genoese traders (line D) influenced the development of commercial institutions in the Mediterranean trading region in the eleventh and twelfth centuries (Greif 1994:912–950).

What are the implications of the theory of induced innovation for the research agenda on the economics of institutional change? It has been possible to significantly advance our knowledge about the rate and direction of technical change by treating it as endogenous—that is, induced primarily by changes in relative resource endowments and the growth of demand. A beginning has been made in developing a theory of induced institutional innovation in which institutional innovation is treated as endogenous. There is now a significant body of evidence suggesting that substantial new insights into institutional innovation and diffusion can be obtained by treating institutional change as an economic response to changes in resource endowments and technical change.

Changes in cultural endowments, including the factors that economists typically conceal under the rubric of tastes and that political scientists include under ideology, are important sources of both technical and institutional change. But our capacity to develop rigorous empirical tests capable of identifying the relative significance of the relationships between cultural endowments and the other elements of the model outlined in Figure 13.1 remains unsatisfactory. Until colleagues in the other social sciences provide more helpful analytical tools, economists will be forced to adhere to a strategy that focuses primarily on the interactions between resource endowments, technical change, and institutional change.

13.3   What Is Institutional Innovation?

Institutions are the social rules that facilitate coordination among people by help- ing them form expectations for dealing with each other. They reflect the conven- tions that have evolved in different societies regarding the behavior of individuals and groups. They include the internal routines and decision rules followed by firms and agencies to facilitate day-to-day operations or the practices espoused by religious organizations.[7] In the area of economic relations they have a crucial role in establishing expectations about the rights to use resources in economic ac- tivities and about the partitioning of the income streams resulting from economic activity (Runge 1981, 1999; Schotter 1981:11; Nelson and Winter 1982:14–21).

To perform their essential role, institutions must be stable for an extended time period. But institutions, like technology, must also change if development is to occur. Anticipation of the latent gains to be realized by overcoming the disequi- libria resulting from changes in factor endowments, product demand, and tech- nical change is a powerful inducement to institutional innovation.[8] Institutions that have been efficient in generating growth in the past may, over time, come to direct their efforts primarily to rent seeking to protect vested interests and thus become obstacles to further economic development.[9] The growing disequilibria in resource allocation create opportunities for political entrepreneurs or leaders to organize collective action and bring about institutional changes.

This perspective on the sources of demand for institutional change bears some similarity to the traditional Marxian view.[10] Marx considered technical change as the primary source of institutional change. The view expressed here is some- what more complex in that it considers changes in factor endowments and product demand as equally important sources of institutional change. This definition of institutional change is not limited to the dramatic or revolutionary changes of the type anticipated by Marx. Rather, I share with Davis and North the view that basic institutions such as property rights and markets are more typically altered through the cumulation of incremental or evolutionary institutional changes such as modifications in contractual relations or shifts in the boundaries between mar- ket and nonmarket activities (Davis and North 1971:9). Very substantial shifts in the demand for institutional services may be required to overcome the transaction costs involved in negotiating changes in institutional arrangements and in over- coming resistance in implementing new institutional arrangements (Williamson 1985:15–42).

There is a supply dimension as well as a demand dimension to institutional change. Collective action leading to changes in the supply of institutional inno- vations often involves intense conflict among interest groups. Clearly, the process is much more complex than the two-class conflict between the property owners and the propertyless assumed by Marx. In this view, the supply of institutional innovations is strongly influenced by the cost of achieving social consensus. The cost of institutional innovation depends on the power structure of vested interest groups. It also depends critically on cultural traditions and ideologies, such as nationalism or religion, that make certain institutional arrangements more easily accepted than others.

Advances in knowledge in the social sciences (and in related professions such as law, management, planning, and social service) can shift the supply and hence reduce the cost of institutional innovation in a manner somewhat similar to the way that exogenous advances in the natural sciences reduce the cost of technical change. Advances in game theory have, during the past several decades, enabled economists and political scientists to bring an increasingly powerful set of tools to bear on their interpretation of institutional and technical change (Schotter 1981; Ostrom 1990; Aoki 1996). In spite of the power of these new tools, application of standard neoclassical microeconomic models involving shifts in the demand for and supply of institutional change remains exceedingly useful.[11]

Insistence that the processes of institutional innovation and diffusion can be understood by treating institutional change as endogenous to the economic system represents a clear departure from the tradition of modern analytical economics.[12] This does not mean that it is necessary to abandon analytical economics. On the contrary, I try to expand the scope of modern analytical economics by treating institutional change as endogenous.

13.4 Demand for Institutional Innovation: Property Rights and Market Institutions

In some cases the demand for institutional innovation can be satisfied by the development of new forms of property rights, more efficient market institutions, or evolutionary changes arising out of direct contracting by individuals at the community or firm level. In this section I draw on agricultural history for examples.

The English agricultural revolutions were associated with the enclosure of open fields and the replacement of a system in which small peasant cultivators held their land from manorial lords with one in which large farmers used hired labor to farm the land they leased from the landlords. The First Enclosure Movement, in the fifteenth and sixteenth centuries, resulted in the conversion of open arable fields and commons to private pasture in areas suitable for grazing. It was induced by expansion in the export demand for wool. The Second Enclosure Movement, in the eighteenth century, involved conversion of communally managed arable land into privately operated units. It is now generally agreed that this conversion was largely induced by the growing disequilibrium between the fixed institutional rent that landlords received under copyhold tenures (with lifetime contracts) and the higher economic rents expected from adoption of new technology, which became more profitable as a consequence of higher grain prices and lower wages. Enclosure was followed by substantial increases in land and labor productivity and by a redistribution of income from farmers to landowners, and the disequilibrium was reduced or eliminated.[13]

In another example, the opening up of nineteenth-century Thailand for international trade and the reduction in shipping rates to Europe resulted in a sharp increase in the demand for rice. The land available for rice production, which had been abundant, became more scarce. Investment in land development, primarily drainage and irrigation for rice production, became profitable. The response was a major transformation of property rights. In the half century after 1850, rights in human property (corvée and slavery) were largely replaced by more precise private property rights in land (Feeny 1982, 1988).[14]

The decollectivization of agriculture in China, beginning in 1979, provides a dramatic contemporary example of the impact of the transformation of property rights (Lin 1987, 1988; Fan 1991). The changes were induced by a productivity disequilibrium (in the range of 30 percent) between crop yields under the collective system and technology frontier yields. A transition to the household system was initiated spontaneously by peasant households, in spite of official sanctions, in a number of collectives in Sichuan province in 1978. By the early 1980s the transition had extended to other provinces and to small-scale industrial and commercial activities at the township level, and by the mid-1980s it had spread to broad sectors of the national economy. In the next section I discuss a recent case in the Philippines. The case is particularly interesting because it represents a carefully researched contemporary example of the interaction between technical and institutional changes in a Philippine village.

A case study of institutional innovation from the Philippines

Research conducted by Hayami and Kikuchi in a village in the Philippines, beginning in the late 1970s, has enabled us to examine in some detail a contemporary example of the interrelated effects of changes in resource endowments and technical change on the demand for institutional change in land tenure and labor relations (Hayami and Kikuchi 1982, 2000). The institutional innovations occurred as a result of private contracting among individuals. The study is unique in that it is based on a rigorous analysis of microeconomic data for a village in East Laguna over a period of several decades.

Changes in Technology and Resource Endowments

Between 1956 and 1976, rice production per hectare in the study village rose dramatically, from 2.5 to 6.7 metric tons per hectare per year. This increase resulted from two major technical innovations. In 1958, the national irrigation system, which permitted double-cropping to replace single-cropping, was extended to the village. And in the late 1960s, high-yielding rice varieties were introduced. The diffusion of modern varieties was accompanied by increased use of fertilizer and pesticides, and by the adoption of improved cultivation practices such as straight-row planting and intensive weeding.

Population growth in the village was rapid. Between 1966 and 1976 the number of households rose from 66 to 109 and the population rose from 383 to 464, while cultivated area remained virtually constant. The number of landless laborer households increased from 20 to 54. In 1976 half the households in the village had no land to cultivate. The average farm size declined from 2.3 to 2.0 hectares. The land was farmed primarily by tenants. Traditionally, share tenancy was the most common form of tenure. In both 1956 and 1966, 70 percent of the land was farmed under share tenure arrangements. In 1963, an agricultural land reform code was passed that was designed to break the political power of the traditional landed elite and to provide greater incentives to peasant producers of basic food crops. A major feature of the new legislation was an arrangement that permitted tenants to initiate a shift from share tenure to leasehold, with rent under the leasehold set at 25 percent of the average yield for the previous three years. Implementation of the code between the mid-1960s and the mid-1970s resulted in a decline in land farmed under share tenure to 30 percent.

Induced Institutional Innovation

The shift from share tenure to lease tenure was not, however, the only change in tenure relationships that occurred between 1966 and 1976. There was a sharp increase in the number of plots under subtenancy arrangements: from 1 in 1956 to 5 in 1966, and to 16 in 1976. Subtenancy represented a new institutional arrangement, rather than the diffusion of an existing institution, within the context of the village. The subtenancy arrangements were usually made without the formal consent of the landowner. The most common subtenancy arrangement was a fifty–fifty sharing of costs and output between subtenant and operator. Hayami and Kikuchi hypothesized that the incentive for the emergence of the subtenancy institution was disequilibrium between the rent paid to landlords under the leasehold arrangement and the equilibrium rent—the level that would reflect both the higher yields of rice obtained with the new technology and the lower wage rates implied by the increase in population pressure against the land.

To test this hypothesis, market prices were used to compute the value of the unpaid factor inputs (family labor and capital) for different tenure arrangements during the 1976 wet season. The results indicate that the share-to-land was lowest and the operators' surplus was highest for the land under leasehold tenancy. In contrast, the share-to-land was highest and no surplus was left for the operator who cultivated the land under the subtenancy arrangement (Table 13.1). Indeed, the share-to-land when the land was farmed under subtenancy was very close to the sum of the share-to-land plus the operators' surplus under the other tenure arrangements. The results are consistent with the hypothesis. A substantial portion of the economic rent was captured by the leasehold tenants in the form of operators' surplus. On the land farmed under a subtenancy arrangement, the rent was shared between the leaseholder and the landlord.

Table 13.1. Factor Shares of Rice Output per Hectare, 1976 Wet Season.

Factor shares[a]	Number of plots	Area (ha)	Rice output	Land Current inputs	Land-owner	Sub-lessor	Total	Labor	Capital[b]	Operator's surplus
Leasehold land	44	67.7	2,889 (100.0)	657 (22.7)	567 (19.6)	0 (0)	567 (19.6)	918 (31.8)	337 (11.7)	410 (14.2)
Share tenancy land	30	29.7	2,749 (100.0)	697 (25.3)	698 (25.3)	0 (0)	698 (25.4)	850 (30.9)	288 (10.5)	216 (7.9)
Subtenancy land	16	9.1	3,447 (100.0)	801 (23.2)	504 (14.6)	801[c] (23.2)	1,305 (37.8)	1,008 (29.3)	346 (10.1)	−13 (−0.4)

[a]Percentage shares are given in parentheses.
[b]Sum of irrigation fee and paid and/or imputed rentals of carabao (water buffalo), tractor, and other machines.
[c]Rents to sublessors in the case of pledged plots are imputed by applying the rate of 40 percent per crop season (a mode in the crop share distribution in the village).
Source: Hayami and Kikuchi (1981).

Table 13.2. Comparison between the Imputed Value of Harvesters' Share and the Imputed Cost of *Gamma* Labor.

	Based on employers' data	Based on employees' data
Number of working days of *gamma* labor (days/ha)[a]		
Weeding	20.9	18.3
Harvesting/threshing	33.6	33.6
Imputed cost of *gamma* labor (pesos/ha)[b]		
Weeding	167.2	146.6
Harvesting/threshing	369.6	369.6
(1) Total	536.8	516.0
Actual share of harvesters:		
In kind (kg/ha)[c]	504.0	549.0
(2) Imputed value (pesos/ha)[d]	504.0	549.0
(2) – (1)	–32.8	33.0

[a]Includes labor of family members who worked as *gamma* laborers.
[b]Imputation using market wage rates (daily wage = 8.0 pesos for weeding, 11.0 pesos for harvesting).
[c]One-sixth of output per hectare.
[d]Imputation using market prices (1 kg rice = 1 peso).
Source: Hayami and Kikuchi (1981).

A second institutional change, induced by higher yields and the increase in population pressure, was the emergence of a new pattern of labor relationship between farm operators and landless workers. According to the traditional system called *hunusan*, laborers who participated in the harvesting and threshing activity received one-sixth of the harvest. By 1976, most of the farmers (83 percent) had adopted a new system called *gamma*, in which participation in the harvesting operation was limited to workers who had performed the weeding operation without receiving wages. The emergence of the *gamma* system can be interpreted as an institutional innovation induced by the disequilibrium between the institutionally determined wage rate and the market rate. In the 1950s, when the rice yield per hectare was low and labor was less abundant, the one-sixth share may have approximated an equilibrium wage level. With the higher yields and more abundant supply of labor, the one-sixth share became larger than the marginal product of labor in the harvesting operation.

To test the hypothesis that the *gamma* system permitted farm operators to equate the harvesters' share of output to the marginal productivity of labor, imputed wage costs were compared with the actual harvesters' shares (Table 13.2). The results indicate that a substantial gap existed between the imputed wage for the harvesters' labor alone and the actual harvesters' shares. This gap was eliminated if the imputed wages for harvesting and weeding labor were added. Those results are consistent with the hypothesis that the changes in institutional arrangements governing the use of production factors were induced when disequilibria between the marginal returns and the marginal costs of factor inputs occurred as a result of changes in factor endowments and technical change. Institutional change, therefore, was directed toward the establishment of a new equilibrium in factor markets.[15]

In the Philippine village case reviewed here, the induced innovation process leading to the establishment of equilibrium in factor markets occurred very rapidly, even though many of the transactions between landlords, tenants, and laborers were less than fully monetized. Informal contractual arrangements or agreements were used. The subleasing and *gamma* labor contract innovations evolved without the mobilization of substantial political activity or bureaucratic effort. Indeed, the subleasing arrangement evolved in spite of legal prohibition. The primary conclusion that I draw from the English enclosure movement, the Thai property rights case, and the Philippines subleasing case is that disequilibrium between institutional rents and economic rents are a powerful source of institutional change.

13.5   The Demand for Institutional Innovation: Nonmarket Institutions and the Supply of Public Goods

Changes such as those described in the previous section are profitable for society only if the costs involved in the assignment and protection of rights do not outweigh the gains from better resource allocation. If those costs are very high, it may be necessary to design nonmarket institutions in order to achieve more efficient resource allocation.[16]

In Japan, for example, although the system of private property rights was developed on cropland during the premodern period, communal ownership at the village level permitted open access to large areas of wildland and forestland, which were used for the collection of firewood, leaves, and wild grasses to fertilize rice fields. Over time, detailed common property rules evolved to govern the use of communal land so as to prevent resource exhaustion.[17] Detailed stipulations of the time and place of use of communal land as well as rules for mobilizing village labor to maintain communal property (such as applying fire to regenerate pasture) were often enforced with religious taboos and rituals. Those communal village institutions remained viable because it was quite costly to demarcate and partition wildland and forestland and to enforce exclusive use.[18] Group action to supply public goods, such as the maintenance of communal land or water resources, may work effectively if the group involved is small (Ostrom *et al.* 1999). If a large number of people are involved in the use of a public good, however, as in the case of marine fisheries, it is more difficult to respond to the demand for more effective resource management by means of voluntary agreements. Action by a higher authority with coercive power, such as government, may be required to limit free riding (Olson 1965).[19]

Agricultural research

The "socialization" of agricultural research has been common in market economies. New knowledge resulting from research is typically endowed with the attributes of a public good characterized by *nonrivalness* or jointness in supply and use, and *nonexcludability* or external economies.[20] The first attribute implies that the good is equally available to all. The second implies that it is impossible for private producers to appropriate through market pricing the full social benefits arising directly from the production (and consumption) of the good—it is difficult to exclude from the use of the good those who do not pay for it. A socially optimal level of supply of such a good cannot be expected if its supply is left to private firms. Because present institutional arrangements are such that much information resulting from basic research is nonexcludable, it has been necessary to establish nonprofit institutions to advance basic scientific knowledge.[21]

A unique aspect of agricultural research in the past, particularly that directed to advancing biological technology, was that many of the products of research—even in the applied area—were characterized by nonexcludability. Protection by patent laws was either unavailable or inadequate. The nature of agricultural production made it difficult to restrict information about new technology or practices. Furthermore, even the largest farms were relatively small units and were not able to capture more than a small share of the gains from inventive activity. Private research activities in agriculture have been directed primarily toward developing mechanical technology for which patent protection is established.[22]

The public-good attributes of the research product together with the stochastic nature of the research production function have made public support of agricultural research socially desirable. It does not necessarily follow, however, that agricultural research should be conducted exclusively in government institutions financed by tax revenue. The social benefit produced by agricultural research can be measured as the sum of increases in consumers' and producers' surpluses resulting from the downward shift in the supply function of agricultural commodities. If the benefit consists primarily of producers' surplus, agricultural research may be left to the self-organizing activities of agricultural producers (i.e., such institutions as agricultural commodity organizations and cooperatives). Research on a number of tropical export crops grown under plantation conditions, such as sugar, bananas, and rubber, is often organized in this manner.

During the past decade, extension of intellectual property rights institutions has induced a dramatic increase in private sector agricultural research. Even today, however, if agricultural research were left entirely to the private sector the result would be serious bias in the allocation of research resources. Resources would flow primarily to areas of technology that are adequately protected by plant variety registration, patents, or trade secrets (such as the inbred lines used in the production of hybrid corn seed). Other areas, such as research on open-pollinated seed varieties, biological control of insects and pathogens, and improvements in farming practices and management, would be neglected. The socialization of agricultural research or the predominance of public institutions in agricultural research, especially in the biological sciences, can be considered a major institutional innovation designed to respond to demand for more profitable technology that could not be embodied in proprietary products.

13.6   Social Science Knowledge and the Supply of Institutional Innovation

The disequilibria in economic relationships associated with economic growth— such as technical change leading to the generation of new income streams—and changes in relative factor endowments have been identified as important sources of demand for institutional change. Institutional innovations are demanded because they enhance the welfare of rational actors. The issue of the supply of institutional innovations has largely been ignored in the economics literature. There is only a limited literature, for example, on how advances in economic knowledge or, more broadly, social science knowledge affect the supply of economic or social policy or institutions (Coleman 1990:61).

Throughout history, improvements in institutional performance have occurred primarily through the slow accumulation of successful precedents or as byproducts of expertise and experience. Institutional change was generated through the process of trial and error, much in the same manner that technical change was generated before the invention of the research university, the agricultural experiment station, or the industrial research laboratory. The institutionalization of research in the social sciences and related professions has opened up the possibility that institutional innovation can proceed much more efficiently, that it will be increasingly possible to substitute social science knowledge and analytical skill for the more expensive process of learning by trial and error.[23]

The research that advanced our understanding of the production and consumption of rural households in less developed countries demonstrates how advances in knowledge increase the supply of more efficient institutions (Schultz 1964; Nerlove 1974). In a number of countries this research has led to the abandonment of policies that viewed peasant households as unresponsive to economic incentives. It has also led both to the design of policies and institutions that make more productive technologies available to peasant producers, and to the design of more efficient price policies for factors and products.

In the examples of the demand for institutional innovation discussed in the previous sections, there has been little attempt to consider the interaction between the supply of and demand for institutional innovation. There is, however, a modest literature that has employed the "political market" metaphor to discuss the market for votes within legislative bodies, the market for the distribution of wealth among constituencies, and exchanges between legislators and bureaucrats and constituencies and interest groups (Keohane *et al.* 1998). In the next section, I present a case study of how shifts in both supply of and demand for SO_2 pollution control led to the design of a "constructed market" for tradable air pollution permits in the United States. Constituents and environmental interest groups were the source of the increase in demand for the regulation of SO_2 emissions. Advances in economic knowledge led to an understanding of the very large cost reductions that could be achieved by utilizing a "constructed market" rather than traditional "command-and-control" methods to reduce SO_2 emissions. Resource economists, the federal bureaucracy, and the US Congress responded by supplying a constructed market in tradable emission permits.

Models that have been widely employed in recent institutional analysis— such as the tragedy of the commons, the logic of collective action, the prisoner's dilemma game, and mechanism design—are profoundly pessimistic about the ability of individuals, acting alone or in cooperation, to achieve common action. The problem of optimal institutional design has not been solved, even at the most abstract theoretical level (Hurwicz 1972, 1998). Producers or consumers, or both, will have an incentive to deviate from the formal rules of the allocation mechanism (a failure of "incentive compatibility"); and they will be able to do so by misrepresenting facts about which (thanks to "information decentralization") they have unique, privileged information (producers about their production functions, consumers about the preferences) (Goodin 1996:32). The research on institutional innovation by Ruttan and Hayami (1984), Ostrom (1990, 2000), and Keohane *et al.* (1998) is much more optimistic. But one can be optimistic only if the objective is to design better rather than optimal institutions. In the next section, I discuss an important recent example of the role of advances in economic knowledge in the design of an institutional innovation. Economics as a discipline came into being and developed during an era of natural markets. It has lagged in its understanding of the behavior and implications of constructed markets.[24]

13.7 Emissions Trading

One of the clear implications of the previous discussion is that continued access to the material sources of production and reduction of the environmental impacts of economic activity will require the design of incentive-compatible institutions. The theory of induced institutional innovation does not assume that these innovations will occur as a simple response to the "invisible hand" of the market. It does assume that changes in both relative prices, as expressed in economic markets, and in the values placed on public goods, as expressed through political markets, will induce institutional change. If market prices do not accurately reflect relative scarcities, the resulting disequilibrium will generate substantial institutional rents, which will, in turn, induce both technical and institutional change.

In this section I discuss the emergence of emissions trading as an example of a public sector institutional innovation—a "constructed market"—induced by the rising economic value of formerly open access resources.[25] The development of constructed markets for emissions trading was one of the most successful institutional innovations in the field of environmental management in the 1980s and 1990s. The concept is relatively simple. It is based on the realization that the behavioral sources of the pollution problem can be traced, to a substantial degree, to poorly defined property rights in environmental resources such as air and water.

The appeal of market-based systems to manage air pollution can be understood to a substantial degree in terms of the deficiencies of the "command-and-control" regulatory system established after the passage of the 1970 Clean Air Act in the United States. The objective of that act was to "protect and enhance the quality of the nation's air." In response to it, the US Environmental Protection Agency (EPA) defined air quality standards to be implemented by the individual states. Emission standards were established for each emitter based on what could be achieved by the use of "best practice," or best available technology.[26] Any emissions above these standards placed the emitter in noncompliance, making it subject to sanctions. In principle, the Clean Air Act regulatory approach involved the specification of emission standards or legal ceilings on all major emission sources at specific emission points—stacks, vents, storage tanks, etc.[27]

A system of property rights and tradable permits for the management of pollution was first proposed in the late 1960s (Crocker 1966; Dales 1968a, 1968b). This institutional innovation did not emerge from its inventors in a fully operational form. The early proposals were followed by a large theoretical and empirical literature (Bohm 1985). Implementation has involved an extended process of "learning by doing" and "learning by using."

Proposals by President Lyndon B. Johnson for effluent fees and by President Richard M. Nixon for a tax on lead in gasoline had been dismissed as impractical and characterized by environmental activists as a "license to pollute." Beginning in the mid-1980s, however, a series of events conspired to make a more market-oriented approach to reducing SO_2 emissions politically feasible (Taylor 1989:28–34; Hahn and Stavins 1991; Stavins 1998). One was President George Bush's predilection for a market-oriented approach to environmental policy. Another was the enthusiasm of EPA administrator William K. Reilly and a number of key staff members in the Executive Office for validating President Bush's desire to be known as "the environmental president." Congress also provided high-profile, bipartisan support for a variety of market-based approaches, including SO_2 allowance trading.

Within the environmental community, the Environmental Defense Fund (EDF) began to support market-based approaches as early as the mid-1980s. In 1989, the EDF worked closely with the White House staff in drafting an early version of proposed legislation. The credibility of the effort was enhanced by the fact that EPA Administrator Reilly, formerly president of The Conservation Foundation, was a "card-carrying" environmentalist. The business community displayed a curious ambiguity toward the emissions trading proposal. Executives of several major corporations, influenced by subtle lobbying by the EDF, commented favorably on the emissions trading proposal. At the same time, lobbyists representing several major business associations opposed the proposed reforms.

The design of the SO_2 emissions trading system under the 1990 amendment to the Clean Air Act drew on earlier EPA experience. The EPA began experimenting with emissions trading permits in 1974. The early programs included the elimination of lead in gasoline, the phaseout of chlorofluorocarbons and halons in refrigeration, and the reduction of water pollution from nonpoint sources. The early programs had a mixed record. They were typically grafted to existing command-and-control programs. The difficulty of converting from command-and-control requirements to tradable emission programs encountered substantial transaction costs. These experiences did, however, provide important lessons for the development of more market-oriented trading programs in the 1990s.

The Clean Air Act created a national market for SO_2 allowances for coal-burning electrical utilities. The commodity exchanged in the SO_2 emissions trading program is a property right to emit SO_2 that was created by the EPA and allocated to individual firms. A firm can make the allowances issued to it available to be traded to other firms by reducing its own emissions of the pollutant below the baseline level.

In 1995, the first year of the program, 110 of the nation's dirtiest coal-burning plants were included. The affected plants were allowed to emit 2.5 pounds of SO_2 for each million British thermal units (Btu) of energy that they generated. During Phase II, which began in 2000, almost all coal-burning electric power plants will be included and allowances for each plant will be reduced to 1.2 pounds per million Btu. Utilities that "overcomply" by reducing their emissions more than required may sell their excess allowances. Utilities that find it more difficult, or expensive, to meet the requirement may purchase allowances from other utilities.

The evidence available suggests that emissions trading has been even more cost-effective than initially anticipated. Prior to initiation of the program, the utility industry had complained that reducing SO_2 in amounts sufficient to meet the projected target (8.95 million tons in 2000, down from about 19 million tons in 1980) might cost as much as US$1,500 per ton. By the mid-1990s allowances were being sold in the US$100–125 range. The decline in the cost of abatement has been due in part to technical changes in coal mining and deregulation of rail transport that have lowered the cost of low-sulfur coal. It has also been due to technical changes in fuel blending and SO_2 scrubbing technology induced by the introduction of performance-based allowance trading. Benefits have exceeded early estimates (Joskow *et al.* 1998).

As of the late 1990s, other emissions trading programs were being implemented. One of the most ambitious is the RECLAIM program developed by the South Coast Air Quality Management District in the Los Angeles area. Emitters of sulfur and nitrogen oxides have been issued annual allowances that decline each year. Any new emission sources must be accommodated within the cap by acquiring allowances from existing emitters. Some oil refiners have met part of their obligations by purchasing and destroying automobiles made before 1971. In 1998, a group of 12 northeastern states were considering an emissions trading system for reducing ozone levels. The Chicago Board of Trade was exploring the possibility of a futures market in carbon dioxide emission certificates (Fialka 1997). In the spring of 1998, the EPA proposed an emissions trading program, modeled on the SO_2 program, to reduce emissions of nitrous oxides. The successful experience with SO_2 emissions trading illustrates a very important principle in inventing new property rights institutions to manage access to formerly open access resources. In a now classic paper, Coase (1960) argued that when only a few decision makers are involved in the generation of externalities and only a few consumers are affected by the externality, the two parties, if left to themselves, will voluntarily negotiate a set of payments (or bribes) that result in a reduction of the externalities to an acceptable level. However important analytically, the Coase theorem has little relevance to most externality problems. The important externality problems that concern society today, such as SO_2 pollution, typically involve large numbers of polluters and even larger numbers of persons affected by the externalities. Direct negotiation would involve unacceptably large transaction costs.

In 1968 Dales argued that "there exists no economically optimum division between amenity and pollution uses of water" (1968a:799). Regardless of political or economic considerations, the decision must be arbitrary. An implication is that, in contrast to a "natural" market, the government must make the decision in order to establish the conditions necessary for a "constructed" market to function (Coggins and Ruttan 1999). In the SO_2 case it was necessary for an outside principle, the US Congress, to define the size (or the boundaries) of the private resource, in this case the maximum tons of SO_2 emissions, and to establish trading rules. In the absence of public intervention there would have been no private market in emissions.

13.8   Perspective

In my research the theory of induced innovation has been used as a primary organizing concept for interpreting historical processes leading to changes in the rate and direction of technical and institutional change. This does not mean that I insist that either the rate or direction of technical or institutional change is entirely endogenous. There is an autonomous or exogenous element in advances in knowledge and in technical and institutional changes. I do insist, however, that changes in relative factor endowments, interpreted through both market and nonmarket institutions, have altered the rate and direction of both technical and institution change.

Three of these changes in relative resource endowments have represented exceedingly powerful driving forces in inducing both technical and institutional change:

- The closing of the land frontier has been a major driving force inducing both technical and institutional change in agriculture in both traditional and modern societies.

- Increases in the price of labor relative to capital have been a pervasive force of technical and institutional change in both industry and agriculture for at least the past two centuries.

- During the last quarter of the twentieth century, the rising value of open access and common property environmental resources became an important driving force in inducing institutional change.

The world is well on its way toward a transition that will lead to a closing of its remaining commons—in establishing more carefully defined property rights in its remaining open access and common property resources. This institutional change will, in turn, have a profound impact on both the rate and direction of technical change.

Acknowledgments

This paper was originally presented at a conference on Induced Technology Change and the Environment, held at the International Institute for Applied Systems Analysis, Laxenburg, Austria, on 21–22 June 1999. I draw heavily on Ruttan (1978, 1997) and on my earlier work with Yujiro Hayami. See, in particular, Ruttan and Hayami (1984), and Hayami and Ruttan (1985). For a very useful review, see Lin and Nugent (1995). Runge (1999) has traced the evolution of thought from induced technical to induced institutional change. I am indebted to Robert E. Evenson for comments on an earlier draft of this paper.

Notes

    1. Schotter (1981:3–4) notes that in economics there have historically been two distinct interpretations of the rise of social institutions—"collectivist" and "organic." He identifies the collectivist view with the work of Commons and the organic view with that of Menger and Hayek. What Schotter terms the organic view is similar to the endogenous or induced innovation view employed in this paper. What he terms the collectivist view is similar to what Hurwicz (1972, 1998) terms the "designer" perspective. I employ a design perspective, informed by an induced innovation perspective, in my discussion of institution design and redesign. Thus, I reject the need to choose among these two perspectives. They are complementary rather than competitive. Further, the objective of the approach that I employ is not to "liberate" economics from its fixation on the market (Schotter 1981:1). Rather, my objective is to apply the tools of neoclassical microeconomics to the analysis and design of institutional change.

    2. Fusfeld uses the terms *pattern model* or *gestalt model* to describe a form of analysis that links the elements of a general pattern together by logical connections. The recursive multicausal relationships of the pattern model imply that the model is always "open"—"it can never include all of the relevant variables and relationships necessary for a full understanding of the phenomena under investigation" (Fusfeld 1980:33). Ostrom (1990) uses the term *framework* rather than *pattern model*: "The framework for analyzing problems of institutional choice illustrates the complex configuration of variables when individuals … attempt to fashion rules to improve their individual and joint outcomes. The reason for presenting this complex array of variables as a framework rather than a model is precisely because one cannot encompass the degree of complexity within a single model." (Ostrom 1990:214)

    3. In economics, the concept of cultural endowments is usually subsumed under the concept of tastes, which are regarded as given, that is, not subject to economic analysis (Stigler and Becker 1977). I use the term *cultural endowments* to capture those dimensions of culture that have been transmitted from the past. Contemporary changes in resource endowments, technology, and institutions can be expected to result in changes in cultural endowments (Ruttan 1988).

4. A major limitation of the Marxian model is the emphatic rejection of a causal link between demographic change and technical and institutional change (North 1981:60, 61). This blindness to the role of demographic factors, and to the impact of relative resource endowments, originated in the debates between Marx and Malthus. An attempt to correct this deficiency represents a major innovation of the "cultural materialism" school of anthropology (Harris 1979).

5. For a critical perspective on the North–Thomas model, see Field (1981:174–198). Field is critical of the attempt by North and Thomas to treat institutional change as endogenous.

6. For a critical review of Olson's work, see North (1983:163, 164).

7. There is considerable disagreement regarding the meaning of the term *institution*. A distinction is often made between the concepts of *institution* and *organization*. The broad view, which includes both concepts, is most useful for our purpose and is consistent with the view expressed by both Commons (1950:24) and Knight (1952:5). This definition also encompasses the classification employed by Davis and North (1971:8, 9). The more inclusive definition is employed to allow consideration of changes in the rules or conventions that govern behavior within economic units, such as families, firms, and bureaucracies; among economic units, as in the case of the rules that govern market relationships; and between economic units and their environment, as in the case of the relationship between a firm and a regulatory agency. Thus, organizations are defined as a subset of institutions involving deliberate coordination (Vanberg 1994).

8. See North and Thomas (1970, 1973) and Schultz (1975).

9. The role of special interest "distributional coalitions" in slowing society's capacity to adopt new technology and reallocate resources in response to changing conditions is a central theme in Olson (1965, 1982) and in the rent-seeking literature (Krueger 1974; Tollison 1982).

10. "At a certain stage of their development, the material forces of production in society come in conflict with the existing relations of production, or—what is but a legal expression for the same thing—with the property relations within which they had been at work before. From forms of development of the forces of production these relations turn into their fetters. Then comes the period of social revolution. With the change of the economic foundation the entire immense superstructure is more or less rapidly transformed" (Marx 1913:11–12). For a discussion of the role of technology in Marxian thought, see Rosenberg (1982:34–51).

11. The microeconomic approach to understanding the process of institutional change is similar to that employed by Becker in analyzing institutions such as the family (Becker 1991, 1993). A major difference is that I focus on the changes in the environment, such as changes in relative factor and product prices, that are exogenous to the institution being studied and that induce institutional change over time.

12. The orthodox view was expressed by Samuelson (1948): "The auxiliary [institutional] constraints imposed upon the variables are not themselves the proper subject of welfare economics but must be taken as given" (221–222). Contrast this with the more recent statement by Schotter: "We view welfare economics as a study ... that ranks the system of rules which dictate social behavior" (1981:6). There are now five fairly well-defined "political economy" traditions that have attempted to break out of the constraints imposed by traditional welfare economics and treat institutional change as endogenous. These include the theory of property rights, the theory of economic regulation, the theory of rent-seeking interest groups, the liberal-pluralist theories of government, and the

neo-Marxian theories of the state. In the property rights theory, government plays a relatively passive role; the economic theory of regulation focuses on the electoral process; the rent-seeking and liberal-pluralist theories concentrate on both electoral and bureaucratic choice processes; and the theory of the state attempts to incorporate electoral, legislative, and bureaucratic choice processes. For a review and criticism, see Rausser *et al.* (1982).

13.   There has been continuing debate among students of English agricultural history about whether the increases in rents that landowners received after enclosure were because enclosed farming was more efficient than open-field farming or because enclosures redistributed income from farmers to landowners. See Chambers and Mingay (1966), Dahlman (1980), Allen (1982), and Overton (1996).

14.   For a similar interpretation of the evolution of property rights in pre-colonial Hawaii, see Roumasset and La Croix (1988).

15.   A second round of technical and institutional changes occurred in the 1990s. Nonfarm employment opportunities expanded as a result of better transport to the metropolitan Manila area and the location of a small metal craft industry in the village; wage rates rose and threshing by small portable threshing machines largely replaced manual threshing. The labor share for harvesting declined and a new form of labor contract, referred to as *new hunusan*, emerged. As a result of nonfarm employment, incomes of former farm labor households have risen (Hayami and Kikuchi 2000).

16.   Demsetz has pointed out that the relative costs of using market and political institutions are rarely given explicit consideration in the literature on market failure. An appropriate way of interpreting the "public goods" versus "private goods" issue is to ask whether the costs of providing a market are too high relative to the cost of nonmarket alternatives (Demsetz 1964). A similar point is made by Hurwicz (1972).

17.   For an explanation of the distinction between open access and common property, see Ciriacy-Wantrup and Bishop (1975). In the case of open access, use rights have not been fully established. In the case of common property, rules have been established that govern joint use. Common property is therefore a form of land use that lies between the extremes of open access and fully exclusive private rights. The problem of resource exhaustion in open access properties is elaborated in Demsetz (1967) and Alchian and Demsetz (1973).

18.   The term *public economies* has been suggested to describe the economic behavior of institutions "that provide services by arranging for the production, regulation, access, patterns of use, and appropriation of collective good" (Ostrom, 1998:6–7).

19.   Several students of institutional change have emphasized that coordinated or common expectations, resulting from the assurance provided by traditional institutions or common assumptions about equity or ideology, have permitted much larger groups to engage in either implicit or explicit voluntary cooperation than is implied by Olson's model (see Runge 1981:595–606). North notes that "the premium necessary to induce people to become free riders is positively correlated with the perceived legitimacy of the existing institution" (1981:54).

20.   For a characterization of the nonrivalness and nonexcludability attributes of public goods, see Samuelson (1954, 1955, 1958) and Musgrave (1959). Nonrivalness is an essential attribute of information. The use of information about a new farming practice (e.g., contour plowing) by a farmer is not hindered by the adoption of the same practice by other farmers. Nonexcludability, in contrast, is not a natural attribute of information but rather is determined by institutional arrangements. In fact, patent laws are an institutional arrangement that makes a certain form of information (called an "invention") excludable, thereby creating profit incentives for private creative activities. Retention of trade secrets

is another legally sanctioned method of retaining control over inventions or other forms of new technical knowledge.

21. For a history of the establishment of public sector agricultural research systems in a number of developed and developing countries, see Ruttan (1983).

22. In a number of countries, "breeders' rights" and "petty patent" legislation have induced rapid growth in private sector agricultural R&D (Ruttan 1982; Evenson and Evenson 1983). Advances in biotechnology have been associated with rapid extension of intellectual property rights (Ruttan 2001).

23. For a review of the role of social science research in social policy, see Coleman (1990:610–649).

24. For further discussion of the distinction between natural markets and constructed markets, see Coggins and Ruttan (1999).

25. This case is discussed in greater detail in Ruttan (2001).

26. The 1970 Clean Air Act recognized two main types of air pollutants. *Criteria pollutants* are relatively ubiquitous substances. They included sulfur dioxide, suspended particulates, carbon monoxide, nitrogen oxides, hydrocarbons, ozone, and lead. "Criteria documents" that summarized the existing research on the health and environmental effects associated with these pollutants were required as a step in the formulation of emission standards. *Hazardous pollutants* included a number of airborne substances, particularly heavy metals, that had been implicated in cancer or showed other serious health effects. The act required the EPA to list and regulate any pollutants that fit this description. These included asbestos, beryllium, mercury, vinyl chloride, benzene, radionuclides, and arsenic (Tietenberg 1985:2, 3).

27. Stavins (1998) points out that, while there are several alternative market-based instruments available, US experience has been dominated by grandfathered trading permits in spite of the fiscal advantages to the government of emissions taxes or auctioned permits. One explanation is that both environmental advocates and politicians have a strong incentive to avoid policy instruments that make the costs of environmental protection highly visible to consumers and voters. A second is that industry prefers grandfathered permits because they minimize the cost of compliance for existing firms (Keohane *et al.* 1998).

References

Alchian, A.A., and Demsetz, H., 1973, The property right paradigm, *Journal of Economic History*, **33**:16–27.

Allen, R.C., 1982, The efficiency and distributional consequences of eighteenth century enclosures, *Economic Journal*, **92**:937–953.

Aoki, M., 1996, Toward comparative institutional analysis: Motivations and some tentative theorizing, *Japanese Economic Review*, **47**:1–19.

Becker, G.S., 1991, *A Treatise on the Family*, Harvard University Press, Cambridge, MA, USA.

Becker, G.S., 1993, The economic way of looking at behavior, *Journal of Political Economy*, **101**:385–409.

Bohm, P., 1985, Comparative analysis of alternative policy instruments, in A.V. Kneese and J. Sweeney, eds, *Handbook of Natural Resource and Energy Economics*, Vol. I, North Holland, Amsterdam, Netherlands.

Chambers, J.D., and Mingay, G.E., 1966, *The Agricultural Revolution, 1750–1880*, B.T. Batsford, London, UK, and Schocken Books, New York, NY, USA.

Ciriacy-Wantrup, S.V., and Bishop, R.C., 1975, "Common property" as a concept in natural resource policy, *Natural Resources Journal*, **79**(October):713–727.

Coase, R.H., 1960, The problem of social cost, *Journal of Law and Economics*, **3**(October):1–44.

Coggins, J., and Ruttan, V.W., 1999, US emissions permit system, *Science*, **284**(April):263–264.

Coleman, J.S., 1990, *Foundations of Social Theory*, Cambridge University Press, Cambridge, UK.

Commons, J.R., 1950, *The Economics of Collective Action*, Macmillan, New York, NY, USA.

Crocker, T.D., 1966, The structure of atmospheric pollution control systems, in H. Wolozin, ed., *The Economics of Air Pollution*, W.W. Norton, New York, NY, USA.

Dahlman, C.J., 1980, *The Open Field System and Beyond: A Property Rights Analysis of an Economic Institution*, Cambridge University Press, Cambridge, UK.

Dales, J.H., 1968a, Land, water and ownership, *Canadian Journal of Economics*, **1**(November):791–804.

Dales, J.H., 1968b, *Pollution, Property and Prices*, University of Toronto Press, Toronto, Canada.

Davis, L.E., and North, D.C., 1971, *Institutional Change and American Economic Growth*, Cambridge University Press, Cambridge, UK.

Demsetz, H., 1964, The exchange and enforcement of property rights, *Journal of Law and Economics*, **7**(October):11–26.

Demsetz, H., 1967, Toward a theory of property rights, *American Economic Review*, **57**(May):347–359.

Evenson, D.D., and Evenson, R.E., 1983, Legal systems and private sector incentives for the invention of agricultural technology in Latin America, in M.E. Piñeiro and E.J. Trigo, eds, *Technical Change and Social Conflict in Agriculture: Latin American Perspectives*, Westview Press, Boulder, CO, USA.

Fan, S., 1991, Effects of technological change and institutional reform on production and growth in Chinese agriculture, *American Journal of Agricultural Economics*, **73**(May):266–275.

Feeny, D., 1982, *The Political Economy of Productivity: Thai Agricultural Development, 1880–1975*, University of British Columbia Press, Vancouver, Canada.

Feeny, D., 1988, The demand and supply of institutional arrangements, in V. Ostrom, D. Feeny, and H. Picht, eds, *Rethinking Institutional Analysis and Development*, International Center for Economic Growth, San Francisco, CA, USA.

Fialka, J.J., 1997, Breathing easy: Clear skies are goal as pollution is turned into a commodity, *Wall Street Journal*, October 3:A1, A5.

Field, A.J., 1981, The problem with neoclassical institutional economics: A critique with special reference to the North/Thomas model of pre-1500 Europe, *Explorations in Economic History*, **18**:174–198.

Fusfeld, D.R., 1980, The conceptual framework of modern economics, *Journal of Economic Issues*, **14**(March):1–52.

Goodin, R.E., 1996, Institutions and their design, in R.E. Goodin, ed., *The Theory of Institutional Design*, Cambridge University Press, Cambridge, UK.

Greif, A., 1994, Cultural beliefs and the organization of society: A historical and theoretical reflection on collectivist and individualist societies, *Journal of Political Economy*, **102**:912–950.

Hahn, R.W., and Stavins, R., 1991, Incentive based environmental regulation: A new era from an old idea?, *Ecology Law Quarterly*, **18**(1):1–42.

Harris, M., 1979, *Cultural Materialism: The Struggle for a Science of Culture*, Random House, New York, NY, USA.

Hayami, Y., and Kikuchi, M., 1981, *Asian Village Economy at the Crossroads: An Economic Approach to Institutional Change*, University of Tokyo Press, Tokyo, Japan (1981), and The Johns Hopkins University Press, Baltimore, MD, USA.

Hayami, Y., and Kikuchi, M., 2000, *A Rice Village Saga: The Three Decades of Green Revolution in the Philippines*, Barnes and Noble, New York, NY, USA, and Macmillan Press, London, UK.

Hayami, Y., and Ruttan, V.W., 1985, *Agricultural Development: An International Perspective*, 2nd ed., The Johns Hopkins University Press, Baltimore, MD, USA.

Hayami, Y., Akino, M., Shintani, M., and Yamada, S., 1975, *A Century of Agricultural Growth in Japan*, University of Minnesota Press, Minneapolis, MN, USA.

Hurwicz, L., 1972, Organized structures for joint decision making: A designer's point of view, in M. Tiute, R. Chisholm, and M. Radnor, eds, *Interorganizational Decision Making*, Aldine, Chicago, IL, USA.

Hurwicz, L., 1998, Issues in the design of mechanisms and institutions, in E.T. Loehman and D.M. Kilgour, eds, *Designing Institutions for Environmental and Resource Management*, Edward Elgar, Cheltenham, UK.

Joskow, P.L., Schmalensee, R., and Bailey, E.M., 1998, The market for sulphur dioxide emissions, *American Economic Review*, **88**:669–685.

Keohane, N.O., Revesz, R.L., and Stavins, R., 1998, The positive political economy of instrumental choice in environmental policy, *Harvard Environmental Law Review*, **22**:313–367.

Knight, F.H., 1952, Institutionalism and empiricism in economics, *American Economic Review*, **2**(2):45–55.

Krueger, A.O., 1974, The political economy of the rent-seeking society, *American Economic Review*, **64**:291–303.

Lin, J.Y., 1987, The household responsibility system reform in China: A peasants' institutional choice, *American Journal of Agricultural Economics*, **36**(3):410–425.

Lin, J.Y., 1988, The household responsibility system in China's agricultural reform, *Economic Development and Cultural Change*, **36**:S199–S224.

Lin, J.Y., and Nugent, J.B., 1995, Institutions and economic development, in J. Behrman and T.N. Srinavasan, eds, *Handbook of Development Economics*, Vol. IIIA, Elsevier Science, B.V., Amsterdam, Netherlands.

Marx, K., 1913, *A Contribution to the Critique of Political Economy*, Charles H. Keff & Co., Chicago, IL, USA.

Musgrave, R.A., 1959, *The Theory of Public Finance*, McGraw-Hill, New York, NY, USA.

Nelson, R.R., and Winter, S.G., 1982, *An Evolutionary Theory of Economic Change*, Harvard University Press, Cambridge, MA, USA.

Nerlove, M., 1974, Household and economy: Toward a new theory of population and economic growth, *Journal of Political Economy*, **82**(Part 2):S200–S218.

North, D.C., 1981, Ideology and the free rider problem, in *Structure and Change in Economic History*, W.W. Norton, New York, NY, USA.

North, D.C., 1983, A theory of economic change, *Science*, **219**:163–164.

North, D.C., and Thomas, R.P., 1970, An economic theory of the growth of the western world, *Economic History Review*, **23**:1–17.

North, D.C., and Thomas, R.P., 1973, *The Rise of the Western World*, Cambridge University Press, Cambridge, UK.

Olson, M., Jr., 1965, *The Logic of Collective Action: Public Goods and the Theory of Groups*, Harvard University Press, Cambridge, MA, USA.

Olson, M., Jr., 1982, *The Rise and Decline of Nations: Economic Growth, Stagflation, and Social Rigidities*, Yale University Press, New Haven, CT, USA.

Ostrom, E., 1990, *Governing the Commons: The Evolution of Institutions for Collective Action*, Cambridge University Press, Cambridge, UK.

Ostrom, E., 1998, *The Comparative Study of Public Economics*, P.K. Seidman Foundation, Memphis, TN, USA.

Ostrom, E., 2000, Collective action and the evolution of social norms, *Journal of Economic Perspectives*, **14**:137–158.

Ostrom, E., Barger, J., Field, C.B., Norgaard, R.B., and Policansky, D., 1999, Revisiting the commons: Local lessons, global challenges, *Science*, **284**:278–282.

Overton, M., 1996, *Agricultural Revolution in England: The Transformation of the Agrarian Economy: 1500–1850*, Cambridge University Press, Cambridge, UK.

Rausser, G.C., Lichtenberg, E., and Lattimore, R., 1982, Developments in theory and empirical applications of endogenous governmental behavior, in G.C. Rausser, ed., *New Directions in Econometric Modeling and Forecasting in U.S. Agriculture*, Elsevier, New York, NY, USA.

Rosenberg, N., 1982, *Inside the Black Box: Technology and Economics*, Cambridge University Press, New York, NY, USA.

Roumasset, J., and La Croix, S.J., 1988, The coevolution of property rights and political order: An illustration from nineteenth century Hawaii, in V. Ostrom, D. Feeny, and H. Picht, eds, *Rethinking Institutional Analyses and Development: Issues, Alternatives, and Choices*, International Center for Economic Growth, San Francisco, CA, USA.

Roumasset, J.A., and Smith, K.R., 1990, Exposure trading: An approach to more efficient air pollution control, *Journal of Environmental Economics and Management*, **18**:276–291.

Runge, C.F., 1981, Common property externalities: Isolation, assurance, and resource depletion in a traditional grazing context, *American Journal of Agricultural Economics*, **63**(November):595–606.

Runge, C.F., 1999, Stream, River, Delta: Induced Innovation and Environmental Values in Economics and Policy, University of Minnesota Department of Applied Economics, 7 May, St. Paul, MN, USA.

Ruttan, V.W., 1978, Induced institutional change, in H.P. Binswanger and V.W. Ruttan, eds, *Induced Innovation: Technology, Institutions and Development*, The Johns Hopkins University Press, Baltimore, MD, USA.

Ruttan, V.W., 1982, Changing roles of public and private sectors in agricultural research, *Science*, **216**(2 April):23–29.

Ruttan, V.W., 1983, *Agricultural Research Policy*, University of Minnesota Press, Minneapolis, MN, USA.

Ruttan, V.W., 1988, Cultural endowments and economic development: What can economists learn from anthropology?, *Economic Development and Cultural Change*, **36**:S247–S271.

Ruttan, V.W., 1997, Induced innovation, evolutionary theory and path dependence: Sources of technical change, *Economic Journal*, **107**:1520–1529.

Ruttan, V.W., 2001, *Technology, Growth and Development: An Induced Innovation Perspective*, Oxford University Press. Oxford, UK.

Ruttan, V.W., and Hayami, Y., 1984, Toward a theory of induced institutional change, *Journal of Development Studies*, **20**:203–223.

Samuelson, P.A., 1948, *Foundations of Economic Analysis. Harvard Economic Studies*, Harvard University Press, Cambridge, MA, USA.

Samuelson, P.A., 1954, The pure theory of public expenditure, *Review of Economics and Statistics*, **36**:387–389.

Samuelson, P.A., 1955, Diagrammatic exposition of a theory of public expenditure, *Review of Economics and Statistics*, **37**:350–356.

Samuelson, P.A., 1958, Aspects of public expenditure theories, *Review of Economics and Statistics*, **40**:332–338.

Schotter, A., 1981, *The Economic Theory of Social Institutions*, Cambridge University Press, Cambridge, UK.

Schultz, M.W., 1975, The value of the ability to deal with disequilibria, *Journal of Economic Literature*, **13**(September):822–846.

Schultz, T.W., 1964, *Transforming Traditional Agriculture*, Yale University Press, New Haven, CT, USA.

Stavins, R., 1998, What can we learn from the grand policy experiment? Lessons from SO$_2$ allowances trading, *Journal of Economic Perspectives*, **12**:69–88.

Stigler, G.J., and Becker, G.S., 1977, De gustibus non est disputandum, *American Economic Review*, **67**(2):76–90.

Taylor, R.E., 1989, *Ahead of the Curve: Shaping New Solutions to Environmental Problems*, Environmental Defense Fund, New York, NY, USA.

Tietenberg, T.H., 1985, *Emissions Trading: An Exercise in Reforming Pollution Policy*, Resources for the Future, Washington, DC, USA.

Tollison, R., 1982, Rent seeking: A survey, *Kyklos*, **35**:575–602.

Vanberg, V., 1994, *Rules and Choice in Economics*, Routledge, London, UK.

Williamson, O.E., 1985, *The Economic Institutions of Capitalism: Firms, Markets, Relational Contracting*, Collier Macmillan, New York, NY, USA.

Wittfogel, K.A., 1957, *Oriental Despotism: A Comparative Study of Total Power*, Yale University Press, New Haven, CT, USA.

Author Index

Subject Index

About the Editors

Arnulf Grübler is a senior research scholar in the Transitions to New Technologies Project at the International Institute for Applied Systems Analysis (IIASA). His books include *The Rise and Fall of Infrastructures* and *Technology and Global Change*. Nebojsa Nakicenovic is leader of IIASA's Transitions to New Technologies Project. He is coeditor with Arnulf Grübler of *Diffusion of Technologies and Social Behavior*; with Arnulf Grübler and Alan McDonald of *Global Energy Perspectives*; and convening lead author of the IPCC's *Special Report on Emissions Scenarios*. William D. Nordhaus is Sterling Professor of Economics at Yale University and a Director on the Resources for the Future Board. His books include *Invention, Growth, and Welfare* and *Managing the Global Commons*. He is coauthor with Joseph Boyer of *Warming the World* and with Paul A. Samuelson of *Economics*, seventeenth edition.